CHEMICAL ENGINEERING
DESIGN PROJECT

CHEMICAL ENGINEERING DESIGN PROJECT

A CASE STUDY APPROACH

(Production of Phthalic Anhydride)

2ND EDITION

Martyn S. Ray and Martin G. Sneesby
Curtin University of Technology, Western Australia

GORDON AND BREACH SCIENCE PUBLISHERS
Australia • Canada • China • France • Germany • India • Japan
Luxembourg • Malaysia • The Netherlands • Russia • Singapore
Switzerland • Thailand

First Edition published 1989
Second Edition published 1998

Amsteldijk 166
1st Floor
1079 LH Amsterdam
The Netherlands

British Library Cataloguing in Publication Data

Ray, Martyn S., 1949–
    Chemical engineering design project : a case study
    approach. – 2nd ed.
    1. Chemical engineering   2. Engineering design
    I. Title   II. Sneesby, Martin G.
    660

    ISBN 90-5699-137-X (softcover)

1005713652

Contents

Preface to Second Edition

The main difference between this edition and the original (1989) is the inclusion of a new case study – the production of phthalic anhydride. Although the design process is essentially similar for most chemicals, no two designs are ever the same. A comparison between the original case study and this new one should emphasise the need for originality and flexibility in process design.

As with the earlier edition, the case study has been incorporated throughout the book so that application of the principles and ideas which are discussed in the main text can be illustrated sequentially. The case study should not be viewed merely as a design blue-print but as an integral part of the book. Two items of equipment have been desinged in detail in this new case study (Part II) whereas only the first item was considered to that depth in the first edition.

All sections of the text material have been revised and new material has been added, e.g. loss prevention and safety, economic evaluation and environmental considerations. The new material reflects important developments and shifting emphasis in chemical engineering over the intervening years. However, the focus is again on the approach of learning by doing. It is necessary to consider a broad range of topics in a design problem, hence the retention of the Technical and Economic Feasibility Study.

The advice is intended to be practical and the reader is directed to an extensive list of useful references – fully updated in this edition. This book is not intended to be a design handbook, it should be considered as a 'road map' for performing a design project.

Acknowledgements

The material included in Section IV(A) is reproduced with permission of the Institution of Chemical Engineers (UK).

I would like to express my thanks and appreciation to all my past project students who have contributed in unseen ways to the evolution of this book, and to the ideas presented here, by our discussions and their hard work and enthusiasm over the years (*MSR*).

This book is dedicated to Creenagh (*MGS*) and to my wife, Cherry (*MSR*), for their understanding, encouragement, and patience during the many hours of preparation.

INTRODUCTION

I HOW TO USE THIS BOOK

(A) The Case Study Approach

This book uses a case study between the lecture-type material to enhance the teaching and appreciation of the work involved in a chemical engineering design project. All undergraduate chemical engineering students are required to perform a design project, usually in the final year of the course. It may be the last piece of work that a student completes (after all other subjects have been examined) prior to graduation, carried out over a period of between 6 to 10 weeks (depending upon departmental policy). Alternatively, the design project may be performed during the entire final year of study. No doubt, variations on these alternatives occur in certain faculties.

Courses that are accredited by the *Institution of Chemical Engineers (IChemE)* UK, must include a design project unit conforming to their specifications (see Section IV). All UK chemical engineering degree courses are accredited by the IChemE; courses in territories having strong historical links with Great Britain, e.g. Africa, Australia, West Indies, etc., also usually aim for IChemE recognition.

In the United States, most engineering courses are accredited by the *Accreditation Board of Engineering and Technology (ABET)*, of which the *American Institute of Chemical Engineers (AIChE)* was a founding organisation. The requirements of the AIChE regarding the teaching of chemical engineering design and the design project are different from those laid down by the IChemE, although all US accredited courses are expected to include some form of design project work to be performed by their students. Only graduates from courses accredited by the IChemE are admitted to professional membership of that institution (or graduates from non-accredited courses who can subsequently fulfil the IChemE requirements).

This book is intended to provide guidance specifically to those students who are enrolled in IChemE accredited courses, and are about to commence the design project. Those same students will also find this book useful when they are studying earlier units in Plant and Process Design. Reference to this book will illustrate the application of certain topics from other taught units during the design project. However, students in courses not accredited by the IChemE (specifically in the USA) should also find this text useful when studying equivalent course units.

The approach adopted here is to provide brief notes and references for a wide range of topics which need to be considered in the design project. Case study material concerning *The Manufacture of Phthalic Anhydride* is presented, and illustrates what is required in a student design project. The case study material is adapted from the design project performed by Martin Sneesby at Curtin University of Technology, Perth, Western Australia, in 1991. This project was awarded the CHEMECA Design Prize for the best Australian university design project in 1991, and the CHEMECA medal was presented to the author (MGS) at the 20th Australian Chemical Engineering Conference. The Curtin University chemical engineering course is accredited by the IChemE, and the design projects performed at the university conform to the Institution requirements.

A coherent view of the design project requirements is obtained by using one typical design example to provide all the case study material in this book. Some of the supporting information and calculations have been abbreviated due to space restrictions. However, the key results are all reproducible, and the validity of the stated results can be checked if desired. Design projects are seldom (if ever) perfect and this book, and the case study material, is no exception.

It was decided that a realistic appreciation of the stages in a design project, and the sequence of tasks that the student performs, would be best obtained by including the descriptive notes in serif typeface 'between' the case study material, which appears in sans serif typeface. This was in preference to presenting all the notes followed by the complete 'typical' student design project. Project work usually proceeds on a week-by-week basis as aspects are investigated, memos and project reports written, decisions are made and changed, and prior work is revised, and hence our choice of presentation.

The aspects of the design that were considered in this project are more comprehensive than those required by the IChemE in their design problem for external students (see Section IV). Topics such as market appraisal, site selection, plant layout, etc., are considered here. The detailed requirements and particular emphasis on certain topics, e.g. control and instrumentation, economic analysis, HAZOP, etc., often depends upon the experience and philosophy of the supervisor and the departmental policy. However, we feel that the aspects of design presented in this book cover a wide and comprehensive range of possible topics, although it is expected that most lecturers would prefer a more detailed coverage in certain areas. Ultimately this book is intended to provide guidance *to the student*, not to be a complete text on all aspects of plant design.

(B)   A "Road Map"

1. Read this preliminary material (Sections I to IV, before Chapter 1), it is there to help you understand what is required in a design project!
2. Identify the difference between the 'case study' material (related to phthalic anhydride) and the 'text' material in each chapter. The case study is in a different type face, and it is *not* all presented together at the end of the book!
3. The 'text' material is intended to provide only *minimum* guidance for a topic, not a comprehensive coverage. It is intended to provide reference to useful and detailed sources for further reading on a particular topic.
4. The book has two main parts. *Part I covers the Technical and Economic Feasibility Study for the Process.* At this stage, all aspects that will impact upon the economic success of the project should be considered. Initially there are many

options and alternatives, all of which should be evaluated. Gradually the number of options is reduced and the scheme of work for the project becomes defined. *Part II is the Detailed Equipment Design Study*, and two major pieces of equipment are designed here. The topics are presented in a logical order and should be studied that way. However, a 'real' project would require consideration of many factors concurrently rather than the convenient academic sectioning and sequencing.

5. Do not assume that the case study is the only way to tackle the problem, or even that it is the best way! It represents one student's approach and decisions within the limitations of an undergraduate exercise. Do not copy this work or the approach, use it as a guide. Make your own evaluations and decisions, make your own design project a far superior piece of work, and know that it is your own work.

II SOME ADVICE

(A)   General Advice to the Student

As a student faced with a chemical engineering design project, you probably have two immediate feelings. First, excitement at finally beginning the project that has been talked about so often in your department. This excitement is enhanced by finally being able to undertake a piece of work that is both challenging and satisfying, and which will enable you to contribute your own ideas. After so much formal teaching it provides the opportunity to create a finished product that is truly your own work.

The second feeling will probably be apprehension about how this daunting task is to be achieved. Will you be able to do what is required? What level of detail is required? Do you already possess the necessary knowledge to complete the project? Other similar questions probably come into your mind. The simple answer is that design projects have been performed by students in your department since the course began, very few students fail this unit and most produce at least a satisfactory project, and often a better than expected report. Previous students have started the project with the same basic knowledge that you possess and, by asking the same questions, they have completed it using the same resources available to you.

The project supervisor should provide information, assistance and advice. Do not stand in awe of this person, ask what you want/need to know, ask for guidance, and persist until you know what is expected. However, you should understand that a supervisor only provides guidance, and will not (and should not) perform major parts of the design project for you. This is the time for you to show initiative, and to impress the lecturers with your knowledge of chemical engineering and your own ability to solve problems.

My main advice to the student undertaking a chemical engineering design project is: 'Don't work in a vacuum!'. By this I mean obtain information and help from as many sources as you can find. Do not assume that you alone can, or should, complete this project unaided. Talk to the project supervisor, other lecturers in your department, research students, lecturers in other departments and at other universities and colleges, other students, technicians, librarians, professional engineers, officers of the professional institutions, etc. Some of these people may not be able to help, or may not want to; however, it is usually possible to find some helpful and sympathetic persons who can offer advice. The most obvious people to approach are the design project supervisor, other lecturers in the same department, and other chemical engineering students (your peers and research students).

Valuable information can often be obtained from chemical companies and contractors (locally and overseas). The information provided may range from descriptive promotional material, press releases, published technical papers, patents and company data sheets, through to detailed advice and information from company employees. Some of this information, especially the latter, may be provided on a confidential basis. A company may refuse to disclose any information, particularly for new products or processes benefiting from recent advances in technology. The older processes used to produce 'traditional' chemicals are usually well documented in the technical literature. Information concerning new project proposals may have been deposited with government departments, particularly concerning the Environmental Impact Assessment report. Some of this information may be available to the public and can provide valuable data for feasibility studies. It is usually necessary to plan well in advance to obtain company information, particularly from overseas sources.

The completed project should be a testimonial to the student's abilities as a chemical engineer, soon to be employed in industry and eventually to become a recognised professional engineer. The work should demonstrate a breadth of knowledge relating to chemical engineering in general, and an appropriate depth of knowledge in relation to particular chemical engineering design problems that have been tackled. The project should be the student's own work, and must represent an achievement in terms of the application of chemical engineering principles.

In my experience, the 'best' projects are usually produced by those students who are widely read and are interested not only in chemical engineering but also in a wide range of subjects. In this case, 'best' means a competent or satisfactory design and a project that includes consideration of a wide range of relevant factors, not only the technical aspects of equipment design. However, I find that most students, even those with a previously poor academic record, are inspired by the prospect of being able to work on a relatively open-ended problem with the opportunity to produce work that is truly their own. Students in general tend to rise to the challenge rather than merely 'going through the motions'.

(B)   Advice from a Former Design Project Student

Throughout the undergraduate engineering course, you will have heard other students (maybe even yourself) exclaim the catch-cry that: "We will never use this in the real world". The design project is probably as close as you will get to 'real world' engineering while you are still at university. You may be surprised at just how much of the knowledge from previous units is pertinent to the design project. During the design project, you will inevitably encounter aspects of chemical engineering that appear new or puzzling. However, if you review your notes from previous units in chemical engineering fundamentals, unit operations or process control, it is likely that you will find some relevant information that will explain the 'unknown' concept or enable you to locate an appropriate reference which will enlighten you.

The ability to apply knowledge is a very important (maybe essential) attribute for graduate engineers, and one which potential employers will value highly. Take this opportunity to demonstrate to yourself, your supervisor, and potential employers that you have this ability. If you don't (or can't) apply knowledge from previous units and choose another approach to completing your design project (e.g. hoping to find a reference where someone else has done something similar), you are probably not doing a good job.

Four points of advice:

(a) try to do things yourself,
(b) try to be a 'real world' engineer,
(c) work hard, and
(d) enjoy the experience.

It will be challenging, and you will probably lose a few nights sleep, but almost everyone who completes the design project comments that it was the most rewarding and satisfying task that they undertook at university. Go for it!

(C)   To the Lecturer

This book is not intended to be a recommended text for a taught unit in Plant and Process Design: there are several books which already satisfy those requirements, although it will provide useful background reading for that subject. This textbook helps the student performing the chemical engineering design project. It provides only essential notes for a range of associated topics, and the case study material (taken from an actual student design) provides a detailed example of the contents and format of the project report.

Many students are overwhelmed, apprehensive and unsure how to proceed when faced with the design project. It is unlike any assignment they have previously been given and represents a true test of their abilities and initiative. However, too often students spend this initial phase wondering what is actually required and viewing past students' projects, these serve merely to emphasis the enormity of the task ahead rather than providing a detailed analysis of what is needed and a plan of action. This book should satisfy the students' need for guidance, and provide a useful case study example as the project proceeds through each stage.

The case study included (production of phthalic anhydride) is just one particular example of the way in which the project can be performed and presented. Each department (and supervisor) will define their own requirements, but our approach and presentation should not be too different. The emphasis in our course at Curtin University is for effective communications. In the design project report this means presenting only essential information for immediate attention and confining additional information and numerical calculations to appendices (although this book contains more substantial material in the case study sections to make it useful as a teaching text). Summaries are required at the beginning of each sub-section, and as an introduction to each of the two major parts of the report.

In this book we present the design project in two parts. Part I describes the *Preliminary Design* related to aspects of the *Technical and Economic Feasibility Study* for the project. During this stage of the project it is still possible to change the earlier major decisions such as production rate, process route, etc., if certain factors indicate particularly adverse conditions or a more economic alternative. The feasibility study should make recommendations such that the *Detailed Equipment Design* can be performed in Part II. Students sometimes assume that the design is (almost) wholly concerned with the design of equipment (i.e. Part II). However, without a thorough feasibility study to preceed these designs, the project becomes more of an academic teaching (rather than learning) exercise.

Included in our preliminary work for week one of the Design Project (at Curtin University) are the initial technical calculations, mainly the detailed mass and energy balances around the entire process (making appropriate estimates of unknown flows, etc., at this stage — to be checked after the detailed equipment design has been performed). Also included is a pumping system study, which includes a pump specification and selection of an industrial model (this may be included as part of the detailed equipment designs). Other aspects include a plantwide overview of the process control strategy, and environmental and safety considerations.

Part II in this book contains the design of two major items of equipment. They are a **Catalytic Tubular Reactor (Chapter 10), and a Shell & Tube Desuperheater–Cooler–Condenser/After-Cooler (Chapter 11)**. Each design includes the chemical and mechanical design aspects, fabrication considerations, materials specifications, detailed engineering drawing, HAZOP study, control scheme and associated instrumentation. In summary, as complete and professional a design as possible is presented within the time available, while recognizing the student's experience and abilities. Part II includes the design of the vessel shell in accordance with an appropriate pressure vessel code or standard, e.g. AS1210, BS5500, ASME Section VIII. (If the design of one of the two specified items of equipment is not classified as a pressure vessel, then the student is also required to design a pressurized-storage tank or similar item.)

The piping design and pump specification/selection and the pressure vessel design are included because they are common tasks given to young graduate engineers in industry, and they emphasize a practical dimension of the project.

In summary, the design project consists of a detailed technical and economic feasibility study for the process, followed by the detailed design of selected plant items and associated equipment. The production rate, selling price, etc., are determined from a detailed market analysis and economic forecasts. An appropriate process route is selected, followed by the site selection, plant layout, site considerations, detailed economic analysis, mass and energy balances, as outlined in the contents list for Part I.

In some cases, the student is required to accept decisions made by the supervisor, e.g. selection of an older process for which design data is readily available in preference to a newer (secret) process, or choice of production capacity (assuming future export markets) in order to design a plant of significant size which is economically feasible. Although these choices may mean that the design no longer represents the optimum or 'best' design possible, it is the learning experience obtained by performing the project that is important.

(D)   The Designer or Project Engineer

Although design can be taught by a traditional lecturing approach like any other topic, the graduating engineer will only become a 'good' designer if they:

(a) can apply the basic knowledge of chemical engineering;
(b) understand the broad constraints placed on chemical plant design, e.g. economics; environmental, social, etc.;
(c) are widely read, think about the ideas encountered, and use the knowledge and ideas in a design study.

In terms of personal qualities, a good design project student must be:

(i) enthusiastic;
(ii) positive;
(iii) realistic;
(iv) self motivated;
(v) a problem-solver;
(vi) an accurate, careful and logical worker;
(vii) superhuman!

Using these notes and the case study material, and the books and papers published on plant design, the designer or project engineer must *apply* what is known in order to produce a good design. Consider the notes included in each section of this book only as a useful reference source (or a bibliography of essential reading), *not* as a condensed version of everything there is to know or study!

The case study material is based upon the work performed in the Design Project at Curtin University in Western Australia. The project was oriented towards local markets and conditions, and the case study material presented here (mainly for the Feasibility Study) has retained that emphasis.

III   PRESENTATION OF DESIGN PROJECTS

(A)   Effective Communications

Written communications need to be effective. They must convey the intended message in a clear and concise manner. In order to achieve this objective, it is necessary to consider both the audience that will receive the information (and act upon it) and the nature of the information itself. In some situations a formal, fully detailed report is required; however, quite often a condensed form of communication (e.g. memorandum) is satisfactory.

Peters and Waterman (1982) identified several factors that were common to successful American companies. One of these factors was the implementation of a system of effective communications within an organisation. Two of the most successful companies at that time, United Technologies and Proctor & Gamble, required that all communications were in the form of a 'mini-memo' of one page *maximum* length.

In most chemical engineering departments, the length of student design projects tends to increase each year or to have stabilized at a rather voluminous 'norm'. Students refer to previous projects and usually assume that their length is acceptable *and* required. Quite often student projects are unnecessarily lengthy and much of the 'extra' information is attributed to other sources, e.g. Perry (1997), Kirk-Othmer (1978–84; 1991–), etc., and could be replaced by an appropriate reference!

We believe that all student projects, including the design project, should contain only *necessary* information. Extensive background information for a project should be reviewed, summarized and referenced, whereas only new mathematical developments and relevant design equations should be included and referenced to the original source. Essential information should be included in the main body of the report and all additional information, data, calculations, etc., presented in appendices. The design project report should be presented so that it can be assessed by someone with a background in chemical engineering, but without any particular knowledge of the chosen process. The following features are suggested for the written report to facilitate the assessment of the proposed chemical plant design.

(a) A one-page summary at the beginning of the project detailing the project specification, the work performed, major decisions, conclusion, etc. This summary may include both Parts I and II. (An overall case study summary of Part I only is included here, and separate summaries in Chapters 10 and 11.)

(b) A one-page summary for each of Parts I and II, to be included at the beginning of the relevant part of the report.

(c) A summary at the beginning of each chapter or major section of the report (or for a particularly significant topic).

(d) Brief conclusions and recommendations at the end of each chapter or major section.

(e) Information that is not essential for an assessment of the project (but which provides useful or necessary background data) is included in appendices. Company literature, materials specifications, trade statistics, etc., are all presented in appendices, whereas conclusions drawn from this information are presented and discussed in the report itself. Calculations relating to the mass and energy balances are also detailed in an appendix, but the basis of all calculations and the results of these balances are presented as ledger balances within the report.

(f) *Reference rather than reproduce* — the use of appropriate referencing rather than reproducing large sections of readily available information.

(g) Guidelines should be given for the expected length of the report and for the design sections contained in Part II. These guidelines should refer only to the main body of the report; appendices can be as long as is required (within reason!).

The important principle is for clear and concise presentation of the design project report. This approach should make the marking and assessment as easy as possible, and the report should truly reflect the student's own work.

Reference

Peters, T.J., and Waterman, R.H., *In Search of Excellence: Lessons from America's Best-Run Companies*, Harper and Row, New York (1982).

(B) General Comments on Preparation of Literature Surveys

The following notes are intended to provide ideas and suggestions — they do not provide the 'correct' presentation or the only way of presenting literature surveys.

Summary

The literature survey is a review of the important information available which is *relevant* to the particular project topic. It may include information available in books, encyclopedias, lecture notes, journal articles, reports (both government and company), doctoral thesis, standards, patents, and personal communications. It is important to identify and include only relevant information, and to state at the beginning of the survey what type of information is required.

1. Ensure that a literature survey is appropriate for:

   (a) the amount and type of information available;
   (b) the intended audience (find out what the supervisor/marker wants!).

   The literature survey must conform to any formal requirements provided by a company or institution (e.g. for the presentation of a graduate thesis). Find out if any requirements exist, and study previous reports.

2. The literature survey of a report is *not* an essay. It must provide clear and concise bibliographical information about the publications themselves, *and* summarize the relevant information they contain.

3. The literature survey *must* include an introductory statement making clear what type of information is required for the particular project.

4. The presentation of the literature survey will depend upon the type of project being undertaken, e.g. laboratory-based or a theoretical study. The presentation will also depend upon the number of publications to be discussed, i.e. a few papers can be reviewed in detail whereas a large number of publications (usually) require more selective discussion.

5. Suppose that you have a 'reasonable' number of worthy references (say 20–30). These are often discussed in chronological order, perhaps from the most recent back to the earliest (the most recent publications are often the most relevant — but not necessarily so!). A chronological survey is usually presented if general background information is required.

6. Alternatively, the 'best' (i.e. containing the most useful/relevant information) 5 to 10 publications could be discussed first, and then the others possibly in less detail.

7. Sometimes one author is prominent in a particular field and these publications can be reviewed separately as a group.

8. In certain situations it is appropriate to review the available literature under relevant headings, e.g. historical developments, textbooks, etc. This would usually be done when a large amount of information is available.

9. If specific factual or numerical material is required, it is useful to organize the discussion of available literature around the specific points or items in question.

What should the literature review review?

10. The literature review is an *evaluation* of the published material in the field of interest. It is *not* merely a list of references — although that is also needed.
    Select material for your literature review such that it includes:

    (a) sufficient background information;
    (b) the essential publications in the area (i.e. the 'definitive works');
    (c) the major fields of investigation that are relevant to the particular project.

    The review should refer to what is of importance/interest for the particular project. Each evaluation of a publication should detail what the publication contains that makes it worthy of inclusion, e.g. a theoretical development, experimental data, details of equipment design. It may be appropriate to add a subjective judgment of the publication, e.g. good/poor; useful/background; overview/detailed study; etc.

11. A summary of what has been revealed by the literature survey must be included. This may take the form of an extraction of the 'best' data, a summary of pertinent opinion, or a judgment on the state of knowledge in the field. This summary must concur with the objective stated at the outset (see point 3 above).

12. The literature review *must* be supported by a complete and accurate reference list. Each reference (in the reference list) *must* include full publication details. Suggestions regarding the presentation of reference lists, and references to material in a report, are included in points 14 and 19 below.

13. Publications that are not referred to in the literature review (or in the report itself), but which provide useful background information, should be detailed in a separate list entitled 'Further Reading' or 'Additional Bibliography' or similar title.

14. Do *not* include full reference details in the review itself, or as footnotes — use a separate and complete reference list.
    Reference to a publication within the literature review (and in the project report) is usually by means of *either*:

    (a) a reference number (in ascending order throughout the report) and a supporting numbered reference list; *or*,
    (b) reference to the first and second authors names and the year of publication, e.g. Ray (1995) or Ray and Tade (1996) — use Ray *et al.* (1997) for three or more authors. If an author has published several papers in one year, it is necessary to distinguish between them, e.g. Ray (1996A).

    The supporting reference list is prepared in alphabetical order, do *not* use the abbreviation "*et al.*" here — include full details of all authors (including *all* initials). Refer to the suggestions for preparation of references in point 19 below.

15. The best advice is to adopt a referencing method and to present a literature review and the reference list as the supervisor wants them to be — go and ask what is required!

16. Examples of literature reviews/surveys are not included here. Students often assume that by using a particular example as a blue-print, they can produce a near-perfect review. This is seldom true, all literature reviews can be improved. Always prepare a literature review so that is it appropriate for the information to be presented, and acceptable to the person who will mark the report!

17. Only say what is important! Be clear and concise, do not copy or quote extensively from the publications. Excessive verbiage often loses marks.

18. Present your report in an impersonal and technical manner, pay attention to your spelling and English 'style'. Avoid using 'humorous comments', the report should be interesting because of the information it contains and the way it is presented. Let someone read and criticize your first draft, and then make corrections. Present it neatly, legibly and correctly. It may seem unfair, but reviews that are typed correctly (without numerous mistakes) are usually initially better received. Aim the survey at the level of the graduate student. Avoid excessive use of jargon.

Additional Detailed Notes for the Presentation of References and Bibliographies

19. The Reference List (sometimes called Literature Cited or Bibliography) should only contain references that are cited in the text. A list of supporting literature and background or further reading contains literature not cited in the text but which is also relevant to the project.

The Format of References

Any reference quoted must be so specific that the reader could easily locate the information. References to **books** must include: all authors and all initials; the full title and edition; the volume (if more than one); publisher, place and date of publication; any pages cited (either given in the text or included in the full reference). The ISBN is optional but sometimes useful. References to **journal articles** must include: all authors and all initials; the full title of the article, the journal name (or accepted standard abbreviation — see below); volume number (underlined or in italics); part number (in brackets); inclusive page numbers for the article; year of publication.

For standard journal name abbreviations refer to: Alkire, L.G., *Periodical Title Abbreviations; Volume 1: By abbreviation,* 9th edn, Gale Research Inc., Washington DC (1994).

Reference to lecture notes should include the course and unit titles and designation numbers, the lecturer, the particular topic, the date, and any other relevant information. References to anonymous authors, personal communications, or where incomplete information is available, should be clearly stated and as much information as possible included, e.g. contact names and telephone numbers. Material taken from electronic sources (e.g. the World Wide Web) should be clearly referenced such that the information could be traced by the reader.

References in the Reference List should be presented as follows:

Journal Articles:

> Ray, M.S., Brown, A.B., and Jones, C.D., Adsorptive and membrane-type separations: A bibliographical update (1995), *Adsorpt. Sci. Technol.*, 13(6), pp. 433–460, 1996.

Note: Authors initials may be written before the surname; the abbreviation '*et al.*' should not be used in the reference list — all names and initials should be included. Only the standard journal abbreviations should be used, not made-up versions.

Books:

> Ray, M.S., *Elements of Engineering Design: An Integrated Approach*. Prentice-Hall International, Hemel Hempstead, UK (1985); pp. 16–58.

Note: For books, the relevant pages or chapter must be cited in the text if pages are not included with the reference details; author's initials may be written first, i.e. M.S. Ray; this is the first edition.

It is essential to be able to clearly identify specific information referred to in the review. For short publications, usually journal articles where information is presented in a concise form, it may not be necessary to refer to specific pages or sections. For books and other lengthy publications, it is essential to quote the particular pages, chapters, figures, appendices, etc. This may be achieved by a note either in the text or in the Reference List. The former method is illustrated in the three examples given below.

Reference to a specific section of a publication in the Reference List may be achieved in the following ways:

In the review: refer to Ref. 8

In the Reference List, for Ref. 8: Include all publication details and pp. 39–46, or see Chapter 6, etc.

If a separate part of this publication (i.e. Ref. 8) is referred to later in the survey, it is then assigned an appropriate (new) number, say Ref. 15, and the Reference List includes:

15. Refer to pp. 51–55 in Ref. 8 above.

Full publication details should not be repeated within the Reference List.

In the Reference List, if two *consecutive* references are from the same source (book or journal) then the second reference can use the abbreviation '*Ibid.*' rather than repeating the reference details. However, different authors or page numbers must be given.

Numbering and Collating References

References within the report may be cited in one of two ways, either *a numerical system* or alphabetical by first author *surname*. The method chosen should be adhered to throughout the report. In the first method (by assigning numbers), references are numbered in the order in which they are cited throughout the report. In the text, use the following reference notation as preferred:

— as previously demonstrated [5];
— as previously demonstrated (Ref. 5; Figure 3)

If it is useful to identify the author (maybe an expert in the field), Then:

— has previously been demonstrated [5; Ray, pp. 39–44]
— as demonstrated by Ray (5; Chapter 6)

It is probably better to use brackets [8] or (8) for the reference number rather than a raised superscript[8].

For the numerical system, the Reference List is presented in numerical order with the appropriate number before the first author's name. The disadvantage of this referencing system is that any reference included in the report at a late stage requires a renumbering of all subsequent references, and of the Reference List.

In the second method, the reference may be cited by using the first author's surname; any second author is also included, and thereafter '*et al.*' for three or more authors; the date of publication is included, and a suitable designation if the author has published several works. Specific pages, etc., are quoted in the text as shown below. The notation in the text may be:

— as supported by Ray (1995: pp. 18–24)
or
— as supported by Ray (1995B: pp. 18–24)
or
— as supported by Ray and Tade (1996: pp. 19–23)
or
— as supported by Ray *et al.* (1997, ref. B: Chapter 5)

An alphabetical reference list is prepared and included. For the examples given above, the list would contain a reference by Ray and Tade in 1996 and other years, or by Ray *et al.* with two references (at least) in 1997, designated Ref. A and Ref. B in the list. The disadvantage of alphabetical referencing is the extra 'bulk' required in the text compared with use of a number system. However, inclusion of an additional reference does not disrupt the numbering system (although the additional reference must be inserted in the list). The choice of referencing method may be a personal decision, or it may be specified by the company funding the work.

Final Comments

20. Prepare and present material in a manner that is acceptable to the supervisor — go and ask what is required!

Only say what needs to be said — be brief, concise and informative.

Do not quote or reproduce extensive amounts of published material.

Present the review in a clear, structured and logical manner.

Your work will be judged by its presentation, i.e. spelling, legibility and English 'style'. Obtain a critical appraisal of your written work and always aim to improve upon previous reports.

The literature review is an important part of a written project. It contains useful information and is a valuable aid to future readers — *if* it is correctly structured and presented.

The notes presented in this section are intended to provide guidance to the student preparing a literature review for a project report. They should not be considered as a definitive set of rules to be followed in all situations. These notes should be applicable to the presentation of any type of report, e.g. design project, laboratory investigation, theoretical study, etc.

Action: Consult a range of project reports and read the literature review sections. Identify their good and bad points. Prepare a draft literature review for your project. Consider carefully the objectives and the intended use of this review. Let someone who is suitably qualified(?) provide a critical assessment of your review. Re-write the draft review.

IV DETAILS OF PARTICULAR DESIGN PROJECTS, AND INFORMATION SOURCES

(A) IChemE Design Projects

Each year, the IChemE set a design project for external candidates. A copy of the detailed regulations is available, and also, *Notes for the Presentation of Drawings* with an accompanying example of a process flow diagram. A list of the design projects set by the Institution from 1959 to 1986 is included here. Full details are also given below (as an example) for the nitric acid production problem set in 1980. More details of selected projects can be found in Coulson and Richardson (eds), *Chemical Engineering*, Volume 6 (2nd edn, 1993; Appendix G, pp. 885–910). Copies of the information provided for particular projects can be obtained from the IChemE (for a small charge).

Students at Curtin University are provided with a set of guidelines for the design project, including requirements for oral presentations, and a booklet: *Presentation of Literature Surveys* (the main points are reproduced here in Section III).

Table 1 contains a summary of design projects set by the IChemE from 1959 to 1986, more details are included in the actual papers, e.g. process description, design data, references, etc. It should be possible to design these plants using information that is freely available in the technical literature.

Table 1

Year	Chemical	Production (per year)	Process/Specifications
1959	Acetic anhydride (95% in acetic acid)	20,000 longtons	Thermal cracking of acetone
1960	Monomeric styrene (95% purity)	17,500 tons	Catalytic dehydrogenation of ethyl benzene
1961	Methyl chloride (<50 ppm water, only impurity)	6000 longtons	Hydrochlorination of methanol
1962	Butadiene (>98% w/w)	20,000 longtons	Catalytic dehydrogenation of n-butenes; feedstock of liquid mixed hydrocarbon stream containing 80.5 mol% n-butenes, 11.5 mol% n-butane, and 1 mol% of higher hydrocarbons
1963	Crystalline ammonium nitrate (<0.1% moisture)	26,250 longtons	Reaction of nitric acid (47.5% w/w) and liquid ammonia
1964	Commercial formaldehyde (formalin): 37% formaldehyde, 10% methanol, 53% w/w water	8750 longtons	Vapour-phase autothermal catalytic oxidation of methanol
1966	Acetone (99.5 wt%)	26,250 tonnes	Vapour-phase dehydrogenation of isopropanol (85.9 wt% plus 12.8% water)
1968	Aniline (99.9% w/w min)	20,000 longtons	Hydrogenation of nitrobenzene (copper on silica gel catalyst)
1969	Amine penicillin salt	10,000 kg	From a fermentation broth containing 5000 units/ml whole broth
1970	Chlorine	10,000 longtons	Catalytic oxidation of HCl gas
1971	Hydrogen (95% purity)	20 million standard cubic feet per day (0.555×10^6 standard m^3/day)	Partial oxidation of a heavy oil feedstock
1972	Urea	100,000 tonnes (metric)	Reaction of ammonia and carbon dioxide at an elevated temperature and pressure, using total-recycle process and CO_2-feed stripping
1973	Styrene butadiene rubber (SBR) latex	25,000 tonnes	Continuous isothermal reaction (5°C) in a series of reactors (33 m^3 capacity each)
1974	Methyl ethyl ketone	10,000 tonnes	Catalytic oxidation of secondary butyl alcohol
1975	Acrylonitrile	100,000 tonnes	Fixed-bed catalytic reactor for ammoxidation process for propylene and ammonia reaction
1976	Monochlorobenzene: Dichlorobenzene:	20,000 tonnes 2000 tonnes	Direct chlorination of benzene
1977	2-Ethylhexanol	40,000 tonnes	Reaction of propylene and synthesis gas
1979	Gas oil	100,000 tonnes	Catalytic hydrogenation process for sulphur removal
1980	Nitric acid (62 wt% HNO_3)	100,000 tonnes	Oxidation of ammonia
1981	Vinyl acetate	50,000 tonnes	Vapour-phase reaction of acetylene and acetic acid
1982	Trichloroethylene: Tetrachloroethylene: Hydrogen chloride:	40,000 tonnes 7000 tonnes by-product	Chlorination and cracking stages

Table 1 (*Continued*)

Year	Chemical	Production (*per year*)	Process/Specifications
1983	Sulphuric acid (98%)	400,000 tonnes	Sulphur-burning process, followed by catalytic oxidation of SO_2 (vanadium pentoxide catalyst)
1984	Substitute natural gas	*Feed:* 600 tonnes per day of coal	Gasification reactor for processing of bituminous coal
1985	Ethylene	30,000 tonnes	Vapour-phase catalytic dehydration of ethanol
1986	Methanol	2000 tonnes	From natural gas by steam reforming and low-pressure synthesis

Instructions for the IChemE Design Project, 1980

The following information is reproduced by permission of The Institution of Chemical Engineers (UK).

Before starting work read carefully the enclosed copy of *The Regulations for the Design Project* in conjunction with the following details for the Design Project for 1980.

In particular, candidates should note that all the questions should be answered in the section headed 'Scope of Design Work Required'.

The answers to the Design Project should be returned to The Institution of Chemical Engineers, 165–171 Railway Terrace, Rugby, CV21 3HQ, by 17.00 hours on December 1st, 1980. In the case of overseas candidates, evidence of posting to the Institution on November 30th will satisfy this requirement. The wrappings must be marked on the OUTSIDE with the candidate's name and words: 'DESIGN PROJECT'.

The Design Project will be treated as a test of the ability of the candidate to tackle a practical problem in the same way as might be expected if he/she were required to report as a chemical engineer on a new manufacturing proposal. The answers to the Design Project should be derived by the application of fundamental principles to available published data, they should on no account include confidential details of plant or processes which may have been entrusted to the candidate. Particular credit will be given to concise answers.

References must be given to all sources of published information actually consulted by the candidate.

The answers should be submitted on either A4 or foolscap paper, but preferably on A4. Squared paper and drawing paper of convenient size may be used for graphs and drawings respectively. The text may be handwritten or, preferably, typewritten; in the latter case it is permissible for another person to type the final copies of the answers. Original drawings should be submitted. Copies, in any form, will not be accepted.

Each sheet and drawing must be signed by the candidate and this signature will be taken to indicate that the sheet or drawing is the candidate's unaided work, except typing. In addition, the declaration forms enclosed must be filled in, signed, witnessed and returned with the answers. The manuscript, drawings and any other documents should be fastened in the folder supplied, in accordance with the instructions appearing thereon.

Answers to the Design Project itself must be written in the English language and should not exceed 20,000 words excluding calculations. The use of SI units is compulsory.

Candidates may freely utilize modern computational aids. However, when these aids are employed, the candidate should clearly indicate the extent of their own contribution, and the extent of the assistance obtained from other sources. For computer programs which have been prepared by the candidate personally, a specimen print-out should be appended to the report. Programs from other sources should only be used by the candidate provided adequate documentation of the program is freely available in recognised technical publications. The candidate must demonstrate clearly that he/she fully understands the derivation of the program, and the significance and limitation of the predictions.

The answers submitted become the property of the Institution and will not be returned in any circumstances.

This following problem was used in the 1st edition of this book and is reproduced here for example, even though the case study material has now been changed to phthalic anhydride (which has not been used by the IChemE).

1980 Design Project (IChemE)

Design a plant to produce 100,000 tonnes/year of nitric acid assuming an operating period of 8000 hours per year on stream.

Process Description

The process consists essentially of the oxidation of ammonia followed by the absorption of nitrogen oxides in water.

A gaseous mixture of ammonia and primary air at approximately atmospheric pressure is preheated before passing to a catalytic reactor where the ammonia is oxidized to nitric oxide. The gases leaving the reactor are cooled in a low-pressure heat-exchange system (which also serves to generate steam and preheat the ammonia/air feed stream) before passing to a water-cooled condenser. The condensate is mixed with a process water stream before entering the top of an absorption column (see below). Secondary air is added to the residual gases and the combined gaseous stream enters a compressor.

Using waste gases and cooling water, the compressed gases are cooled in a high-pressure heat-exchange system, where most of the nitric oxide is oxidized to nitrogen dioxide and most of the water vapour condenses to produce weak aqueous nitric acid.

The residual gas and condensate streams pass, separately, to a tray column where the remaining oxides of nitrogen are absorbed in water. The waste gases leaving the top of the absorption column are used to cool the compressor discharge stream before undergoing expansion in a turbine and final discharge as stack gases. The liquid nitric acid product leaves from the bottom of the column.

Feed Specification

(i) Pure gaseous ammonia is available from storage at atmospheric pressure and 303 K.

(ii) Atmospheric air, suitably dried, is available at 303 K.

Product Specification

(i) Nitric acid: 62% HNO_3 (by weight) solution in water.

Operating Parameters

(i) *Excess air.* The 'total air' used by the process is 11% in excess of that based on the overall equation: $NH_3 + 2O_2 = HNO_3 + H_2O$
'total air' = primary air + secondary air.

(ii) *Ammonia oxidation reactor.* Primary air : NH_3 ratio = 10 : 1 (molar).
Temperature = 1120 K; pressure = atmospheric.
Catalyst: platinum-rhodium gauze.
Conversion: 98% to nitric oxide, 2% to nitrogen.

(iii) *Low-pressure condensation.* Condensate and residual gas streams leave at 303 K.

(iv) *Compression.* Discharge pressure = 8.0 bar.
It may be assumed that no oxidation of nitric oxide occurs during compression.

(v) *High-pressure cooling.* Oxidation of nitric oxide occurs according to the reaction: $2NO + O_2 \rightarrow 2NO_2$ and the nitrogen dioxide is in equilibrium with its dimer, N_2O_4. The condensate may be assumed to be a 50 wt% aqueous solution of nitric acid consistent with the overall equation: $4NO_2(g) + 2H_2O(l) + O_2(g) \rightarrow 4HNO_3(l)$.
Both condensate and residual gas streams leave at 303 K.

(vi) *Absorption.* Isothermal operation at 303 K. The total amount of oxides of nitrogen in the gas stream is to be reduced to less than 1500 parts per million by weight.

Utilities

(i) Saturated steam at 18 bar and 2 bar.

(ii) Cooling water at 293 K and 4.5 bar (at ground level). Maximum allowable discharge temperature is 320 K.

(iii) Process water, boiler feed water, instrument air, inert gas, electricity and liquid ammonia refrigerant are all available at the conditions required by the plant.

Scope of the Design Work Required

Answer *all* the following sections. Candidates will be expected to show full calculations in support of Sections 1 and 2.

1. *Process Design*

   (a) Prepare material balance and energy balance flow diagrams for the entire process showing a tabulated summary of the process stream flowrates (kg) and compositions (wt%) on *a basis of one hour*. Also indicate all heat exchanger duties (kW), stream pressures (bar) and temperatures (K).

   (b) Prepare a process flow diagram for the plant showing all major items of equipment, in approximately correct elevation relative to each other, together with a suitable control and instrumentation scheme. Indicate all utilities requirements and the nominal size of all major pipelines. Detailed design of the compressor and turbine are not required.

2. *Chemical Engineering Design*
   Prepare a detailed chemical engineering design of the absorption column assuming that sieve trays are used.

3. *Mechanical Design*
   Prepare mechanical design sketches of the absorption tower suitable for submission to a draughtsman, paying particular attention to the tray layout and any associated cooling equipment.

4. *Loss Prevention*
   A full-scale Design Project would include an operability study followed by appropriate hazard analysis. For the purpose of this examination, candidates are only required to make recommendations to minimize environmental pollution during operation of the plant.

Data

(i) *Velocity constant for nitric oxide oxidation* (k_1)

$$2NO + O_2 \xrightarrow{k_1} 2NO_2$$

There is a discrepency between References 1 and 2 in the value given for k_1 and the following expression should be used in this design project:

$$\log_{10} k_1 = \frac{641}{T} - 0.725$$

(ii) *Heat of solution of nitric acid in water at 298 K* (ΔH_S^0)

$\dfrac{mol\,H_2O}{mol\,HNO_3}$	1	2	3	4	5	6	8	10	15	20	30	40	50	100
(ΔH_S^0) (kJ/mol HNO$_3$)	13.1	20.1	24.3	27.0	28.8	29.9	31.2	31.9	32.5	32.7	32.8	32.8	32.8	32.8

References

1. Chilton, T.H., *Chem. Eng. Prog. Monograph Series*, **55**, No. 3 (1960).
2. Solomon, C.H. and Hodges, A.W., *Brit. Chem. Eng.*, **8**, 551 (1963).
3. Bump, T.R. and Sibbitt, W.L., *Ind. Eng. Chem.*, **47**, 1665 (1955).
4. Nonhebel, G., *Gas Purification Processes for Air Pollution Control*, Chapter 5, Part B: Absorption of Nitrous Gases, Newnes-Butterworths (1972).
5. Sherwood, T.K. Pigford, R.L. and Wilke, C.R., *Mass Transfer*, Chapter 8, McGraw-Hill (1975).

(B)   Information Sources

Selected Books

The following books are arranged alphabetically by author, they all have their good points but all could be improved in several aspects — as could even the best design! These books are mainly concerned with a range of aspects of chemical engineering plant and process design, rather than particular operations such as distillation, reactor design, etc. Books describing specific chemical engineering topics are included in Chapters 8 and 9.

Aerstin, F. and Street, G., *Applied Chemical Process Design*, Plenum Press, New York (1978).

Austin, D.G. and Jeffreys, G.V., *The Manufacture of Methyl Ethyl Ketone from 2-Butanol (A Worked Solution to a Problem in Chemical Engineering Design)*, The Institution of Chemical Engineers (UK) and George Godwin Ltd, London (1979).

Basta, N., (Ed.), *Shreve's Chemical Process Industries*, 6th edn, McGraw-Hill, New York (1994).

Baasel, W.D., *Preliminary Chemical Engineering Plant Design*, 2nd edn, Van Nostrand Reinhold, New York (1990).

Backhurst, J.R. and Harker, J.H., *Process Plant Design*, Elsevier, New York (1973).

Briggs, M. (ed.), *Decommissioning, Mothballing and Revamping*, IChemE, UK (1996).

Cook, T.M. and Cullen, D.J., *Chemical Plant and its Operation (Including Safety and Health Aspects)*, 2nd edn, Pergamon Press, Oxford (1980).

Coulson, J.M., Richardson, J.F. and Sinnott, R.K., *Chemical Engineering, Volume 6: An Introduction to Chemical Engineering Design*, 2nd edn, Pergamon Press, Oxford (1993).

Douglas, J.M., *Conceptual Design of Chemical Processes*, McGraw-Hill, New York (1988).

Edgar, T.F. and Himmelblau, D.M., *Optimization of Chemical Processes*, McGraw-Hill, New York (1987).

Felder, R.M. and Rousseau, R.W., *Elementary Principles of Chemical Processes*, 2nd edn, John Wiley, New York (1986).

Horsley, D. and Parkinson, J. (eds), *Process Plant Commissioning: A user guide*, IChemE, UK (1990).

Hussain, A., *Chemical Process Simulation,* Halstead Press, New York (1986).

Kirk-Othmer Encyclopedia of Chemical Technology, 4th edn in production (1991–); 3rd edn, 25 volumes (1978–84), John Wiley, New York.

Landau, R. and Cohan, A.S., *The Chemical Plant*, Van Nostrand Reinhold, New York (1966).

Luyben, W.L. and Wenzel, L.A., *Chemical Process Analysis: Mass and Energy Balances*, Prentice-Hall, New Jersey (1988).

Mecklenburgh, J.C., *Process Plant Layout*, Halstead Press, New York (1985).

Mill, R.C. (ed.), *Human Factors in Process Operations*, IChemE, UK (1992).

Perry, R.H. and Green, D.W. (eds), *Perry's Chemical Engineers' Handbook*, 7th edn, McGraw-Hill, New York (1997).

Peters, M.S. and Timmerhaus, K.D., *Plant Design and Economics for Chemical Engineers*, 4th edn, McGraw-Hill, New York (1991).

Raman, R., *Chemical Process Computations*, Elsevier, New York (1985).

Ray, M.S., *Elements of Engineering Design*, Prentice-Hall, UK (1985).

Ray, M.S., *The Technology and Applications of Engineering Materials*, Prentice-Hall, UK (1987).

Ray, M.S., *Engineering Experimentation: Ideas, Techniques and Presentation*, McGraw-Hill, New York (1988).

Ray, M.S. and Johnston, D.W., *Chemical Engineering Design Project: A Case Study Approach*, Gordon and Breach Publishers, UK (1989).

Ray, M.S., *Chemical Engineering Bibliography (1967–1988)*, Noyes Publications, New Jersey, USA (1990).

Ray, M.S., *Chemical Process and Plant Design Bibliography (1959–1989)*, Noyes Publications, New Jersey, USA (1991).

Rudd, D.F., *Process Synthesis*, Prentice-Hall, New Jersey (1973).

Scott, D. and Crawley, F., *Process Plant Design and Operation: Guidance to Safe Practice*, IChemE, UK (1992).

Scott, R. and Macleod, N., *Process Design Case Studies*, IChemE, UK (1992).

Townsend, A. (ed.), *Maintenance of Process Plant*, 2nd edn, IChemE, UK (1992).

Ulrich, G.D., *A Guide to Chemical Engineering Process Design and Economics*, John Wiley, New York (1984).

Van den Berg, P.J. and De Jong, W.A. (eds), *Introduction to Chemical Process Technology*, Reidel-Holland, New York (1980).

Vilbrandt, F.C. and Dryden, C.E., *Chemical Engineering Plant Design*, McGraw-Hill, New York (1959).

Wells, G.L., and Rose, L.M., *The Art of Chemical Process Design*, Elsevier, New York (1986).

Both the American Institute of Chemical Engineers (AIChE), New York, and the Institution of Chemical Engineers (IChemE), UK, publish a wide range of symposium series, books, design guides, pocket guides, user guides, standards, directories, procedures for equipment testing, etc. It would be useful to obtain a catalogue of the publications from each institution.

Chemical Engineering (published by McGraw-Hill, New York) publishes reprints of particular articles on selected topics, e.g. distillation, absorption, design, etc. In addition to these reprints of small groups of related published papers, bound volumes containing usually 100 or more relevant papers are also available, titles include Physical Properties, Process Technology and Flowsheets (Volumes I and II), Capital Cost Estimation, Process Heat Exchange, Modern Cost Engineering, etc. More recent volumes include Instrumentation and Process Control, Heat Transfer Technologies and Practices, Safety and Risk Management Tools and Techniques (all published in 1996).

Journals

The following journals are useful to the process engineer, and the first five are probably the best sources of data for the student undertaking a chemical engineering design.

Chemical Engineering (New York) — published monthly, the 'feature articles' provide excellent updates and overviews of particular topics. Single article reprints and bound volumes of selected papers are also available (see above).

Hydrocarbon Processing — excellent flowsheets, thermodynamic data series, and major articles.

Chemical Engineering Progress — published by the AIChE (New York), articles concerning engineering and technical subjects. Also produces a symposium series of volumes on selected topics.

The Chemical Engineer — published monthly by the IChemE (UK).

Chemical Week and *Chemical and Engineering News* and *Chemical Marketing Reporter* — provide facts and figures for the chemical industry.

Ammonia Plant Safety

Chemtech

Chemistry and Industry

Environmental Progress

Gas Separation and Purification

Industrial and Engineering Chemistry Research (this journal replaced the journals: *Industrial and Engineering Chemistry Process Design and Development* and *Industrial and Engineering Chemistry Product Research and Development* which ceased publication in 1986).

Journal of Chemical and Engineering Data

Journal of Physical and Chemical Reference Data

Oil and Gas Journal

Powder Handling and Processing

Process Engineering

Process Safety Progress

Separation and Purification Methods
Separation Science and Technology
Separations Technology

The following journals feature papers concerned with research studies:

AIChE Journal
Biotechnology Progress
Canadian Journal of Chemical Engineering
Catalysis Reviews in Science and Engineering
Chemical Engineering and Processing
Chemical Engineering Communications
The Chemical Engineering Journal
Chemical Engineering Research and Design (formerly the Transactions of the IChemE)
Chemical Engineering Science
Computers and Chemical Engineering
Developments in Chemical Engineering and Mineral Processing
Energy and Fuels
Food and Bioproducts Processing (IChemE)
Fuel
Industrial and Engineering Chemistry Fundamentals (ceased publication in 1986)
International Chemical Engineering (ceased publication in 1994)
International Journal of Heat and Mass Transfer
Journal of Chemical Technology and Biotechnology
Journal of Food Engineering
Journal of Heat Transfer (Transactions of the ASME)
Journal of the Institute of Energy
Journal of Loss Prevention in the Process Industries
Powder Technology
Process Safety and Environmental Protection (IChemE)
Solvent Extraction and Ion Exchange
Journals dealing with specific subjects are also available, e.g. *Cost Engineering, Plastics World*, etc.

Other Sources

Chemical Abstracts and *Engineering Index* provide a useful source of data from the technical literature (chemistry and engineering) published worldwide. Other indexes associated with metals, mining, etc., are also available. The chemical engineering journal literature is accessible through the *CHERUB* database on the *Engineering & Applied Science CD-ROM* (see References after Section 2.5). Information can also be obtained from symposium series, conference proceedings, company literature, and patents.

As discussed later in Section 2.3, there is a vast quantity of information available on the Internet and through the *World Wide Web*. No doubt this will only expand with time. However, the reliability of this information needs to be considered *before* it is used, and also the time required to search for information and then to analyse large amounts of data. Perhaps this time could actually be better spent in performing design calculations with reference to the more traditional sources? Many previously inaccessible sources of information will become available for personal searching, e.g. world patents, and this resource should be used — *if* it will allow time to be spent more effectively and *if* a better design can be produced.

PART I

TECHNICAL AND ECONOMIC
FEASIBILITY STUDY

(References are included within each chapter. References for general information sources are in Section IV, and design books for various topics are included in Chapters 8 and 9.)

1. THE DESIGN PROBLEM

1.1 INITIAL CONSIDERATIONS AND SPECIFICATION

A chemical engineering design project *does not* follow a set of standard steps similar to the familiar undergraduate textbook problems, nor does it have a single 'correct' solution. The considerations in a design project are many and varied. The solution that is finally accepted is (usually) the 'better' solution (often based upon economic considerations) from several alternatives. *The important feature of a design study is that decisions must be made at every stage, and compromises are frequently required.* It is also necessary to reconsider various decisions in the later stages of a project, and to re-work solutions and re-evaluate proposals.

1.1.1 The Feasibility Study

The feasibility study for a design project involves mainly the technical and economic evaluation of the process in order to determine whether the detailed design (and subsequent construction) stages should be undertaken. There are many aspects that need to be considered, such as legal implications, political influences, etc., and these should be considered before the expensive detailed design stage commences.

An undergraduate design project is not unlike the work of a project engineering team in industry. There is a problem to be solved, there is a project manager (the lecturer-in-charge), a team of project engineers (the students, or perhaps one student), and various deadlines to be met. The initial task is to define the *actual* problem; depending upon the information provided by the lecturer (see later), the problem may be well-defined or it may be quite unclear initially exactly what is required. The first task in any project is to consider the options and alternatives and to gradually eliminate those which are not feasible or not required, and then evolve a plan and hence a project definition. The approach taken is often to try and define the project as quickly as possible in order to proceed with the "real work". While time management is an essential attribute, the rush to eliminate options can lead to problems and revisions later in the project and it is worthwhile keeping the number of useful options open as long as possible. Students often complain initially that they do not know what is required or what is possible, the answer in both cases is **everything**! Consider all possibilities, and even options that don't seem possible! Decide when decisions will need to be made, and at that time eliminate the least attractive options and eventually arrive at a feasible project definition.

1.1.2 Time Management

An essential ingredient of all successful projects is realistic time management. Once a project is defined it is necessary to identify the associated tasks, the required deadlines, and the resources available. Undergraduate projects have the same requirement; and without appropriate time allocations to specific tasks then the project will either not be completed on time or it will be an inferior piece of work. It is necessary to look to the end of a project at the beginning of the work, rather than focussing on the day-by-day or week-by-week tasks. An overall view is required in the early stages of a project proposal, and the essential/key tasks, decisions and bottlenecks need to be identified. Appropriate actions and planning can then be incorporated in the time-plan for the project. It is also necessary to be flexible and to modify the timing of the project stages as the work proceeds. However, it is the unforeseen and unexpected events that are usually most crucial!

1.1.3 Stages in a Design Problem

Many textbooks have described and analysed the design process (e.g. Ray, 1985), and whatever approach is taken there are several common steps. The following six steps in the design of a chemical process have been identified (see Ulrich, 1984):

1. Conception and definition.
2. Flowsheet development.
3. Design of equipment.
4. Economic analysis.
5. Optimization.
6. Reporting.

This is obviously rather cursory and in need of further consideration. The student would benefit from reading the introductory chapters in one or more of the popular texts on plant and process design (e.g. Peters and Timmerhaus, 1991; Ulrich, 1984; Baasel, 1990). The importance of a clear (but open) project definition and consideration of the constraints and limitations cannot be over-emphasised. There are many aspects and stages in the technical and economic feasibility study which preceed the detailed equipment design work. There is often the need for compromise, and for optimization of decisions at both the local (or individual problem) level, and in terms of the overall (or big picture) view of a project. Some of these appreciations come with experience and some can be gained by wider reading, and some by keeping an open mind! Further consideration of the stages in a project and the approach to design will be left to the student and the lecturer to explore.

1.1.4 The Search for Information

The first step in a design project is to try and identify (all?) the relevant information that is available. The second step is to identify the information *required*, decided after initial consideration of the problem. The aim should be for quality rather than quantity of information, i.e. a few useful/relevant references rather than hundreds of general references. Two possibilities exist:

(a) Not enough information is available — a search is required (see Sections IV and 2.3 for ideas regarding sources of information).
(b) Too much information is available — the task is to assess the reliability of conflicting information.

It is unlikely that *all* the required information is initially available, or that the data is sufficient to complete the project (except for university teaching assignments!). *Remember*: Published information is not necessarily correct!

The assessment and definition stage of a project is often either rushed, overlooked or postponed by eager students, such an approach usually leads to wasted time and effort later in the design. Make it a rule to *know* what you have and where you are going, rather than simply *thinking* that you know.

1.1.5 Scope of the Project

Undergraduate design projects are often well-defined, although this need not be the case. The design project set by the *IChemE* (UK) is for the design of a process for the production of a particular chemical at a given production rate, and the design of particular items of equipment (see Section IV). In universities, the chemical(s) to be produced is (are) specified and the following information may be given (depending upon the philosophy of the supervisor):

(a) production rate (plant throughput);
(b) purity of final product;
(c) raw materials to be used/available;
(d) utilities available;
(e) site location;
(f) expected markets;
(g) the process route.

Some or all of this information may be expected to be obtained by the student as part of the design project. Industrial process designs are usually (although not always) more clearly defined — 'this is what we want, this is what we know, now decide how to achieve it (and make a profit)'.

1.1.6 Evaluating the Alternatives — Making Decisions

Probably by the end of the first (or second) week of a project a wide range of options and alternatives will have been considered, some will have been rejected quite quickly. There may still be several alternatives under consideration but the general scope and constraints of the project should have been identified. A detailed plan should evolve within the first three weeks, and details of the main tasks to be completed within the time allocations should be prepared.

Some Questions to Ask for the Chemical to be Produced	
How is it made?	Alternative processes
Uses	New applications
Current sales/production — expanding or declining?	Raw materials
Main product or by-product?	Environmental problems?
Political influences?	Utilities
Special site location requirements	Storage/handling
Etc., etc.	

Further Reading

Brennan, D.J., Chemical engineering design and undergraduate education, *Chem. Eng. Australia*, 20(3), 19–22 (1995).

Crowther, J., Feasibility studies, *Bull. Proc. Australasian Inst. Min. Metall.*, August, 14–20 (1985).

Harding, J.S., A crash course in project engineering, *Chem. Eng. (N.Y.)*, July, 118–126 (1995).

Nowicki, P.L., Rapid application planning for project success, *Chem. Eng. (N.Y.)*, November, 123–124 (1996).

Ray, M.S., *Elements of Engineering Design: An Integrated Approach.* Prentice-Hall International, Hemel Hempstead, UK (1985); pp. 16–58.

Sloma, M., Process design: A winning service industry, *Chem. Eng. Australia*, 20(3), 10–18 (1995).

Spitz, P.H., Handling your process engineering, *Chem. Eng. (N.Y.)*, 23 October, 175–180 (1967).

Ulrich, G.D., *A Guide to Chemical Engineering Process Design and Economics*, John Wiley, New York (1984).

Action: *Identify useful sources of information and their location. Obtain personal copies of essential/useful information. Initiate a filing and reference system for useful information (preferably on a computer database). Prepare a complete design specification (as detailed as possible at this stage, but leave some options open). Identify the essential information that is available. Identify the information that is required. Obtain all necessary information. Prepare a detailed plan for the work to be completed over the next eight(?) weeks.*

Note: The case study material in this book is based upon the work performed for the Design Project at Curtin University in Western Australia. The project was oriented towards local markets and conditions, and the case study material presented here (mainly for the Feasibility Study) has retained that emphasis. The Case Study provides an example, not a blue-print to be followed, each project should be approached on the basis of its own particular needs and problems and not seen as a copy of an existing report.

CASE STUDY: PRODUCTION OF PHTHALIC ANHYDRIDE

Overall Summary for the Technical and Economic Feasibility Study

Phthalic anhydride (PAN) is a white crystalline solid which is the first in the series of cyclic anhydrides. In the molten state, it is a clear liquid which resembles water. Although its normal boiling point is 280 °C and its normal melting point is 131 °C, the vapour has a strong tendency to exist at temperatures below the boiling point and to sublime. It is toxic at low levels (2 ppm), a severe skin irritant and can form explosive mixtures with air at concentrations of 1.7–10.5%.

Ortho-xylene is now the primary feed material, having replaced naphthalene in the 1970s. A fixed-bed, catalytic oxidation process using air is preferred by most current manufacturers. The principal uses for PAN include plasticisers, polyester resins, alkyd resins and dyes. Global production of PAN has grown only slowly in recent years.

Phthalic anhydride is currently produced in Australia by one company, at two small plants in New South Wales. Both of these plants were built in the 1960s and use naphthalene as the feed and outdated technology. Several large producers of PAN operate in Japan, Korea and Taiwan. These plants, and several others in the region, satisfy the local and regional demand for PAN and most plants are operating at 20–40% below capacity.

The LAR process for PAN production is the most recent technology and is capable of producing PAN at less than the current market price of A$0.95–A$1.00 per kg. However, after capital repayments and a realistic profit margin is considered, the likely selling price for a new plant based in Western Australia could be as high as A$1.45. A plant capacity of 30 kT/yr should satisfy the Australian demand for PAN and provide some margin for exports. However, a capacity of up to 60 kT/yr could also be viable if appropriate markets are found. The capital costs of a 30 kT/yr plant was estimated to be A$25–A$70 million while the capital cost of a 60 kT/yr plant was estimated to be A$41–A$105 million. The overall viability of a new PAN plant is likely to depend on the establishment of downstream processing near the development.

Within Western Australia, a site near Kemerton in the state's south-west was preferred economically and politically. The development will be well placed to receive government support to assist the decentralisation of major industry. Highly developed infrastructure and transport facilities are either available or planned for the selected area, and the local boat building industry is a potential market for the product.

The environmental impact of a new plant is anticipated to be small although there will be certain hazards involved in the day-to-day operation of the plant. Some air and water emissions are expected. A PAN plant which uses the LAR process is also able to export electricity to the local grid or surrounding industries.

Note: *All the references for the Technical and Economic Feasibility Case Study are included after Section 8.4.6.*

1.2   CASE STUDY — DEFINING THE PROBLEM AND BACKGROUND INFORMATION

Summary

Phthalic anhydride is an intermediate petrochemical used for the manufacture of plasticisers (especially for PVC), alkyd resins, polyesters and dyestuffs. It is traded in either molten form (requiring heated tankers) or as a white powder ('flake'). *Ortho*-xylene (or *o*-xylene) is the primary raw material although naphthalene has been used widely in the past. The manufacture of phthalic anhydride consumes almost all *o*-xylene produced so that *sourcing a suitable supply of high purity o-xylene is a critical consideration* in the early stages of the feasibility study. Fixed bed and fluidised bed reactors are both viable alternatives as both can produce high yields by using appropriate specific catalysts. Care must be taken in the process handling stages as explosive mixtures form easily between phthalic anhydride dust and air, and spontaneous reaction occurs with many organic compounds including skin tissue.

1.2.1   Background and Objectives

The first stage of a feasibility study for the design of a chemical plant is the identification of key chemical properties, salient market issues and potential production difficulties associated with the chemical being produced. The main applications and uses of the chemical should be summarised and current market demand determined. Possible process routes may also be identified at this stage and assessed at an elementary level for viability. This section is intended to establish a general 'feel' for the chemical.

1.2.2   Chemical Structure and Physical Properties

Phthalic anhydride (PAN) is an anhydrous acid-derivative formed by partial oxidation of *o*-xylene:

At room temperature PAN is a white, crystalline solid; it melts at 131.6 °C and boils at 284.5 °C. It has a tendency to sublime when condensing which causes problems in recovery systems. In the molten state it is 10–20% heavier than water and has a similar viscosity (0.55–1.2 cP) [1–3].

It is a dangerous chemical and forms explosive mixtures with air at concentrations of 1.7–10.5 vol% at temperatures above 140 °C [2,4]. It is toxic at levels above 2 ppm and is a severe irritant to the eyes, respiratory tract, skin and moist tissue [1].

1.2.3 Applications and Uses

About 50% of all phthalic anhydride produced is used in the manufacture of plasticisers for PVC processing. Several other uses are also significant: unsaturated polyester resins (15–20% of all PAN produced); alkyd resins (10–15%); and dyes including phenolphthalein and anthraquinone [1,5], as shown in Figure 1.1. Plasticisers are formed by reaction with alcohols. Chemical combination with maleic anhydride and glycols forms the prepolymer for unsaturated polyester resins [3].

In Western Australia, phthalic anhydride would be used mainly in the form of plasticisers for the building industry and a potential automotive industry. Alkyd resins could be used in the boat building industry. There is unlikely to be a sufficient market for phthalic anhydride-based dyes in Western Australia.

Worldwide production has steadily, but slowly (3–5% p.a.), increased since the 1950s with current production now approaching the estimated capacity of 3500 kT per year after a long period of surplus capacity. The selling price has varied widely due to fluctuating demands and production rates, but is currently around US$0.80–$1.00 per kg [1,6].

1.2.4 Basic Chemistry

The oxidation (under controlled conditions) of either *o*-xylene or naphthalene will produce phthalic anhydride with the co-evolution of water. The pathway is via

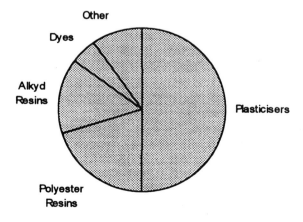

Figure 1.1 Major uses of phthalic anhydride.

ortho-phthalic acid (*o*-PA). If naphthalene is used, then carbon dioxide is a co-product (thus reducing the maximum yield per kg of feed).

The synthesis reaction can be conducted in either the liquid phase or gas phase and must utilise a catalyst to achieve high product selectivity. *The reaction is highly exothermic and a means of dissipating the evolved heat must be incorporated into the reactor design.* The maximum stoichiometric yields are 1.39 kg of phthalic anhydride per kg of *o*-xylene, and 1.15 kg of PAN per kg of naphthalene. Commercial yields are about 1.09 and 0.95 kg, respectively [1,7,8].

Phthalic anhydride can be reacted to form esters, salts and acid chlorides. The benzene ring will undergo halogenation and sulphonation. Other important reactions occur with benzene (to form anthraquinone derivatives by condensation), phenol (to form phenolphthalein, also by condensation) and urea and metal diacetates (to form metal phthalocyanines) [3].

1.2.5   Evaluation of Alternative Processing Schemes

There are three possible feeds for use in a phthalic anhydride plant: *o*-xylene; naphthalene; or *n*-pentane. The traditional source is coal-tar naphthalene but the most common feed is now *o*-xylene. The first patent (BASF in 1896) used concentrated sulphuric acid in the presence of mercury salts to perform the oxidation. The first vapour-phase process was patented in 1917. Since then, there have been great improvements in the catalysts used, both in terms of increased yield and reduced catalyst aging. Potassium-modified vanadium pentoxides (with promoters such as molybdenum and manganese oxides) are now the most common catalysts used in PAN production. The *n*-pentane process is the most recent development but requires a cheap supply of raw material in order to be a viable option [1–3,5,7,9–13].

The main factors affecting the choice of feed stock are the yield (*o*-xylene is higher yielding as all carbon atoms appear in the product), availability (naphthalene sources are becoming rarer) and cost (naphthalene is only available at a competitive cost in an impure form which is less suitable). Catalyst aging, utility costs (*o*-xylene normally requires higher temperatures) and product quality (a wider range of by-products are possible using naphthalene) also have potential to influence the final choice. *If o-xylene is preferred, a suitable source must also be determined.* As nearly all the *o*-xylene produced is used for phthalic anhydride manufacture there is unlikely to be an adequate supply in the general market for a new plant. Therefore, a xylene separation plant, using a mixed xylene feed, is often constructed on the same site as a PAN plant.

Apart from raw material selection, *the main design consideration is the type of reactor.* Both fixed-bed and fluidised-bed processes are viable for the gas phase process. The oxidation reaction can be conducted at both high and low air ratios, which has significant implications for the design of the reactor and downstream equipment. Plants throughout the world utilise a range of technologies, including the liquid phase process which appears to be less viable.

The product typically needs *to be purified to around 99.7%* [1,3]. The major contaminants are usually maleic anhydride and single-ring aromatics. Both the raw materials and the product are skin, eye and lung irritants. Appropriate safety equipment (overalls, face shield, long gloves) should be worn by plant operators whenever they are likely to be in contact with any of the process hazards.

1.2.6 Conclusions

- Phthalic anhydride is a white, crystalline solid at ambient conditions which forms explosive mixtures with air and is an irritant to human tissue requiring appropriate personal safety equipment to be worn.
- The main uses for PAN are as a chemical intermediate for the manufacture of plasticisers, alkyd resins and unsaturated polyesters.
- PAN is predominantly synthesised from o-xylene via a catalytic, gas-phase, strongly exothermic, oxidation reaction with air.
- Worldwide PAN demand is slowly growing (3–5% p.a.) and is now approaching capacity.
- *The o-xylene supply is a critical feasibility study consideration for a proposed plant.*
- *The reactors can either be fixed bed or fluidised bed but must incorporate an appropriate mechanism for removing the liberated heat.*

1.2.7 Recommendations

- Identify PAN producers in the region and provide an overall assessment of the current and prospective PAN supply.
- Identify end users of PAN in the region.
- Find the most economic new plant size for the region.
- Determine the preferred processing route and type of reactor.

2. FEASIBILITY STUDY AND MARKET SURVEY

2.1 INITIAL FEASIBILITY STUDY

The feasibility study for a chemical process design investigates both the technical and economic feasibility of the proposed project. The initial feasibility study is only an introductory assessment, at this stage the process route has not yet been finalized although a preferred route may be apparent. Part of the work in preparing the feasibility study is to obtain information regarding the alternative process routes, and to provide an assessment of the suitability of these routes for the particular project.

The technical part of the feasibility study considers the alternative processes, and the equipment that constitutes the chemical plant in each case. At this stage it is necessary to identify any items of equipment that pose particular or unusual design problems, or which are very expensive or hazardous. The feasibility study should determine whether it is possible to design and build a chemical plant for a particular manufacturing process. Any external factors that may influence the operation of the plant should be noted, e.g. discharge levels, stability of raw materials supply, etc.

The economic feasibility of the process should be established at this stage. Again, this is only an introductory assessment performed more to establish that the plant is not a definite loss-maker from the outset, rather than deciding that it is a particularly attractive proposition. A full and detailed economic evaluation of the plant and process is performed later in the design study (see Chapter 6) after the process route has been finalized, the detailed equipment listing prepared, and preliminary equipment designs have been performed.

2.2 PRELIMINARY MARKET SURVEY/ECONOMIC ANALYSIS

The following steps need to be performed to establish the initial economic feasibility of the process:

(a) Determine the **cost of *all* raw materials** (per tonne or kg) used in the process, these should be *delivered* prices to the anticipated plant location including shipping and handling.

Determine the **current selling price for the proposed chemical** (per tonne or kg) to be produced at the plant location, or at the market location if these are very different (and in this latter case, estimate the transportation costs).

Estimate the quantities of raw materials required to produce (say) one tonne of product. **Determine whether the combined purchase costs of all the raw**

37

materials is *more* **than the selling price of product** for the production process (based on one tonne of product).

If it is, why bother to build the plant?

(b) **Identify the markets for the chemical,** location and size, both locally and for exports. Determine whether these are contracting or growth markets, *analyse the sales and production trends over the last five years.* Identify and explain any anomolies, i.e. large increases in selling price or production. Establish the *minimum* sales level that would be expected: Is this sufficient to make a profit or operate at breakeven during short-term adverse conditions? Local or national sales are usually more predictable and stable than overseas markets. Compare this expected sales figure with estimates of the economic throughput for such a plant — based on available data for similar plants.

(c) **Identify the main (traditional) uses for the product.** Then concentrate on those main uses which account for (say) 60–80% of production. Remember that markets can disappear quite quickly, e.g. adverse toxicology data, ozone destruction effects! Therefore, it is necessary to survey any new uses or applications for the product and see these not only as developing sales opportunities but also as a support in the event of declining sales or competition. A report or memo on the market evaluation should not only present the available information but it should also analyse and interpret the value of this data to the specific project. Regular articles and product reports in *Chemical & Engineering News* may provide useful data, e.g. 'The Top 50 Chemicals and Top 100 Chemical Producers' and 'Facts and Figures of the Chemical Industry' are regularly published updates; also Product Reports on Fine Chemicals, or Soaps and Detergents, etc. Journals such as *Hydrocarbon Processing* and *Process Engineering* publish regular updates on chemical plant projects currently under construction (Construction Boxscore Supplement) including locations, throughput, and related plants.

Local information relating to the markets and demand for the chemical, or even a list of current producers of the chemical, may not be readily available and it may be necessary to use any worldwide data (often from the USA or Europe) that is available. However, it is important when presenting this data to relate the analysis back to your particular project, and to identify any specific factors which may override a global trend and outlook. Global data provides an overview but the local markets are where you will operate. To quote the environmentalists: Think global, act local!

(d) At this stage it is necessary (and probably possible) to **establish the production capacity throughput** for the proposed plant, so that preliminary calculations can be performed.

(e) **Obtain total capital cost data for an existing plant** producing the same chemical, preferably a recently commissioned plant of approximately the same required production rate. However, any data is (usually?) better than none, so if no data for such a plant is available try to obtain information for a plant producing a similar type of chemical with comparable plant equipment and throughput. Use one of the factored cost estimation methods (see Section 6.2) to estimate the present capital cost of the proposed facility.

Remember, this is only an estimate (±40% at best!). An accurate capital cost figure requires detailed designs of the plant equipment to be performed and a large amount of time to be spent. All that is required at this stage is a 'ball-park figure' to establish the feasibility of the project. Note any major differences between the existing plant and the proposed facility, e.g. cost of land, local availability of raw materials, existing transportation networks, etc. In particular, be careful about the choice of exchange rates. If the exchange rate has fluctuated widely over the previous 12 months, it will be preferable to use the highest and lowest rates to obtain a range for the projected capital cost estimate, rather than using the current or average exchange rate.

(f) Estimate the number of years of possible plant operation (longer for a "basic" chemical commodity) and **determine the pay-back period for the plant capital cost**. Determine the likely rate of interest charged (almost as difficult as estimating exchange rates!). Determine the yearly capital cost and interest repayments, and also the repayment cost per tonne of product. Estimate (very approximately) the operating costs, e.g. labour, utilities, maintenance, etc., for the plant (is 10% of the capital cost a suitable 'first' estimate?).

(g) Itemize all costs either annually or per tonne of chemical produced, including raw materials, capital cost repayments, operating costs, etc. Compare this final total figure with the expected (maximum and minimum) revenue from sales of the chemical.

Does the plant appear to be economically feasible at this stage of the project?

(h) Present all the findings of the feasibility study in a clear and concise written form. Identify all assumptions made and any factors that significantly affect this study, and any specific local considerations. Consider various scenarios, e.g. changes in selling price of product and/or costs of raw materials, increased export markets, etc, and present alternative strategies. Do not present one set of data or conditions and assume this is the only possibility. Make recommendations and present alternative approaches and conditions.

With most stages of project work, the key is to identify the main points or factors upon which critical decisions need to be based. There is often much data and many aspects appear to be important, but once the key element(s) are identified the path to be followed becomes apparent. Each project has a different set of important factors and constraints, for example this may be the need to establish (or significantly expand) the market for a project, or it may be the influence of political factors, or the need to sell or reuse a by-product produced in large quantities. The general factors which affect the success of most projects are: environmental considerations, safety issues, and the most obvious need to operate the plant as economically as possible. Which of these three factors (or any others) has most influence needs to be determined as soon as possible, then the focus of the project work becomes clear and the work will proceed more smoothly.

Note: Students sometimes complain that a project is not feasible at this stage and should be abandoned. However, the design team may not be fully aware of external developments, e.g. anticipated oil price rises, shortages of key commodities, etc. If a project does not appear feasible, it may be necessary to complete the design study in readiness for possible market changes and more favourable economic conditions. In this situation, it is preferable to perform the detailed economic evaluation (see Section 6.4) in terms of the changes in raw materials costs, increase in product price, etc., which are necessary to make the project profitable and feasible, rather than as a detailed statement of loss per annum. Present alternative scenarios and analyses.

Action: *Decide whether the design project should proceed to the next stage. If not, list all the reasons (in order of priority) and indicate what changes (if any are possible) which would reverse this decision. If yes, list all the assumptions/ restrictions upon which this decision is based (in priority order).*

References

Baasel (1990; Chapter 3); Peters and Timmerhaus (4th edn, 1991; Chapter 2); Ulrich (1984; Chapter 2); (see Section IV here). Extensive reference lists are included at the end of the quoted chapters in Baasel (1990) and Peters and Timmerhaus (1980, 3rd edn not 4th!).

2.3 INFORMATION SOURCES

By the time undergraduates begin the Design Project they should be familiar with the usual methods of searching for information, and the various sources of useful data. If this is not the case, then an information tutorial in the library is what is needed — not a chapter in this book! However, some pointers and tips will probably be welcome. Despite the ever-expanding number of electronic information sources, the first step for most students is still probably the university library and a search of the paper journals. Surveying the chemical engineering literature can be done relatively quickly (and painlessly) by using the bibliographies produced by the author (MSR), see references after Section 2.5. At the outset of a literature search it is essential to ask (and answer!) the question: *What type of information do I actually want, and need?* Unless this is clear and an appropriate search strategy is applied, then the outcome may be numerous references that are of little help in advancing the project.

Most detailed process data can be found in the journal issues from the 1960s, especially articles published in *Chemical Engineering (McGraw-Hill, New York)*. Obviously process improvements have been made since then, but these improvements are often related to energy efficiency, plant safety aspects, and reducing or processing plant discharges. Detailed design data on process plants usually is not published in the open literature for 10–20 years. Patents and licencing agreements are often the only source of recent developments. Most papers published now on process design tend to be fairly general and descriptive. A chemical engineering journal search for process information can be performed quite easily by reference to the '*Chemical Process and Plant Design Bibliography (1959–1989)*' and subsequently from the

CHERUB CD-ROM available with six-monthly updates (see References at the end of Section 2.5). The most useful journal sources of process data are *Chemical Engineering (N.Y.)*, *Hydrocarbon Processing*, *Process Engineering*, *Chemical & Engineering News*, *Chemical Engineering Progress* and *The Chemical Engineer (UK)*. The 'academic' journals such as *Chemical Engineering Science*, *AIChE Journal*, etc., rarely provide this type of information, although they often contain detailed analytical papers related to particular equipment design such as reactors, heat exchanger networks, etc. There are also several specialist journals, such as *Plastics World*, *Oil and Gas Journal*, and *Catalysis Today*, which are useful for specific types of information. Refer to Section IV for more details.

After chemical engineering sources and the closely-related literature has been explored, then the search may need to extend to *Chemical Abstracts*, the patent literature, foreign journals, and direct requests to companies that currently produce the chemical. On-line library searches are available, but are only useful if the keywords are carefully chosen (a librarian can be a useful asset here). Several years experience teaching design has shown that although a patent may appear to be the panacea for all gaps in knowledge, this is often not so. Although patents are supposed to divulge all information relating to the process, often the most useful data or operating details are conspicuously absent! Also an article written in a foreign language may seem essential (from a translation of the abstract) but the cost and time of translation is usually extremely high. Furthermore, the information may not be that useful when translated. However, it is usually possible to extract useful numerical data quite easily from any tables or equations, and this may actually be the extent of the value of the paper!

Students are probably far more adept than their lecturers at 'surfing the Net' for sources of information. There are numerous bulletin boards, newsgroups (see *sci.engr.chem*, *sci.chem*, etc.), newsletters (try *chenews*, information from *majordomo@cae.wisc.edu*), and home pages on the Internet and the World Wide Web (WWW). These resources are (and will) undergo rapid expansion and changes and any attempt at detailed description here will soon become outdated. It is now possible to conduct restricted patent searches on the WWW (see http://www.ibm.com/patents) and this is a very useful resource which will undoubtedly grow and develop, as will other data sources currently not generally available. However, the main point to be considered is that there is no regulation or control of most of these sources, and hence no guarantee that the information is accurate or that it does not have copyright elsewhere! Information on the Internet can disappear without trace and depending upon the source it can also be altered quite easily, which is not the case with paper sources. The Net and WWW can also absorb much time in searching and discovery, sometimes for scant rewards! The Internet is possibly a useful source of information but one that needs to be 'handled with care'.

2.4 EVALUATION OF AVAILABLE LITERATURE

After finding some (useful?) data, it is quite surprising that only a few students manage to do anything particularly interesting or informative with it! Most times the data is used just as that — data, often without any attempt to work with the numbers to provide particular insights. The reader of reports and memos usually wants some interpretation and evaluation of the information. Most times the information is readily available and contains no great surprises or secrets. But the job of the graduate engineer (as distinct from the technician) is to evaluate and make reasonable decisions based on the data. Many student reports are surprisingly dull because what is presented is a set of published facts rather than a set of ideas for further investigation. Many times, reports are presented without Recommendations for further work, sometimes even without Conclusions! Project work is always leading to the next stage and the graduate should be proposing directions, not just collecting information. Descriptive work is not the stuff that good projects are made of, it is the food of followers rather than leaders.

The inital rule of project work can be summarised by the *3-Ds*, namely *Drive, Direction* and *Decisions*. The students need to exhibit self-motivation and *Drive* to obtain information, this needs to be analyzed (not just reproduced) in order to identify the *Direction*, and this should lead naturally into appropriate *Decisions*.

2.5 CONSIDERATIONS FOR LITERATURE SURVEYS

Detailed notes regarding the presentation of literature surveys are included in the preliminary pages (see Section III) at the beginning of this book. The following are a few points for consideration before starting a survey.

Is a Literature Survey necessary, or required?

The Literature Survey should be *a review* of the important information available *relevant* to the particular project topic.

It is important to identify and include *only relevant information*, and to state at the start of the survey what type of information is required.

It is hard to do a comprehensive survey before the subject has matured. Once the project is well developed much of the literature survey may be of little interest (apart from providing a Reference Listing).

There are many stages and aspects to be covered in a Design Project, and a single survey may not be appropriate. An evaluation of the useful and available literature within particular sections may be what is needed.

References

Anon., A wealth of information online, *Chem. Eng. (N.Y.)*, June, 112–127 (1989).

Chowdhury, J., Taking off into the world of Internet, *Chem. Eng. (N.Y.)*, March, 30–35 (1995).

Graham, M.H., Information storage and retrieval in the 70s, *Chem. Eng. Prog.*, February, 74–75 (1973).

Kleiner, K., What a tangled Web they wove..., *New Scientist*, 30 July, 35–39 (1994).

Krieger, J.H. and Illman, D.L., Internet offers alternative ways for chemists to hold conferences, *Chem. Eng. News*, 12 December, 29–40 (1994).

Ray, M.S., *Chemical Engineering Bibliography (1967–1988)*, Noyes Publications, New Jersey, USA (1990).

Ray, M.S., *Chemical Process and Plant Design Bibliography (1959–1989)*, Noyes Publications, New Jersey, USA (1991).

Ray, M.S., *CHERUB* chemical engineering database, included on *Engineering and Applied Science Database* on CD-ROM, published by Royal Melbourne Institute of Technology, Australia. Updated every 6 months, full details available from the author (MSR).

Ray, M.S., Electronic Communications and the Internet, *Chem. Eng. Australia*, 20(2), 7–12 (1995).

Reid, R.C., The archive journal, *Chem. Eng. Prog.*, August, 105–108 (1973).

Rosenzweig, M. and Gardner, G., Searching for information, ChEs see significant gains and changes, *Chem. Eng. Prog.*, November, 7, 54–61 (1994).

Sondak, N.E. and Schwartz, R.J., The paperless journal, *Chem. Eng. Prog.*, January, 82–83 (1973).

2.6 CASE STUDY — FEASIBILITY STUDY AND MARKET ASSESSMENT

Summary

A preliminary feasibility study for phthalic anhydride production was performed which included:

(a) market assessment to determine the scale of production and demand on a global, regional, national and local basis (where appropriate);
(b) initial estimates for capital, operating and raw material costs to determine whether the product can be sold competitively.

At this stage, *production appears to be marginally viable*. The data gathered here will provide the basis for future decisions regarding plant capacity and site location.

2.6.1 Market Assessment

2.6.1.1 Production: Worldwide

Phthalic anhydride production rates have cycled since the 1950s due to fluctuating demands, variable feedstock prices and availability. The estimated worldwide production of phthalic anhydride was 2700 kT in 1993 [1], although worldwide capacity (in 1993) was much higher at around 3500 kT [1–3]. Therefore, the utilisation of capacity was only about 77% in 1993 and has typically been only 65–75% since the early 1970s [3]. Growth has been slow in North America and Europe (less than 3% per year since 1981) but high in the Far East, Eastern Europe and South America (around 15% per year) [1,2]. However, more recently, growth has increased and demand is now approaching capacity [14–16].

The most significant trend over the last 30 years has been the move away from the traditional raw material, naphthalene, to *o*-xylene. In 1968, only around 30%

of production was o-xylene based. By 1987, this figure had increased to 88% in the United States [3]. *Since 1971, no new plant has been constructed which uses naphthalene as a feedstock.*

The number of active producers has decreased significantly as older plants become uneconomical and they are replaced by larger capacity plants. No major phthalic anhydride projects were initiated during the 1970s and 1980s because demand was relatively constant and current facilities were not operating at full capacity. However, towards the end of the 1980s feasibility studies of several projects were investigated and several new plants were built including some in the Western Pacific region. The United States and Europe appear to be relying on existing facilities, at least until 2000.

Most modern plants have an annual capacity of between 75 and 120 kT/yr. Only five operating plants remain in the US (compared with eight in 1980) and they have an average capacity of 91 kT/yr (compared with 77 kT/yr in 1980) [3]. Currently, the largest phthalic anhydride producer has a capacity of 150 kT/yr [1,17].

2.6.1.2 *Production: Regional*

Production of phthalic anhydride in the Western Pacific region is dominated by Japan. In 1995, Japanese plants had a combined capacity of 323 kT/yr and production was increasing at 5–10% per year. However, these plants are still only being operated at 60–70% capacity [15]. Current and proposed major regional producers are listed in Table 2.1 [15,17] by location, feedstock and current production rates. All producers in the region, except the dual-feed ICI plants at Rhodes and Mayfield, NSW, use o-xylene exclusively. The Australian plants are also the smallest in the region, with a combined capacity of only 21.5 kT/yr [17].

Traditionally, East Asia has imported a major proportion of its phthalic anhydride however, more recently, existing plants have been upgraded and new projects have been launched in a drive for self-sufficiency. Two large plants in Taiwan (90 kT/yr) and Korea (110 kT/yr) have both come onstream since 1990, and they have the potential to flood the market with a cheap supply of phthalic anhydride. Indonesia, Australia's closest neighbour, have also recently completed an o-xylene plant and they are now constructing a PAN facility. *These developments will significantly reduce the export potential of a new Australian plant.*

2.6.1.3 *Production: National*

Australian production of phthalic anhydride is restricted to the twin ICI plants in NSW [17,19]. However, there are several traders and importers of phthalic anhydride scattered around the Eastern States [20]. Within Western Australia, ICI can supply phthalic anhydride if it is required but they do not have local production facilities.

Table 2.1 Operating and proposed phthalic anhydride plants in the Western Pacific region.

Location	Type	Feed	Annual Production (kT/yr)
Australia	operating	naphthalene	21.5
Indonesia	operating	o-xylene	30
Japan	operating	o-xylene	86
	operating	o-xylene	72
	operating	o-xylene	65
	operating	o-xylene	40
	operating	o-xylene	30
	operating	o-xylene	30
Korea	operating	o-xylene	110
	operating	o-xylene	60
	operating	o-xylene	40
	construction (due end 1996)	o-xylene	150
Malaysia	planned (due end 1998)	o-xylene	20
Philippines	operating	o-xylene	16
Singapore	construction (due end 1996)	o-xylene	30
Taiwan	operating	o-xylene	90
	planned	o-xylene	50

2.6.2 Current and Future Prices

Limited production rates for phthalic anhydride allow the majority of the product to be sold on the spot market. The current price for molten phthalic anhydride is 75–80 US¢/kg (34–36 US¢/lb) [6]. At current exchange rates, this equates to A$0.95–1.00 per kg. Prices have varied quite widely in recent years, for example, in 1987, typical prices were 65–75 US¢/kg [21]; in 1989, prices were 88–97 US¢/kg [22]; and, in 1994, prices were 70–75 US¢/kg [6]. Reduced operating costs are mainly due to changes in raw material prices and improved energy usage.

Prices are strongly dependent on general economic circumstances, and also on consumer demand as the end-product uses are dominated by housing and automotive applications. With current prices and demand, there are few phthalic anhydride plants worldwide that are being operated profitably, at least after depreciation and reinvestment margins are considered [21].

2.6.3 Demand

There are three principal markets for phthalic anhydride: phthalate plasticisers for flexible PVC and other resins (55%); unsaturated polyesters for fibre-reinforced plastics (20%); and alkyd resins for oil-based coatings (15%) [1,3]. Demand for phthalate plasticisers has remained stable because of their relatively low cost, even though higher molecular weight substitutes perform better. Unsaturated polyester remains a growth industry and increased usage in the automotive and construction industries is promising. Alkyd-based resins are being phased out in favour of water-based products. The overall assessment is that end uses were declining slightly in 1989 [22].

The phthalic anhydride market has grown slowly and US growth in the ten year period to 1989 was only 0.1% p.a. [22]. Export markets are also shrinking as plants in SE Asia increase their capacity. Indonesia is already a net exporter of phthalic anhydride [23]. Local demand is similarly slow with much of Australia's production being consumed in-house at ICI, Rhodes, NSW, for plasticisers and alkyd resin production.

2.6.4 Australian Imports and Exports

Most of the phthalic anhydride that is produced internationally is bought and sold on the spot market. Prices are based on short-term trends and can fluctuate dramatically. Therefore, imports and exports can vary significantly from year to year. However, over the last decade, imports to Australia have typically been only 1–5 kT/yr. This low volume and the large number of trading partners (at least 21 countries) suggests that there is no permanent import market. There was a small export market in phthalic anhydride but this has evaporated in recent years with increasing self-sufficiency in the Asian region [24–27].

Although trade in PAN is weak, there have been substantial quantities of polyester and alkyd resins imported during the last decade. A new phthalic anhydride plant may find local markets if this import demand for PAN end-products is replaced with local manufacturing capacity.

2.6.5 Plant Capacity

The import market is currently weak and, if acquired, is unlikely to provide more than 2 kT/yr at present. However, there is potential for this to increase if local users of PAN increase their production. The export market is also weak and appears to have little prospect of strengthening in the medium to long term due to new developments in the region. Australia's current facilities are old and outdated and should prove to be uncompetitive if new facilities come onstream. Therefore, the base demand is likely to be around 25 kT/yr. Allowing some surplus capacity for expansion, *a plant capacity of 30 kT/yr is suggested*. A second option of 60 kT/yr, based on the co-development of end-use markets and favourable development of export and spot markets (15–20 kT/yr) is also considered.

2.6.6 Product Value and Operating Costs

2.6.6.1 Capital Costs

Two types of capital estimate were considered: one based on a nomograph, and scaled for location and inflation; and, one based on recent construction costs, scaled for capacity, location and inflation.

Nomograph Estimate

For a Western European location in mid-1982, costs for a 30 kT/yr plant are estimated at US$28 million and for a 60 kT/yr at US$43 million [28]. Using an exchange rate of US$0.79 per A$1, a location factor of 1.4 (Western Europe to Australia) [29] and an inflation index of 1.15 to 1990 (the plant cost index [30]) and 1.19 from 1990 to 1996 (average 3% p.a.), these costs are equivalent to A$70 million and A$105 million, respectively.

Recent Construction Estimate

A 60 kT/yr plant was completed in Japan in January 1983 for ¥4030 million, excluding license and engineering fees or catalyst charging. If the additional costs are estimated at 20% of the overall cost, applying current exchange rates equates the total cost to A$23 million. Indexing for location and inflation gives A$41 million [28,30]. Using a size correction exponent of 0.70 [31], a 30 kT/yr plant would cost A$25 million.

2.6.6.2 Operating Costs

The total operating cost for a phthalic anhydride plant in 1987 using *o*-xylene as a feedstock was estimated at 46.3 US¢/kg [21]. Primary components of the estimate are 33.3 ¢/kg for raw materials (yield of 1.01 kg/kg), 1.2 ¢/kg for utilities, 0.5 ¢/kg for catalyst, 4.5 ¢/kg for fixed costs (salaries, maintenance, administration, taxes, insurance, etc.) and 6.6 ¢/kg for plant depreciation (10% of investment). This distribution is shown in Figure 2.1. Adjusting for location, inflation and raw material cost fluctuations, the production cost (i.e. total operating cost) of PAN in Australia in 1996 is estimated at A$0.87 per kg.

2.6.6.3 Approximate Selling Price

The final selling price for phthalic anhydride is determined by the operating costs, capital cost repayments and other fixed costs, transportation costs and the required profit margin. Location in Western Australia is distant from potential markets and incurs a penalty in transportation costs for both raw materials and product. Initial estimates of these costs are A$0.03 per kg of feed and A$0.07 per kg of product (including insulation and heaters to maintain phthalic anhydride in the molten phase). Capital cost repayments depend upon the size of the plant but could be expected (as a first estimate) to be between A$0.08 and A$0.24 per kg,

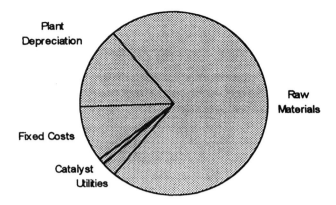

Figure 2.1 Distribution of operating costs.

at 10% interest rates. Therefore, the break-even price is around A$1.10 to A$1.20 per kg of phthalic anhydride.

With current market prices for phthalic anhydride of approximately A$0.95–$1.00 per kg, there appears to be little margin for profit. Allowing a 20% profit margin for viability, the final selling price could be as high as A$1.45. Such a price represents the upper limit for commercial sales and would be a major obstacle to successfully competing with other producers on the open market. Without government subsidies, local and overseas users of phthalic anhydride would not consider buying from a Western Australian plant under these conditions.

The overall viability of a local phthalic anhydride plant is dependent on several factors other than the anticipated production costs. The development of a new industrial complex that incorporated several major petrochemical plants may attract government assistance. This could be in the form of subsidies during the early years of the development in order to establish plants with long-term economic benefits for WA and Australia.

The most likely option for increasing the viability of this plant is to incorporate downstream processing options, possibly in the form of a polyester resin plant. If several plants were constructed together as part of a single development, profits from some industries could be used to offset small losses in other areas. A terephthalic acid plant (using a *p*-xylene feed from a xylene plant that also supplied a phthalic anhydride plant) is likely to be very profitable in the current market. Finally, it may be possible to take advantage of Australia's stability compared with some of its regional competitors where continuity of supply was an important concern. Contract negotiations may favour WA over Taiwan on this basis.

2.6.7 Conclusions

- Worldwide production of phthalic anhydride is currently approaching the estimated global capacity of 3500 kT/yr, and is growing at 3–5% annually after a long period of under utilisation of available capacity.

- *Ortho-xylene is now the feedstock for about 90% of all phthalic anhydride produced. No plant constructed since 1971 uses naphthalene.*
- Phthalic anhydride production in the Western Pacific region is dominated by Japan, Korea and Taiwan, with a combined capacity of about 630 kT/yr. An additional 250 kT/yr of PAN capacity is planned or under construction in the region.
- Australia produces 21.5 kT/year from twin ICI plants at Rhodes and Mayfield in NSW. The Australian plants are the smallest in the region and the only ones to use naphthalene as a feedstock.
- No phthalic anhydride is produced in Western Australia and there are no major end-use industries in the state.
- *The price for molten phthalic anhydride on the spot market is A$0.95–1.00 per kg.* Prices are falling in real terms.
- The three major uses for phthalic anhydride (plasticisers, unsaturated polyester resins and alkyd resins) consume about 98% of all phthalic anhydride produced.
- Demand is rising for polyester resins, it is stable for plasticisers, and is falling for alkyd resins. Overall demand is slowly increasing.
- Australian imports of phthalic anhydride are variable but, on average, account for less than 5 kT/yr.
- New processing methods offer the opportunity to reduce costs by 25% by using *o*-xylene as the feed, utilising better catalysts and implementing energy integration techniques. The Australian plants are operating at a significant economic disadvantage.
- *Ortho*-xylene prices are linked to *p*-xylene prices which are increasing more rapidly due to greater demand.
- If a phthalic anhydride plant was constructed *in Western Australia, it should have a capacity of about 30 kT/yr.* The market share would be obtained by the forced closure of the ICI plants in NSW and the development of local end-use industries. If a larger plant were constructed (60 kT/yr) aggressive marketing would be required to obtain some export markets.
- Capital costs estimated from nomograph would be A$70 and A$105 million for 30 and 60 kT/yr plants, respectively. Capital costs estimated from recent construction costs would be A$25 and A$41 million for 30 and 60 kT/yr plants, respectively.
- Operating costs were estimated at A$0.87 per kg of phthalic anhydride produced. The break-even selling price was estimated at A$1.10 to A$1.20 per kg phthalic anhydride.
- Viability of a proposed plant is likely to depend on other factors, such as government assistance, downstream processing or specialist contracts.

2.6.8 Recommendations

- Select the most economically viable phthalic anhydride process.
- Size the major items of equipment and determine other equipment requirements.
- Develop a plant layout and determine space requirements.

- Investigate alternative sites for a potential phthalic anhydride plant in Western Australia.
- Perform more detailed costings to determine the optimum size for a potential plant.
- Decide upon key process conditions and assess the environmental impact.
- Approach Government and industry to promote a consortium of chemical companies to exploit possible synergies.

3. PROCESS SELECTION, PROCESS DESCRIPTION AND EQUIPMENT LIST

3.1 PROCESS SELECTION CONSIDERATIONS

Following the initial feasibility (and market) evaluation, the next stage in the design project is the evaluation and comparison of the alternative process routes for manufacture of the chemical. The selection of an appropriate process is an important decision, as all the subsequent work depends upon this choice. Although the selection can be changed or modified at a later stage, at least before the plant is built, changing this decision results in a serious waste of time and money. However, probably not such a waste as building an uneconomic or unsafe plant!

3.1.1 Flow Diagrams — PFD and P&ID

Obtain a *Process Flow Diagram* (*PFD*, or simple flowsheet) or *Piping and Instrumentation Diagram (P&ID)* for each process route under consideration. The PFD (which includes the main plant items drawn using recognisable symbols rather than blocks and circles) may be a simple representation of the process stages but the P&ID is usually quite complex and includes all the ancillary equipment, utilities, instrumentation and control details. To obtain an appreciation of the alternative processes, and to make an initial comparison, first draw a simple *Block Diagram* from each PFD showing only the main chemical engineering plant items, e.g. reactor, separation columns, purification stages, etc. Omit the instrumentation and the various pumps and valves at this stage. This simple Block Diagram will not provide sufficient detail for final process selection, but it will provide an initial comparison of different processes and a quick familiarisation with the process stages comprising each route. Consider process selection using the simple Block Diagrams in conjunction with the PFDs which includes relevent process data such as temperatures, pressures, flow-rates, and any stream compositions (if known).

3.1.2 The Reactor

The reactor is the heart of most chemical production processes, although many processes such as oil refining and minerals extraction processes do not contain a traditional reaction vessel. Developments in chemical process technology often centre around improvements in the design and operation of the reactor, and this is often the basis of 'new and improved' processes. Subsequent stages in the chemical process are usually concerned with the separation of various by-products and unused reactants from the desired product, followed by the final purification stages.

3.1.3 Product Purity

Previously in the feasibility study (Chapter 2), the appropriate purity of the final product should have been established. The product purity must be dictated by the customer (market) requirements. In some situations it may be possible to remove a low-purity product from the process for a particular application, and then purify the remaining chemical for another customer. The possibility of further purification should be determined in anticipation of new applications, environmental regulations, etc. There is a compromise between increased capital cost of the initial investment and flexibility of future operation. The alternative is to modify the process later, with a possible cost penalty.

3.1.4 Process Conditions

Processes often differ in terms of the process conditions, e.g. high-pressure and low-pressure processes, or the type of reactor that is used, e.g. gas-phase catalytic reactor or liquid-phase CSTR. The differences should be clearly marked on both the simple block diagrams for each process and on the detailed process flowsheets, these differences often determine which process route is ultimately selected. Many older, established processes were operated under conditions of high pressure or high temperature (mainly in the reactor), whereas the newer processes have often been 'improved' by operation under less severe conditions, often through improvements in catalysts.

3.1.5 Process Data

For student design projects, the older and less efficient processes are often the only ones for which detailed process data (suitable for the detailed equipment design stage) are available. Information concerning 'new' technology and the relevant experimental design data may still be secret. Therefore, it is sometimes necessary for students to select a process route and perform a design study for what appears to be a less efficient process than they would ideally choose. The design project should then focus on producing an improved process (and plant equipment designs) based upon the student's knowledge and abilities. The training and experience obtained in plant and process design is not diminished if an older process route is selected, and the real thing is awaiting the student's talents in industry!

3.1.6 Energy Efficiency

Many older processes (developed before the 1970s oil price crisis) were much less energy efficient than those developed more recently. This should be apparent by consideration of the energy conservation features included in new plants, and the subsequent increase in complexity of the associated P&ID. Energy conservation is discussed in Section 8.1.

3.1.7   Factors in Process Evaluation and Selection

The selection of a process route for production of a chemical depends upon the following, not necessarily complete or comprehensive, list of questions and considerations.

(a) Will the process produce what the customer requires?
(b) Is it possible to design, build and operate this plant safely and economically?
(c) Are the necessary design data, technology, fabrication methods and materials, raw materials, etc., available?
(d) Is the plant able to operate in a safe manner, providing an acceptable hazard risk (see Section 8.3) to the plant employees and the public?
(e) Will the plant conform to any environmental protection requirements (see Chapter 5), and any possible future restrictions?
(f) Is the process as energy efficient (and energy self-sufficient) as possible?
(g) Have maintenance requirements been minimized?
(h) Will the plant be able to operate adequately under conditions of reduced throughput (say 50–70%), and for increased production (say + 50%, having already been designed for a 25% overcapacity)? The latter case represents an overdesign of the plant and additional capital costs. In these situations, the energy consumption is greater than the optimum requirement. However, the need for operational versatility usually overrides these considerations. The design basis should be set after consideration of future operational changes, and then minimise any additional overdesign of equipment.
(i) Has the production of any unusable by-products been minimized?
(j) Are all necessary utilities, e.g. electricity, cooling water, etc., available? The utilities specifications must be established, these are dependent upon whether a 'grass-roots' plant is to be built or if the plant is to be situated in an existing chemical complex. These specifications include the steam pressures, cooling water and/or refrigerants available, etc. The unit costs must also be determined.

3.1.8   Choices and Compromises

As with many aspects of design work, the final choice will usually depend upon a compromise between various features of different processes. It is unlikely that one process will possess all the advantages and no disadvantages. Sometimes there will be one overriding factor that influences the selection of a particular process, e.g. availability of a particular raw material, minimum cooling water requirements, etc. However, whichever process route is finally selected, it must fulfill *all* the criteria established in the project brief (see Section 1.1).

3.1.9   The Optimum Design

The aim of process design is to build the 'best' overall plant, although all units within the designated plant may not operate at maximum efficiency or full potential. It is necessary to achieve a balance between the conflicting requirements of the individual units in order to produce the 'best' plant possible. Plant design is an exercise in compromise and optimisation!

3.1.10   Process Control and Instrumentation

The process control and instrumentation requirements (see Section 8.2) must also be considered when selecting the process route. The ability to provide suitable operational control over the process and the availability and cost of necessary instrumentation are major considerations. If these aspects are ignored at this stage, it could well be that the detailed plant design is at best more expensive than is necessary or at worst impossible to operate. The attitude that 'whatever is built can be operated and controlled' is both dangerous and unprofessional. At this stage, the process control and instrumentation for the process will only be specified on the basis of either previously published information (and an available P&ID), or from a basic knowledge of the subject rather than a detailed study of the particular process. Process control for the plant will need to be specifically developed at the detailed design stage. Some additional points regarding process control are discussed in Section 8.2.

Action: *Select the 'best' process route for production of the chemical, and justify this decision.*

    Prepare a detailed process flow diagram and, as far as possible, a P&ID.

References

Kirk–Othmer Encyclopedia of Chemical Technology, 4th edn in production (1991–); 3rd edn, 25 volumes (1978–84), John Wiley, New York.

Basta, N. (ed.), *Shreve's Chemical Process Industries*, 6th edn, McGraw-Hill, New York (1994).

Cavaseno, V. (ed.), *Process Technology and Flowsheets, Volume I* (1980).

Greene, R. (ed.), *Process Technology and Flowsheets, Volume II* (1983).

    These two books contain reprints of papers from *Chemical Engineering*, McGraw-Hill Publications, New York.

Ulrich, G.D., Chapter 3: *Flow Sheet Preparation*, and references at the end of that chapter (1984).

Woods, D.R., *Process Design and Engineering Practice*, Prentice-Hall Inc., New Jersey (1995).

Woods, D.R., *Data for Process Design and Engineering Practice*, Prentice-Hall Inc., New Jersey (1995).

3.2 PROCESS DESCRIPTION

Once the process selection has been made, then it is necessary to provide a detailed description of the particular process and especially the main features and advantages and disadvantages. This description must be accompanied by the process flow diagram (PFD, or the P&ID if this is available from the literature) and the description and PFD must be linked to the Equipment List (see next section) by an appropriate equipment numbering system. The description must be as detailed as possible at this stage and must reference all useful publications and include any available data. Conflicting data must also be reported and commented upon as it will have to be resolved at the detailed design stage (if not before). The detailed description should not be an essay. This is an engineering report and engineers are interested in easily digestible facts, therefore only the important points need to be included, generally in point form. The advantages and disadvantages of the choosen process must be clearly listed and their implications explained, e.g. effects on process economics: capital versus operating costs; process control and safety, etc.

Descriptions of the alternative processes are also required but in less detail, mainly concentrating upon the advantages and disadvantages and explaining clearly why these processes were rejected. If the process selection needs to be changed later, then details included here will make this decision easier to implement. Also the aim in design work is to improve upon the chosen process and it may be useful to start by reviewing the features of the alternative schemes.

3.3 PREPARING THE EQUIPMENT LIST

After the process route has been selected, it is both possible and necessary to prepare an initial *Equipment List* or *Equipment Schedule*. The process description, preliminary design calculations and equipment listing are usually performed at the same time as the preliminary mass and energy balances (for obvious reasons). However, for convenience of presentation and concise discussion, the preliminary calculations are discussed here and the detailed mass and energy balances are considered separately in Chapter 7. The Equipment List should be as detailed as possible, but it will obviously not be complete until the detailed equipment designs are available (see Part II). The reason for preparing a listing at this stage is to start the compilation of the full documentation for the project, and to establish a database of useful information for subsequent stages.

The Equipment List is used in conjunction with the P&ID (as some minor items are not included on the PFD). Each item of equipment on the flowsheet should be assigned a unique reference number, and that number is used to cross-reference items in the Equipment List. Particular letters may be used to identify similar items of equipment, e.g. reactors as Rxxx, pumps as P-001, etc. There is no standardised lettering system for the designation of equipment, although there are some obvious choices and ambiguity or repetition should be avoided.

The Equipment List should include all currently available information that will be useful for the detailed design of the equipment. The Equipment List will be continually updated as more information is acquired and as the various feasibility and design stages are completed, and the results of further calculations become available. This is a dynamic list, not a 'once-off' fixed contingency. The list should include the physical size of an item (if known), operating temperature and pressure, materials of construction, etc. Details such as wall thicknesses and height of packing are unlikely to be known at this stage unless previously published design data is available. The original reference sources for all this information must be included. If one major publication was used, then it may be stated in the introduction to the Equipment List that 'all data were obtained from this source unless stated otherwise'. Although the majority of this information will be for similar plants (i.e. similar process routes), the final design details for the new plant may be very different. However, the Equipment List provides a useful reference source at this early stage of the design, and as a basis for subsequent stages.

The following information must be included in the Equipment List so that the cost of purchasing and installing each piece of equipment can then be estimated.

(a) Specific type of equipment.
(b) Size and/or capacity.
(c) Material of construction.
(d) Operating pressure.
(e) Maximum operating temperature (if above ambient), or minimum temperature if refrigerated.
(f) Corrosion allowances (if large).
(g) Special features, e.g. jackets on heat exchangers, special insulation.
(h) Duplication of plant items (for safety and/or reliability).

3.4 RULES OF THUMB

The heat exchangers and pumps cannot be sized accurately until the piping layout has been prepared and the energy balances for the plant are completed (Chapter 7). The final energy balance depends upon the energy conservation measures to be employed (see Section 8.1), and the plant layout (see Section 4.2). Other equipment should be sized as accurately as possible at this stage, although much of this data will only be 'ballpark' figures. More accurate sizings will become available as the project proceeds and during the detailed design. Many approximate methods of equipment sizing are available, reference can be made to the books by Aerstin and Street (1978); Baasel (1990); Ulrich (1984); Walas (1988) for particular examples (see Section IV here). References for the sizing and design of particular items of equipment are included in Baasel (1990; pp. 134–146). A few other useful references for preliminary design and 'rules of thumb' are given below. It is both adequate and necessary to design only to the appropriate level of detail at each stage as the project proceeds.

Action: Prepare a complete Equipment List for the chosen process route, including references for all information and data.

3.5 SAFETY CONSIDERATIONS AND PRELIMINARY HAZOP STUDY

Once the preliminary P&ID and Equipment List are available, then quantitative consideration can be given to the safety aspects of plant operation. It will be possible (and necessary) to perform a preliminary *Hazard and Operability study (HAZOP)* at least around the major plant items, and preferably around the entire plant. Consideration of the aims of the HAZOP study and the procedures to be followed are given in Section 8.3. A HAZOP is not the only tool to be used in the safety evaluation of a process, but it is quite straightforward to perform and has been well documented. However, specialist advice is required and other safety evaluation methods also need to be applied (see Section 8.3). The preliminary HAZOP will obviously not be as detailed as the final study performed after the detailed design work, but the initial evaluation is necessary to help in the development of the control scheme and also to identify (and avoid) some obvious design and operational problems at an early stage.

A detailed HAZOP study on the phthalic anhydride reactor design is presented in Chapter 10 (see Section 10.16). This is included as a part of the detailed equipment design, but it should be remembered that in some ways this is rather an artificial or academic exercise. The HAZOP is a sequential study performed line-by-line throughout the entire plant starting from the front-end, and it considers not only the operation of a particular item but also the effects of the (mal-)operation of previous units on downstream plant items. The work involved in a HAZOP is considerable and it was decided that a book of this nature (and the case study) would be incomplete without an example, but that two would be sufficient (see also Section 11.14 for the After-Cooler HAZOP study).

The HAZOP has several outcomes, the most obvious is the development of operating procedures for the equipment, but it also leads to the development or re-evaluation of the process control and instrumentation requirements. This in turn leads to the design of alarm systems, where the aim should be to structure a system of responses such that the original alarm (s) can be identified (and easily corrected) rather that a series of independent alarms and actions. The final outcome is the determination of start-up and shut-down procedures, a commissioning programme, and the maintenance requirements.

References

Baasel (1990; Chapter 5) and Ulrich (1984; Chapter 4).
Branan, C., *Rules of Thumb for Chemical Engineers: A manual of quick, accurate solutions to everyday problems*, Gulf Publishing, Houston, Texas (1994).
Fisher, D.J., *Rules of Thumb for the Physical Scientist*, Trans World Publishing, UK (1988).
Fisher, D.J., *Rules of Thumb for Engineers and Scientists*, Gulf Publishing, Houston, Texas (1994).
Korchinski, W.J. and Turpin, L.E., The five-minute chemical engineer, *Hydrocarbon Process.*, January, 129–133 (1996).
McAllister, E.W., *Pipeline Rules of Thumb Handbook*, 3rd edn, Gulf Publishing, Houston, Texas (1993).

Walas, S.M., Rules of thumb: Selecting and designing equipment, *Chem. Eng. (N.Y.)*, 16 March, 75–81 (1987).

Walas, S.M., *Chemical Process Equipment: Selection and Design*, Butterworth, Boston, Massachusetts (1988); pp. xiii–xix.

3.6 CASE STUDY — PROCESS SELECTION AND EQUIPMENT LIST

Summary

Potential PAN processes were identified and evaluated. *Ortho*-xylene is clearly the preferred raw material. A fixed-bed reactor is preferred. A catalyst which allows a low air to *o*-xylene ratio is also preferred as it reduces equipment sizes and the capital cost. The major items of equipment within the PAN plant, requiring the most detailed design, will be the reactor, switch condensers and purification columns.

3.6.1 Trends in Phthalic Anhydride Processing

Ortho-xylene is the clearly preferred modern day feedstock for phthalic anhydride manufacture. Few naphthalene-based plants remain operational, and no new naphthalene plants have been built since 1971. *Ortho*-xylene produces higher yields (all carbon atoms appear in the product), it is cheaper than naphthalene and provides a more efficient process. Converting an existing plant to *o*-xylene can reduce raw material costs by 25% and utility costs by 30% [8].

A second development in processing has been the integration of the energy evolved by the oxidation process. This energy can be used to produce process steam at high pressure and should negate the need to purchase power. New low-energy processes have been developed which recycle the waste gases [32]. Advances are also being made in catalyst selectivity to improve the quality of the product and eliminate side reactions. Dual-use catalysts have been developed to allow both *o*-xylene and naphthalene to be used with only minor alterations to operating conditions [3,21].

3.6.2 Raw Material

Both naphthalene and *n*-pentane can be used for PAN production but both are at an economic disadvantage compared with *o*-xylene, and they must be supplied at prices well below market value to be viable alternatives. Neither naphthalene nor *n*-pentane is available locally and therefore, should not be considered for a new plant located in WA. *Ortho*-xylene is clearly the preferred raw material as it has proven to be cheaper, higher yielding and more selective (fewer impurities are produced by the reaction). Some processes maintain the flexibility to also use naphthalene as protection against sudden large increases in the *o*-xylene price as a result of extra demand for *p*-xylene [1,2]. However, the choice of *o*-xylene as feedstock for a proposed plant would represent a significant technological advantage over the existing Australian plants (ICI at Rhodes and Mayfield, NSW).

Ortho-xylene is produced from three main regions around the world: the USA, Western Europe and East Asia. Of these, East Asia is the closest to Australia and is also the fastest developing. Japan, Korea and Taiwan have markets that a WA-based plant could utilise. However, the most economically attractive option is to use by-products from Australian oil refineries in a local xylene separation plant. A supply of cheap BTX (benzene–toluene–xylene) is likely to become available in the near future as Australian regulations concerning gasoline composition are changed to exclude aromatics in line with worldwide trends.

Xylene separation plants process a mixed xylene feed and include both isomerisation and separation units. The close boiling points of the three isomers require large columns (200 stages) and high reflux ratios (around 10) [33]. Isomerisation units can be used to control the ratios of the three products. *Para*-xylene has the highest demand and production is increasing rapidly. Phthalic anhydride consumes about 98% of all *o*-xylene produced and *o*-xylene production rates reflect the phthalic anhydride trends. *Meta*-xylene has no commercial application and is normally recycled to isomerisation units. The most recent development in this area, which appears set to significantly reduce processing costs, is the reactive distillation of mixed xylenes to terephthalic acid (or another product) and a by-product which is rich in *o*-xylene. Current prices for *o*-xylene are approximately 55–75 US ¢/kg [6] and are likely to increase as future *o*-xylene prices are linked to *p*-xylene demand which is currently strong.

3.6.3 Process Configurations

Phthalic anhydride processes can be classified according to the type of reactor used. Three reactor configurations have been commercially developed: (a) fixed-bed vapour-phase reactor; (b) fluidised-bed vapour-phase reactor; and (c) liquid-phase reactor. Many variations are possible and a range of processes have been licensed [4,7,9–13,34,35]. In recent years, the fixed-bed vapour-phase process has proven to be superior to the alternatives and all plants built in the last ten years have used this process. Fluidised bed processes have proved difficult to maintain and have suffered from erosion problems and excessive catalyst losses [4]. The construction costs of liquid-phase processes have proven prohibitive [1].

Rival companies have established several different versions of the fixed-bed vapour-phase process, these differ mainly in catalyst performance and operating conditions. Of these, two are significantly better in their ability to integrate the energy produced and either the Low Energy Von Heyden (LEVH) or the Low Air Ratio (LAR) process would be preferred for plants built in the 1990s [2,7,11]. The Vent Gas Recycle (VGR) process also appears to have some potential [2,32].

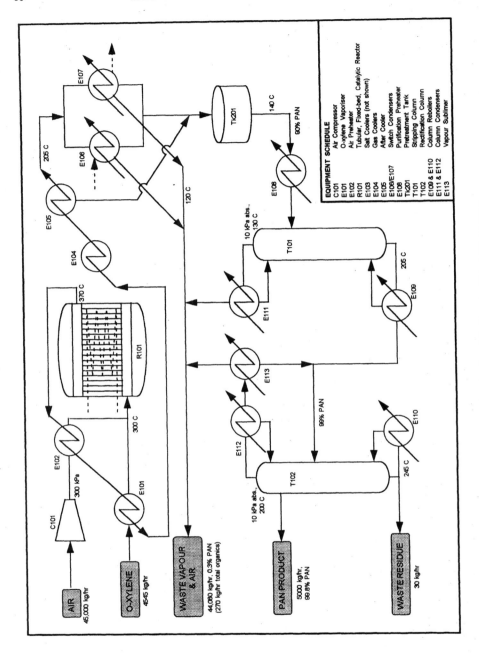

Figure 3.1 PFD for the phthalic anhydride LAR process.

Based on the above information, the following decisions were made regarding the construction of a new phthalic anhydride plant in Western Australia:

* *use o-xylene as the primary or sole feed,*
* *use a fixed-bed vapour-phase reactor,*
* *use either the LEVH or the LAR process.*

3.6.4 Detailed Process Description

The LAR and LEVH processes are based on essentially the same flowsheet and use similar equipment. Figure 3.1 is the process flow diagram (PFD) for the LAR process. It indicates the layout of the major items of equipment used in phthalic anhydride production. The key difference between the LAR and the LEVH process is the presence of an extra condensing unit (the After-Cooler) in the LAR process which allows some of the PAN to be recovered from the reaction gas by condensation rather than sublimation. The higher air to *o*-xylene ratio which is used in the LEVH process necessitates additional capacity in some items of equipment, such as the air compressor, the reactor, the gas-cooler and the switch condensers.

There are three principal steps in the production of phthalic anhydride: (a) reaction; (b) condensation; and (c) purification. The reaction is highly exothermic and must take place at high temperature (based on thermodynamic consider-ations). This requires heat removal via molten salt circulation. The hot molten salt can then be used to generate high-pressure steam. The reaction vapours are. cooled which causes the phthalic anhydride to sublime in switch condensers, where flakes of high-purity phthalic anhydride can be removed. Coolant and steam are fed alternately to the condensers in order to condense the vapour and then melt the solid phthalic anhydride, respectively. The impure product (98% PAN) is then removed to an intermediate storage tank. The effluent gas from the condensers (which still contains various organics) is stripped or incinerated before being vented to atmosphere. Crude PAN is purified by pretreating (to convert any remaining acid to anhydride) and then using vacuum distillation to remove maleic anhydride and other impurities. The product can be stored and sold either in molten form or as flake [3,7], and must meet the specifications shown in Table 3.1 [3]. These targets are normally met or comfortably exceeded by all commercial processes.

Table 3.1 Typical phthalic anhydride product specification.

	Typical Production	*Specification*
Phthalic anhydride purity	99.9%	99.7% min
Maleic anhydride content	0.05% max	0.15% max
Solidification point	130.9°C	130.8°C min
Molten colour	10	30 max
Molten appearance	water white	clear liquid

3.6.5   Advantages of the LAR Process

Major developments in phthalic anhydride manufacture in recent years have been in the area of catalyst activity. The LAR process uses a catalyst that requires an air:o-xylene ratio of only 9.5:1 (compared to around 20:1 for other technologies). This increases the catalyst productivity by 40% and dramatically reduces the reactor size, and lowers the duty of other critical equipment (including the switch condensers and evaporator). Energy requirements for blowers and pumps are reduced by about 60%. The reduced flow of inerts in the reactor allows more sensible heat to be extracted via the cooling salts. The plant can become completely energy self-sufficient and can export considerable quantities of steam to neighbouring plants. An after-cooler can also be added to remove about half of the product phthalic anhydride as a liquid before sublimation. This increases the product quality by reducing the concentration of impurities entering the product stream in the condensers [7,10].

The LAR process has a lower capital cost and a higher operating efficiency compared with other processes. This is a direct result of the low air to o-xylene ratio in the reactor. No electricity or fuel is required and a significant steam credit can be exported. The final selling price for phthalic anhydride produced by the LAR process is likely to be lower than by competitive processes.

3.6.6   Advantages of the LEVH Process

The principal advantage of the LEVH process is that it is proven technology with 65 plants around the world (a combined capacity of 1400 kT/yr) having been constructed to use the VH or LEVH process and most are successfully operated. Maintenance costs have proven to be low (less than 2% of capital per year) and support facilities are likely to be well developed [7].

By using a special catalyst, VH plants can be modified to operate on naphthalene or a mixture of naphthalene and o-xylene. This represents a safeguard against increasing o-xylene prices and flexibility if naphthalene prices fall dramatically. Reactors have been constructed with capacities up to 50 kT/yr.

The Low Energy Von Heyden process is a safe option, suitable for conservative developments where a guaranteed market exists or where the operating company has extensive prior knowledge through operating other plants based on the same process. An LEVH plant will have lower production costs than most older plants but is unlikely to compete effectively with a successfully operated LAR plant operating on an equal basis [11].

3.6.7   Process Selection

Although the Low Air Ratio process is not yet proven technology (only one plant is currently operating worldwide), it is considered to be the best choice for this plant as both operating costs and capital repayments are likely to be lower than for alternative routes. The potential saving is probably sufficient to offset any operational difficulties which may be encountered. However, adopting the same process and technology as other plants in the region also constrains minimum production costs to approximately those of other plants. Given locational disadvantages, this would significantly reduce the ability of a plant in Western Australia to compete effectively.

3.6.8 Initial Equipment Design

A list of major items of equipment, with approximate sizes, is required at an early stage to determine how much land is required so that a suitable site can be selected. Estimates of the equipment sizes have been based on data for a plant using the Low Air Ratio process to synthesise phthalic anhydride from o-xylene. A capacity of 40 kT/yr and an operational factor of 90% (8000 hr/yr on-line) produce a design basis of 4545 kg/hr of o-xylene and 45 T/hr of air (at a ratio of 9.5 [7]).

The major items of equipment (reactor, switch condensers and purification columns) were designed in some detail and are discussed below. Other items were specified by function and capacity, and only very approximate size estimates have been obtained at this time. Rules-of-thumb and approximations were applied, and calculation details (where relevant) are included in Appendix A at the end of this section.

The PAN process has the potential to be a net exporter of steam, but utility steam is required at various points throughout the process. By integrating the PAN plant with other plants on the same site, the steam balance can be satisfied more easily. It is assumed that any boilers required for high-pressure steam will be shared with other users. Similarly, cooling water facilities (treatment and a cooling tower) and a main stack which vents waste gases and combustion products will also be shared with other industries. These items are not specified as part of this study as their capacities will be intimately linked with cocurrent developments.

Phthalic anhydride is moderately corrosive. Consequently, some items of equipment need to be constructed from a stainless steel, such as type 316, as it offers a corrosion rate of only 0.025 mil/yr [36]. Items of equipment operating under less severe conditions can be constructed from plain carbon steel.

3.6.9 Equipment List

1. Reactor: flow 4545 kg/hr o-xylene, 45 T/hr air; temperature 350°C inlet, 500°C max.; pressure 200–300 kPa; internals 13,400 tubes of 3.0 m × 25 mm Ø, catalyst 2.8 m high; shell 5.9 m Ø × 3.5 m; cooling duty 21.4 MW; salt circulation 310 T/hr; stainless steel type 316.
2. PAN Pre-Condenser: total flow 50 T/hr, 2500 kg/hr PAN; duty 450 kW; carbon steel.
3. Switch Condensers (2): total flow 47 T/hr, 2500 kg/hr PAN; duty 450 kW; carbon steel.
4. Tail Gas Scrubber: flow 45 T/hr vapour; duty 250 kg/hr organics; carbon steel.
5. Vaporiser: flow 4625 kg/hr o-xylene, 45 T/hr air temperature 125–350°C; duty 3.9 MW; carbon steel.

6. Gas Cooler: flow 50 T/hr total gases; duty 5.0 MW; carbon steel.
7. Pretreatment Tank: flow 5000 kg/hr PAN, 300 kg/hr other components; temperature 70–270°C; heating duty 290 kW; carbon steel.
8. Stripping Column: flow 5000 kg/hr PAN, 300 kg/hr other components; partial vacuum; 15–20 packed stages (low pressure drop); diameter 900 mm; high reflux ratio (5–10); reboiler temperature 205°C; reboiler duty 150 kW; stainless steel type 316.
9. Rectification Column: feed 5000 kg/hr PAN, 20 kg/hr residue; partial vacuum; 15–20 trays (very low liquid rate); diameter 1200 mm; very low reflux ratio (0.1); reboiler temperature 245°C; reboiler duty 650 kW; stainless steel type 316.
10. Miscellaneous Heat Exchangers (8): air preheater; o-xylene preheater; gas cooler; reboilers (2); condensers (2); sublimator.
11. Tanks and Drums (6): various sizes for temporary storage of feed, crude PAN and product PAN to provide 14 days supply.
12. Pumps (8 + spares): various sizes for transport of feedstocks and intermediate products.
13. Air compressor: flow 45 T/hr air to reactor; pressure 200–300 kPa.
14. Filter: air cleaning prior to reaction.
15. Ejector System: vacuum source (10 kPa abs.) for purification system.

3.6.10   Conclusions

- Naphthalene is being phased out as a feedstock for phthalic anhydride plants as o-xylene is cleaner and produces higher yields.
- *Ortho*-xylene must be separated from a mixed xylenes stream and its price is linked to the overall xylene demand. Of the other isomers, p-xylene markets are growing strongly but m-xylene has no significant use.
- New processing methods offer the opportunity to reduce costs by 25% by using o-xylene as the feed, utilising better catalysts and implementing energy integration techniques. The Australian plants are operating at a significant economic disadvantage.
- *The LAR process for phthalic anhydride synthesis is newer than competitive processes but it is the most attractive as equipment sizes can be reduced.*

3.6.11   Recommendations

- Locate a suitable site within Western Australia for the construction of a new phthalic anhydride plant.
- Develop a plant layout and determine space requirements.
- Perform more detailed mass and energy balances for the plant to confirm equipment sizes and determine other equipment requirements.
- Decide upon key process conditions and assess the environmental impact.
- Approach Government and industry to promote a consortium of chemical companies to exploit possible synergies.
- Determine the costs of the major items of equipment required for a 40 kT/yr PAN plant and conduct an economic evaluation of the process.

APPENDIX A. PRELIMINARY EQUIPMENT SPECIFICATIONS

A.1 Reactor

Configuration: packed bed, tubular reactor; tubes 25 mm $\emptyset \times 3.0$ m [7]
Phthalic anhydride yield $= 1.10$ kg PAN/ kg o-xylene [7]
O-xylene feed rate $= 5000$ kg/hr/1.10 $= 4545$ kg/hr
Hydrocarbon loading $= 0.34$ kg/h/tube [7]
Number of tubes required $= 4545/0.34 = 13,400$
Estimated tube spacing $= 40$ mm c-c
Estimated reactor diameter $= \sqrt{13,400 \times (4/\pi)} \times 40$ mm $= 5.9$ m (say, 6.0 m)
Heat generated $= \{$selectivity $\times \Delta H_{PAN} + (1 -$ selectivity$) \times \Delta H_{CO_2}\} \times$ moles o-xylene
$\qquad = \{0.788 \times 265$ kcal/mol $+ (1 - 0.788 \times 1046$ kcal/mol$)\} \times (4545/0.106)$
$\qquad = 18.5 \times 10^6$ kcal/hr $= 21.4$ MW
Salt circulation rate $= Q/c_p \Delta T = 21,400/(2.5$ kJ/kg/°C $\times 100$°C$) = 86$ kg/s $= 310$ T/hr
Heat transfer area $= 13,400$ tubes $\times (\pi \times 25.4$ mm $\times 3.0$m$) = 3200$ m^2

A.2 Salt Coolers

Reactor cooling duty $= 21.4$ MW
H_v of steam @ 20 bar $= 1880$ kJ/kg
Steam produced $= (21,400/1880)$ kg/s $= 11.4$ kg/s $= 41.0$ T/hr
Typical boiling HTC $= 500$ W/m^2/°C [38]
Estimated temperature difference $= 40$°C
Heat transfer area required $= Q/U \Delta T = 21.4 \times 10^6/(500 \times 40)$ m$^2 = 1100$ m^2
HT area per 6.0 m $\times$ 1.8m $\emptyset$ exchanger shell (max. ASTM size) $= 1040$ m^2 [39]
Number of shells required $=$ at least 2 due to high vapour flow rates

A.3 Switch Condensers

Process objective: sublime 50% PAN vapour to solid [7]
Inlet gas flow $= 45$ T/hr air $+ (0.50 \times 5000$kg/hr PAN$)$
Outlet gas flow $= 45$ T/hr air
Outlet liquid flow $= 2500$ kg/hr crude PAN
Duty $= 2500$ kg/hr PAN $\times 650$kJ/kg $= 450$ kW
Estimate $U = 500$ W/m^2/°C; $LMTD = 25$°C
HT area $= Q/(U \times LMTD) = 450/(0.500 \times 25) = 36$ m^2 (390 ft^2)

A.4 After-Cooler

Process objective: condense 50% PAN vapour to liquid [7]
Inlet gas flow $= 45$ T/hr air $+ (0.50 \times 5000$kg/hr PAN$)$
Outlet gas flow $= 45$ T/hr air
Outlet liquid flow $= 2500$ kg/hr crude PAN

Duty $= 2500$ kg/hr PAN $\times 650$ kJ/kg $= 450$ kW
Estimate $U = 500$ W/m^2/°C; $LMTD = 40$°C
HT area $= 450/(0.500 \times 40) = 22.5$ m^2 (240 ft^2)

A.5 Vaporiser

Process objective: vapourise o-xylene feed and heat to reactor inlet temperature (350°C)
Duty $= 4545$ kg/hr $\times$ (346 kJ/kg $+ 325$°C $\times 0.7$ kJ/kg/°C) $= 725$ kW
Estimate $U = 500$ W/m^2/°C; $LMTD = 50$°C
HT area $= 725/(0.500 \times 50) = 29$ m^2 (310 ft^2)

A.6 Air Preheater

Process objective: heat air to reactor inlet temperature (350°C)
Duty $= 45,000$ kg/hr $\times$ (325°C $\times 1.2$ kJ/kg/°C) $= 4.9$ MW
Estimate $U = 200$ W/m^2/°C; $LMTD = 50$°C
HT area $= 4900/(0.200 \times 50) = 490$ m^2 (5200 ft^2)

A.7 Gas Cooler

Process objective: cool reaction gases to after cooler inlet temperature (205°C)
Duty $=$ (5000 kg/hr PAN $\times 295$°C $\times 0.8$ kJ/kg/°C)
       $+$ (45,000 kg/hr air $\times 295$°C $\times 1.2$ kJ/kg/°C) $= 4.75$ MW
Estimate $U = 200$ W/m^2/°C; $LMTD = 50$°C
HT area $= 4750/(0.200 \times 50) = 475$ m^2 (5100 ft^2)

A.8 Stripping Column

Process objective: remove approx. 270 kg maleic anhydride (light) from PAN product [7]
Pressure: vacuum (10 kPa abs.), therefore packing preferred
Shortcut method: minimum stages $= 6.6$ [39]
Actual stages: 20–30 @ 70% estimated overall efficiency (via simulation)
Low vapour loading, therefore small diameter (750–900 mm); low reboiler duty (approx. 110 kW); high reflux ratio (5–10) [39]
Reboiler temperature: 205°C @ 15 kPa abs.
Condenser temperature: 130°C @ 10 kPa abs.

A.9 Purification Preheater

Process objective: heat crude PAN to column feed temperature (180°C)
Duty $= 5300$ kg/hr $\times 40$°C $\times 0.8$ kJ/kg/°C $= 170$ kW

Estimate $U = 300\,W/m^2/°C$; $LMTD = 25°C$
HT area $= 50/(0.300 \times 25) = 6.3\,m^2$ $(70\,ft^2)$

A.10   Rectification Column

Process objective: remove approx. 20 kg heavy residue from PAN product [7]
Pressure: vacuum (10 kPa abs.)
Shortcut method: minimum stages $= 2.8$ (depending on impurities) [39]
Actual stages: 20–30@60% estimated overall efficiency, with safety margin for unknown impurities
Low liquid loading, therefore trays preferred; very low reflux ratio (0.1–0.15)
Reboiler duty: approx. 650 kW [39]
Column diameter: approx. 1200 mm [39]
Reboiler temperature: 245°C@20 kPa abs.
Condenser temperature: 200°C@10 kPa abs.

A.11   *Ortho*-Xylene Feed Tanks

Process objective: 10 days cover at maximum feed rate
Volume $= 10$ days $\times$ 24hrs/day $\times$ 4545 kg/hr $/$ 850 kg/m^3 (ρ@25°C) $= 1285\,m^3$
Tank dimensions: estimate $H/D = 0.5$; 10% freeboard
   volume $= (\pi D^2/4)\,(0.5\,D)$
   $D^3 = 1285 \times 110\% \times (8/\pi)$
   $D = 15.3\,m$; $H = 7.7\,m$
Use standard dimensions of 15.0 m (50 ft) $\emptyset \times$ 7.5m (25 ft)

A.12   Phthalic Anhydride Product Tanks

Process objective: 10 days supply at maximum product rate
Volume $= 10$ days $\times$ 24hrs/day $\times$ 5000kg/hr $/$ 1210 kg/m^3 (ρ@140°C) $= 990\,m^3$
Tank dimensions: estimate $H/D = 0.5$; 10% freeboard
   volume $= (\pi D^2/4)\,(0.5\,D)$
   $D = 14.0\,m$; $H = 7.0\,m$
Use standard dimensions of 15.0 m (50 ft) $\emptyset \times$ 7.5m (25 ft) to match *o*-xylene feed tanks

A.13   MAN Product Tank

Process objective: 21 days supply due to lower lifting rate
Volume $= 21$ days $\times$ 24hrs/day $\times$ 270kg/hr$/$1100 kg/m^3 $= 125\,m^3$
Tank dimensions: estimate $H/D = 1.0$; 10% freeboard
   volume $= (\pi D^2/4)\,(D)$
   $D^3 = 125 \times 110\% \times (4/\pi)$

$D = 5.6$ m; $H = 5.6$ m
Use standard dimensions of 6.0 m (20 ft) $\emptyset \times 6.0$m (20 ft)

A.14 PAN Pretreatment Tank

Process objective: 24 hrs supply to the purification equipment
Volume $= 24$ hrs $\times 5300$kg/hr / 1210 kg/m^3 (ρ @ 140°C) $= 105$ m^3
Tank dimensions: estimate $H/D = 1.0$; 10% freeboard
   volume $= (\pi D^2/4)\,(D)$
   $D^3 = 105 \times 110\% \times (4/\pi)$
   $D = 5.3$ m; $H = 5.3$ m
Use standard dimensions of 6.0 m (20 ft) $\emptyset \times 6.0$m (20 ft)

A.15 Ejector

Process objective: provide 90 kPa vacuum to the purification columns
Estimate two stages required, 25 kg/hr vapour

A.16 Air Compressor

Flow $= 45$ T/hr (38,600 Nm^3/hr)
Discharge pressure $= 300$ kPa

A.17 Pumps

Typical liquid flow $= 5000 - 5500$ kg/hr $= 4.1 - 4.5$ m^3/hr
Typical head $= 30 - 60$ m
Number required $= 7 + 7$ spares

4. SITE CONSIDERATIONS:
SITE SELECTION AND PLANT LAYOUT

4.1 SITE SELECTION/LOCATION

Some teaching institutions that undertake design projects omit consideration of the site selection for the chemical plant. The IChemE (UK) design project does not specify this aspect of the design to be considered, requiring only that detailed equipment designs for the plant are performed. However, in order to make the design project realistic and to provide the student with experience of as many aspects of design as possible, some consideration should be given to the site selection and plant layout.

4.1.1 Local Industrial Areas

In order to make the acquisition of relevant information easier, it is preferable to consider site selection on a local basis. If no local markets exist for the chemical or there are no areas suitable for chemical plant construction, it will be necessary to consider industrial regions elsewhere within the state or the country. Maps showing areas designated for heavy industrial development and the location of existing plants are usually available from the relevent government department. Even though an area has been designated as suitable for industrial development does not mean that permission will automatically be given to construct a new chemical plant, or that the location is actually suitable for the proposed plant. This comment can also apply to existing chemical plant sites, and in some cases further development of a site may be restricted by the government due to public opinion, environmental impact problems, or concerns regarding the risk of a major accident in a highly industrialised area. Workers usually prefer to live near their employment, and housing usually develops near industrial sites which were originally uninhabited, e.g. Bhopal! This can then lead to strong public opinion against further development.

Assuming that areas for industrial development can be identified, it is necessary to select a particular site. This should not be considered a 'one horse race'. Several sites need to be considered and the merits and disadvantages of each site should be itemized. The final outcome will be a list of suitable sites in decreasing order of preference. To concentrate on one site and neglect all others leaves the project vulnerable to the dictates of government.

4.1.2 Some Important Factors

Several factors influence the selection of a suitable site for the location of a chemical plant. The following list contains a few of the important considerations, but should not be considered exhaustive:

(a) Designation as a heavy industrial development area ('light' industry usually means assembly of electronic components, small metal fabricators, etc., and a major chemical plant would not be acceptable).

(b) Prior existence of similar chemical plants and location of other industrial centres.

(c) Existing infrastructure, e.g. roads, and services such as electricity, gas, water, etc.

(d) Appropriate terrain, sub-surface, drainage, etc.

(e) Suitable access for transportation of raw materials and chemicals, and for the construction of a chemical plant.

(f) Proximity to major transportation networks, e.g. roads, railways, airports, waterways, ports. This is a major consideration in the location of a plant. In some cases direct pipelines for the transportation of chemicals or utilities (e.g. water, gas, oil) may be the most economical method. The cost of transportation by tanker (road, rail or sea) is reduced if a return load can also be carried.

(g) Availability of a local workforce and distance from local communities.

(h) Availability of domestic water and plant cooling water.

(i) Environmental discharge regulations.

(j) Proximity to both the raw materials supply and the market for the product chemical.

(k) Existence of services equipped to deal with a major industrial accident.

(l) Climatic conditions, e.g. humidity, maximum wind velocity and its prominent direction, rainfall, etc.

(m) Proposed or possible government restrictions regarding industrial development or discharge emissions.

(n) Space for expansion.

(o) Price of land.

(p) Public opinion.

(q) Possibility of earthquakes, subsidence, avalanches, etc.

(r) Availability of government regional development grants or tax incentives, subsidies, etc.

Many other factors could be included, the most important factors are those specific to the requirements of a particular process or peculiar to a certain site. Several of these factors are discussed in more detail in Backhurst and Harker (1973; pp. 374–377); Baasel (1990; pp. 21–39); Coulson and Richardson Volume 6 (2nd edn, 1993; pp. 797–799); references are given in Section IV here.

4.1.3 Prioritizing the Factors

The three major site location factors are usually the location of raw materials, location of markets, and transportation. Once these aspects have been determined, other important (but secondary) considerations can be identified and evaluated.

However, even if many of the basic requirements for a chemical process are satisfied by a particular site, the site still may not be acceptable if certain important criteria cannot be fulfilled, e.g. lack of government approval due to adverse public opinion, environmental regulations, availability of labour, etc. Therefore, it is necessary to consider the process of site selection as an overall package of *essential* requirements — all of which must be satisfied by the preferred site location; primary factors (probably 3 or 4) which are highly desirable; and achievement of as many desirable (but non-essential) requirements as possible. Final site selection (i.e. obtaining a listing of site preferences) may be helped by listing all the features (required and existing) for each site, and assigning a subjective numerical value (say 1 to 5, or 9) for each feature. Each location can then be given a final points rating. Some discretion must be exercised because serious disadvantages must not be obscured by this approach. For example, a site that was otherwise ideal (or top of the list of preferences) but which had no local water available would probably be of little use as a chemical plant complex!

Action: *Prepare a list of suitable sites for the construction of the chemical plant, arranged in descending order of preference.*

List all the merits and disadvantages of each site, and explain in detail why the first site is preferred.

State any factors that need to be monitored throughout the project, e.g. changes in government policy.

References

Baasel (1990; Chapter 2) and Peters and Timmerhaus (3rd edn, 1980; Chapter 3). Both chapters include detailed reference lists.

4.2 PLANT LAYOUT

Having selected a suitable site for the chemical plant, it is then possible and necessary to make a preliminary decision regarding the layout of the plant equipment. Although the equipment has not been designed in detail, *preliminary estimates* of the physical size of each item should be available in the equipment list (see Section 3.3). Any sizing differences between the initial and final estimates should not be too excessive, and appropriate areas for access, maintenance and safety (and insurance) should be allowed around the plant items when determining the layout.

A preliminary determination of the plant layout, also called *the plot plan*, enables consideration of pipe runs and pressure drops, access for maintenance and repair, access in the event of accidents and spills, and location of the control room and administrative offices. The preliminary plant layout can also help to identify undesirable and unforeseen problems with the preferred site, and may necessitate a revision of the site selection (Section 4.1). The proposed plant layout must be considered early in the design work, and in sufficient detail, to ensure economical construction and efficient operation of the completed plant. The site layout must be suitable for equipment construction purposes, e.g. movement of cranes, etc. The plant layout adopted also affects the safe operation of the completed plant, and acceptance of the plant (and possibly any subsequent modification or extension) by the community.

4.2.1 Plant Layout Strategies

There are two schemes that can be adopted for determination of the plant layout. First, the *'flow-through' layout (or 'flow-line' pattern)* where plant items are arranged (sequentially) in the order in which they appear on the process flowsheet. This type of arrangement usually minimises process pipe runs and pressure drops, and is often adopted for small plants. Second, the equipment is located on site in *groupings of similar plant items*, e.g. distillation columns, separation stages, reactors and heat exchangers, etc. The grouped pattern is often used for larger plants and has the advantages of easier operation and maintenance, lower labour costs, minimising utility transfer lines and hence reducing the energy required to transfer materials. These two schemes represent the extreme situations and in practice some compromise arrangement is usually employed. The plant layout adopted depends upon whether a new ('grass roots') plant is being designed or an extension/modification to an existing plant. Space restrictions are the most common restraints on plants to be developed on existing sites, however, space limitations are usually imposed even with new sites.

4.2.2 Factors Influencing Plant Layout

Other factors to be considered in the equipment arrangement are:

(a) Location of the control room, offices, etc., away from areas of high accident risk, and upstream of the prevailing winds.
(b) Location of reactors, boilers, etc., away from chemical storage tanks.
(c) Storage tanks to be located for easy access, and a decision made as to whether all tanks (for raw materials and product) should be located together or dispersed around the site perimeter.
(d) Ease of access for plant operators, and distances between equipment.
(e) Elevation of equipment.
(f) Requirements of specific plant items, e.g. pumps, and advantages of static head.
(g) Supply of utilities, e.g. electricity, water, steam, etc.
(h) Minimizing plant piping systems.
(i) Suitable access to equipment requiring regular maintenance or repair.
(j) Plant layout to facilitate easy clean-up operations and dispersion of chemicals in the event of a spillage.
(k) Access to the plant in the event of an accident.
(l) Location of equipment requiring cooling water close to rivers, estuaries, etc.
(m) Location of plant waste collection and water drainage systems (separate or combined?) and treatment tanks.

(n) Adopting a plant layout that will act to contain any fires or explosions.

(o) Spacing between items of equipment. Companies specialising in the insurance of chemical plants have specific recommendations for the distances required between particular items of equipment. Some recommendations are given in national Standards and Codes of Practice.

Particular factors that need to be considered in the plant layout stage of the design are discussed in more detail in Backhurst and Harker (1973; pp. 377–391; Coulson and Richardson Volume 6 (2nd edn, 1993; pp. 799–805); and Baasel (1990; Chapter 6).
The layout of plant equipment should aim to minimise:

(i) damage to persons and property due to fire or explosion;
(ii) maintenance costs;
(iii) the number of plant personnel required;
(iv) operating costs;
(v) construction costs;
(vi) the cost of plant expansion or modifications.

Some of these aims are conflicting, e.g. (i) and (iv), and compromises are usually required when considering the plant layout in order to ensure that safety and economic operation are both preserved. The final layout will depend upon the requirements for energy conservation schemes within the plant (see Section 8.1) and any subsequent modifications, and the associated piping arrangements.

Action: *Prepare a preliminary plant layout, explaining the reasons for the particular proposal and any alternative schemes.*

References

Baasel (1990; Chapter 6); and Peters and Timmerhaus (3rd edn, 1980; Chapter 3); detailed reference lists are also included in these chapters.

House, F.F., An engineers guide to process-plant layout, *Chem. Eng. (N.Y.)*, 28 July, 120–128 (1968).

Jayakumar, S. and Reklaitis, G.V., Chemical plant layout via graph-partitioning; Part 1: Single level; Part 2: Multiple levels, *Comput. Chem. Eng.*, 18(5), 441–458 (1994); 20(5), 563–578 (1996).

Kern, R., Practical piping design (12 part series), *Chem. Eng. (N.Y.)*, published between 23 December, 58–66 (1974) and 10 November, 209–215 (1975).

Kern, R., Plant layout (12 part series), *Chem. Eng. (N.Y.)*, published between 23 May, 130–136 (1977) and 14 August, 141–146 (1978).

Madden, J., Pulford, C. and Shadbolt, N., Plant layout: Untouched by human hand? *The Chemical Engineer (Rugby, Engl.)*, 24 May, 32–36 (1990).

Marshall, V., Consider plant layout in HAZOP studies, *The Chemical Engineer (Rugby, Engl.)*, 28 July, 18–19 (1994).

Mecklenburgh, J.C., *Process Plant Layout*, Halstead Press, New York (1985).

Meyer, M. *et al.*, Optimal selection of sensor location on a complex plant using a graph-oriented approach, *Comput. Chem. Eng.*, 18(supplement), S535–S540 (1994).

Ottino, C., Design and layout of a control room, *Chem. Eng. (N.Y.)*, September, 174–180 (1991).

Parkinson, G., Site selection in Europe, *Chem. Eng. (N.Y.)*, April, 41–48 (1992).

Penteado, F.D. and Ciric, A.R., An MINLP approach for safe process plant layout, *Ind. Eng. Chem. Res.*, 35(4), 1354–1361 (1996).

Realff, M.J., Shah, N. and Pantelides, C.C., Simultaneous design, layout and scheduling of pipelines in batch plants, *Comput. Chem. Eng.*, 20(6), 869–884 (1996).

Robertson, J.M., Design for expansion, *Chem. Eng. (N.Y.)*, 22 April, 179–184; 6 May, 187–194 (1968).

Russo, T.J. and Tortorella, A.J., CAD in plant layout, *Chem. Eng. (N.Y.)*, April, 97–101 (1992).

Thompson, D., Rational approach to plant design, *Chem. Eng. (N.Y.)*, 28 December, 73–76 (1959).

Various, Plant layout (feature report), *Chem. Eng. (N.Y.)*, April, 81–101 (1992).

4.3 CASE STUDY — SITE CONSIDERATIONS: SITE SELECTION AND PLANT LAYOUT

Summary

It is proposed to develop a 40 kT/yr phthalic anhydride plant in Western Australia at the Kemerton Industrial Park, near Bunbury. The development should be a joint venture with several other industries with whom important synergies exist (shared raw materials and end-users). The location was chosen for its existing infrastructure, close proximity to road, rail and sea transport, and minimising overall transportation costs. Government assistance is anticipated in order to encourage the decentralisation of industry within Western Australia.

The plant layout was specified primarily for safety considerations, but also to minimise piping costs and to maximise ease of access and flexibility. Sufficient land is available on the site for the co-development of other related industries which will share some central facilities (steam, cooling water, discharge stack and administration buildings).

4.3.1 Background and Objectives

Based on the previous market survey and preliminary feasibility study, there appears to be a niche for a new phthalic anhydride plant in Western Australia. However, it will be important to use the latest technology and to absorb the market share of Australia's existing, outdated PAN production facilities in New South Wales. Two scenarios are being considered: a production rate of 30 kT/yr; and a production rate of 60 kT/yr. The higher capacity case is the optimistic option and relies on a significant volume of export sales on the spot market. *A 30 kT/yr capacity plant appears to be more viable at this stage.* For the purposes of estimating equipment requirements and for site selection, a 40 kT/yr plant will be assumed. A larger (or smaller) plant may still prove to be more viable after considering more accurate capital cost estimates and a detailed profitability analysis. It will be important to integrate the plant capacity with other industries located at the same industrial site.

Site selection is likely to have a major influence on the overall viability of the process as it has already been determined that a stand-alone development is unlikely to be profitable. Government sponsorship of a new industrial area is probably required and synergies with other industries will need to be sought. Ideally, the selected site must:

(a) meet all process requirements;
(b) satisfy environmental regulations and community aspirations;
(c) minimise transportation costs (both raw materials and product) in order to maximise profitability.

Once the site is chosen, a satisfactory plant layout that incorporates safety and functionality and minimises piping costs, must be designed. The layout should be specific to both the industry and the location.

4.3.2 Potential Sites

Four sites within Western Australia were selected for further consideration: Geraldton, Karratha, Kemerton and Kwinana. Of these, only Kwinana is an existing industrial estate of any significant size although the other sites have available land classified for heavy industrial use. Government assistance during the construction phase and possibly also in the form of subsidies or tariffs is possible for the sites outside Perth (i.e. Geraldton, Karratha and Kemerton). They would be developed as part of a new industrial complex in order to decentralise heavy industry within Western Australia. This is considered essential to reduce the costs of infrastructure development and overcome marginal economics. The following industries have been identified as having significant synergies with a phthalic anhydride plant and would be encouraged in a joint venture development:

- mixed xylene separation plant (to provide cheap and reliable raw material to PAN plant)
- phthalate ester plant (end-user and sink for exported steam and energy)
- polyester resin plant (end-user and sink for exported steam and energy)
- maleic anhydride plant (processes PAN plant by-product)
- terephthalic acid plant (processes p-xylene from xylene separation plant)

Currently, the Australian phthalic anhydride market is based in the Eastern States. End-use industries would need to be relocated to Western Australia (whilst simultaneously updating their ageing facilities to remain competitive) for a WA-based PAN plant to be viable. The only current Australian manufacturer of PAN is ICI in Sydney, and they receive their raw material (naphthalene) directly from a neighbouring industry. This feed arrangement would also need to be replicated for a WA PAN plant, and would require the construction of Australia's first mixed xylene separation plant which would process a benzene–toluene–xylene (BTX) fraction from the BP Oil Refinery at Kwinana. Otherwise, o-xylene will need to be imported from SE Asia. This would create a significant vulnerability for a new industry and may render the entire development uneconomic.

4.3.2.1 Kemerton

Kemerton Industrial Park is already a site for several new industries within Western Australia. Infrastructure development is at a useful level and road, rail and sea access is available. It is the closest of the remote sites to Perth and has a significant amount of available land for future developments. Labour could come directly from Perth or from the developing city of Bunbury (only 15 km away). Construction costs are also likely to be lower than for the other remote sites.

4.3.2.2 Geraldton

Bootenal West, near Geraldton, has good road, rail and sea access, and is located away from local residents so that council approval is probable. The area is not particularly environmentally sensitive and river water could be used for cooling. The terrain is flat and would be suitable for major construction works. Perth is still reasonably close (600 km) for head office facilities, emergency response and raw material supply.

4.3.2.3 Karratha

A new industrial development in Karratha would utilise infrastructure which is being developed for major gas projects on the North-West shelf. Although some condensate production is expected, there is unlikely to be an adequate supply of aromatics (BTX) to support a PAN plant. However, it does offer relatively easy access to Indonesian o-xylene. The downstream industries which would be located on the same site have markets that are distant from Karratha, and this would provide little incentive for their development. Port facilities are well developed and there is a ready supply of sea water for cooling.

4.3.2.4 Kwinana

The Kwinana industrial area is congested and has little room for a new complex containing several plants. Although road, rail and sea transport are available, and construction costs are probably the cheapest of any of the potential sites, land is only available at a premium. Government assistance for another development at Kwinana is unlikely to be forthcoming.

4.3.3 Preferred Site and Layout

Kemerton is the preferred site for a new phthalic anhydride plant in Western Australia. A Kemerton site would satisfy all the requirements for heavy industrial development, and the particular needs of a PAN plant. Figures 4.1 and 4.2 are

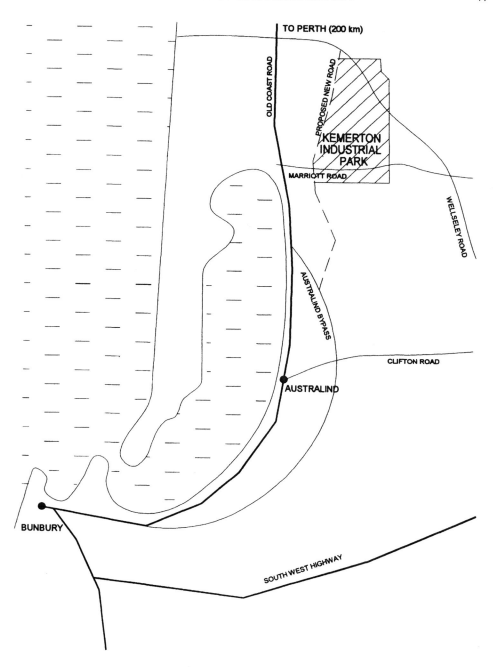

Figure 4.1 Kemerton region.

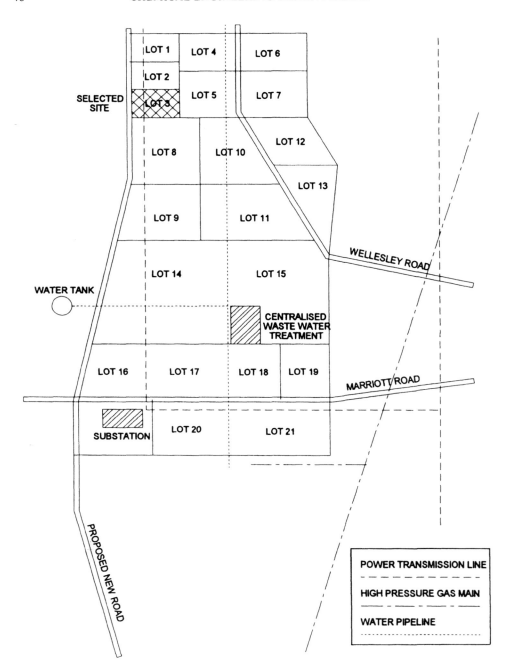

Figure 4.2   Kemerton Industrial Area layout.

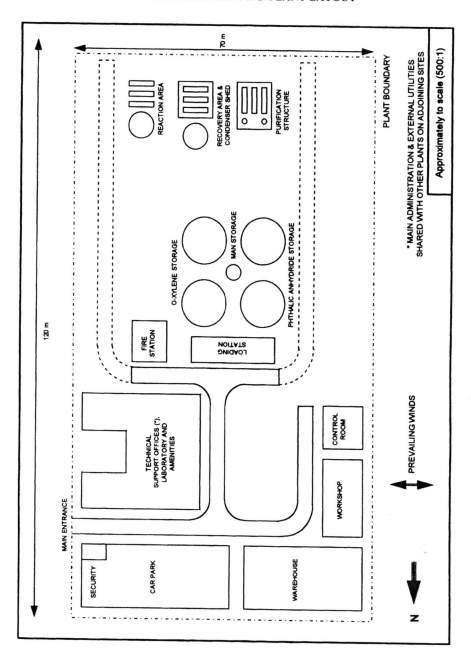

Figure 4.3 Phthalic anhydride plant layout.

maps of the area showing the core industrial zone and a buffer zone for a proposed industrial complex and existing infrastructure.

There are several suitable sites within the Kemerton area. Lot 3 (see Figure 4.2) was selected as having sufficient area to accommodate a PAN plant and several cocurrent developments, and also direct road and rail access. Shared facilities (steam, cooling water and stack) could be located on this lot or an adjoining lot. A phthalic anhydride plant has only a small demand on cooling water and could be situated away from the central cooling water supply if required. Similarly, the PAN plant would be largely self-sufficient in steam. The prevailing wind direction is westerly (south-westerly in summer and north-westerly in winter), but easterly winds are also common in summer. PAN forms hazardous vapour clouds if process leaks occur so that the plant should be located downwind of other plants. The total number and types of other industries to share this site will influence the final choice of individual location and orientation.

A preliminary layout for the major items, including process equipment and buildings, of a phthalic anhydride plant located at Kemerton is shown in Figure 4.3. The plant is based on a north–south axis as winds are common from both the east and the west. With this alignment, personnel and service buildings should almost never be downwind. The control room is also to the north of the plant to protect it in the event of an emergency. The small overall size of the plant allows the control room to be located away from the centre of the main processing equipment. In the event of an emergency, the plant can be viewed (line of sight) from both the office buildings and the control room.

Storage facilities will be located reasonably close to both the feed point and the product discharge point which minimises piping and pumping costs. The loading station can be accessed easily via the main entrance. Clearance around the storage areas is low but should be sufficient in emergency situations.

The process equipment is split into two areas: reaction and purification. This separates the potentially hazardous reactor from other equipment. In the event of an accident, the prevailing winds should carry vapours away from other equipment. Market conditions suggest that future expansion will be very limited. If necessary, there is sufficient space for the de-bottlenecking of major items of equipment (for example, a second reactor could be added), but there is no specific area set aside for a second train.

Medical facilities and other essential services will be shared with other industries on the site. Senior management, secretarial, personnel and accounting staff will also be housed in the same building, away from the process operations. The small size of the phthalic anhydride plant precludes the need for non-operational staff to be on site.

4.3.4 Conclusions

- A 40 kT/yr phthalic anhydride plant is considered to be moderately viable. More detailed feasibility issues will be considered on this basis.
- Sites at Kemerton, Geraldton, Karratha and Kwinana were considered for a PAN plant in Western Australia.

- *The preferred site is located at the Kemerton Industrial Park.* This site will decentralise industrial development from Kwinana, has suitable site conditions, available land, good transport facilities, and the centre is still sufficiently close to Perth (200 km) to utilise those services.
- *A phthalic anhydride plant should be built as part of an industrial complex* with at least one downstream processing industry at the same site. *A polyester resin plant would be most suitable for WA* as it would supply the local boat building industry.
- Government assistance would be important in the initial construction of a new industrial complex and in maintaining competitive prices given the remoteness of the region.
- A plant layout was designed to protect workers and vital buildings, and to separate potentially hazardous process items. The process layout is also designed to minimise piping and pumping costs. Some auxiliary services will be shared with other industries on the same site.

4.3.5 Recommendations

- Investigate the environmental impact of a new phthalic anhydride plant on a Kemerton site.
- Perform more detailed mass and energy balances for the plant to confirm equipment sizes.
- Determine the costs of the major items of equipment required for a 40 kT/yr PAN plant, and estimate transportation costs for a detailed profitability analysis.
- Ensure design modifications are made to minimise process hazards.
- Complete detailed hazard and operability (HAZOP) studies after the P&ID is developed.

5. ENVIRONMENTAL CONSIDERATIONS

5.1 ENVIRONMENTAL IMPACT ASSESSMENT

This very important topic needs to be considered at an early stage in the design, particularly in relation to the site considerations, and reassessed during the detailed design stage (see Chapters 8 and 9). Consideration of the environmental aspects of a project are usually assigned a specific place in an undergraduate course (and in a design textbook), however in an industrial project the environmental factors would be considered from the outset and then at each subsequent stage of the project. A well thought-out environmental plan will be an essential factor in a successful chemical engineering development. Academic convenience means that design units are divided into structured modules to be completed in a logical order, however this should not be considered as a 'real-life' approach. The effect of operation of the chemical plant upon both the environment and the population *must* be considered during both the feasibility study and during subsequent design stages.

The formal *Environmental Impact Assessment (EIA)* has two parts, these relate to:

(a) the treatment of unwanted chemicals (by-products), and the concentrations of liquid discharges and gaseous emissions during normal operation (and during startup and shutdown);
(b) the handling of a major chemical accident or spill, including all chemicals within the plant and any subsequent reaction products, their containment and cleanup.

The EIA will be specific to the plant under consideration, although the comments/suggestions made here are of a general nature.

5.2 GENERAL CONSIDERATIONS

All waste chemicals from the plant must be disposed of in an acceptable manner. Dumping of the waste is not allowed or, if it is, it may be prohibitively expensive. Some form of treatment, e.g. dilution, neutralization, purification, separation, etc., may be necessary prior to disposal. It is necessary to determine whether it is more economical (and preferable for the efficient operation of the plant) to perform this treatment within the chemical plant itself. Consideration should also be given to the installation of separate drainage systems from certain sections of the plant and for particular waste streams, e.g. rainwater, domestic waste, relatively pure process water, acid spills, oils, etc. Separate drainage systems are best installed during the plant construction stage as it will be more difficult and costly to implement when the plant is operational. Therefore, the plant layout needs to be re-considered in terms of effective waste treatment and siting of containment and drainage systems. Separate drainage schemes will be more expensive than a central collection tank for all liquid waste, but will allow separate treatment and possible recycling of liquids.

The concentrations of all chemicals to be discharged (including gaseous emissions), and the flowrates of the particular components, must be determined. For example, the gas composition from a discharge stack should be determined (from a reasonably accurate mathematical model) to allow ground concentrations at various distances from the plant to be estimated. Measures must be taken to ensure that these levels conform to allowable legislative standards. In some cases discharge levels are actually specified, e.g. for SO_x and NO_x, and the allowable discharge for a new plant may have to conform to a maximum level for a particular region where it will also be necessary to consider the amounts expelled by neighbouring plants. In other situations actual allowable discharge levels are not published and a new plant proposal will be judged on its own merits, and best performance in relation to minimising environmental effects will be expected. In general, all wastes and discharges should be (pre-)treated to minimise the amount of material for removal (or further treatment), e.g. mechanical dewatering of solid wastes prior to disposal.

The cost of clean-up can be high, e.g. scrubbing systems, filters, etc., and will affect the economic analysis of the plant (see Chapter 6). It is prudent to ensure than not only are current emission standards observed, but also that the plant could conform to any subsequent legislative reductions in these emission levels (while still operating economically). When considering a new plant proposal it is probably sensible to design for discharge levels (in say 5 or 10 years) which are at least half of the current values.

The effects of all emissions on the environment, and upon company employees and the population must be assessed. The possibility of future litigation, say in 10 years time, should not be ignored, e.g. compensation for prolonged exposure to asbestos dust at the mining town of Wittenoom in Australia. Noise levels from the plant need to be considered, and finally, its aesthetic acceptability.

Why is Waste Treatment Such a Problem?

It is worth asking this question before embarking on an environmental assessment. It should also lead to the reason why environmental clean-ups are so costly. The answer is (usually) quite simple: Wastes from a chemical plant are often dilute, either in gas streams or in aqueous solutions, and the separation of the undesirable components prior to discharge requires energy intensive and, therefore, expensive operations. Traditional unit operations such as distillation require large amounts of heat if the components to be separated/removed are dilute, and this is why newer environmental separations are investigating the use of 'alternative' techniques such as membrane separators or supercritical fluid extraction. If the waste stream were concentrated then it would probably be easier to separate, or would be recycled, or it would have value as a by-product.

The latest catch-cry of the environmental lobby is *'clean technology'* — meaning that wastes are not produced in the first place, thereby avoiding difficult and expensive clean-up. It also means that 'throwing money at the problem' is not the answer, and simply spending more on state-of-the-art equipment means that the philosophy of reinventing process design has not been grasped. This is the message that should be emphasised to engineering graduates as they will be the plant designers of the future. It is also an essential principle to be used in a design project. *Reuse, Recycle, and Replace* is a good motto, but *Avoid Refuse* is equally valid!

The second part of the EIA relates to the effects of a major accident or spill within the plant. The safety aspects of, say, an explosive gas discharge should be considered in conjunction with the loss prevention studies for the plant (see Section 8.3). However, proposals for containment, clean-up and discharge of major chemical spills are required within the EIA report. All proposals should ensure the safety of personnel, minimise the discharge and its effect on the environment, and preserve the integrity of the plant. The worst situation should be evaluated, not just the most likely scenario. Factors to be considered include the quantity and location of chemicals stored within the plant, the prevailing wind direction, the location of plant personnel and the general public, and access to the plant and to particular high-risk areas.

There is a large amount of information available concerning chemical discharges, environmental protection, clean-up, and the applicable legislation. The Environmental Protection Agency (EPA) in the USA has produced a prodigious number of reports concerning a wide range of chemical plants. Much of the available information is specific to particular chemicals and process routes, however so much information has been published that locating a relevant source is usually the main problem. Most countries now require a company to file an EIA report as part of the proposal for a new chemical plant. This document is usually open to public scrutiny and can be a source of much useful information. Copies of the relevant legislation are readily available, and advice can be obtained from appropriate government departments.

Remember that however attractive a particular chemical plant appears (technically and economically), public and political opinion especially in relation to environmental issues can prevent the project from proceeding, or seriously reduce its economic feasibility. This can be seen in Australia where the three-mines (only) policy for uranium extraction was used to appease the environmental lobby (and other groups), despite the prospects of high economic returns for this commodity on international markets.

5.3 EIA POLICY AND SCOPE

Environmental Impact Assessment (EIA) has evolved as a comprehensive approach to project evaluation in which environmental factors, as well as economic and technical considerations (e.g. cost benefit analysis), are given appropriate consideration in the decision-making process. The purpose of an EIA study is to determine the possible environmental, social and health effects of a proposed development. It attempts to define and assess the physical, biological and socio-economic effects, so that logical and rational decisions are made. The identification of possible alternative sites and/or processes may assist in the reduction of potentially adverse impacts.

There is no general or widely accepted definition of an EIA and although many organizations may share common objectives, the contents and scope of an EIA may vary considerably. The conclusions or main points of an EIA are assembled into a document known as an *Environmental Impact Statement (EIS)* which contains a discussion of beneficial and adverse impacts considered relevant to the project, plan or policy. In some countries and states, the draft EIS must be made available for public inspection, debate and comment, and a final EIS may be required to include a response to relevant public opinion and discussion.

An EIA should improve the efficiency of decision-making by ensuring that subjectivity and duplication of effort are minimised, and by avoiding the short-term and long-term consequences of inappropriate decisions. However, in order to be effective an EIA should be implemented in the early stages of project planning and design, and it must be an integral component in project design rather than a technique to be utilised after the design stage is completed. An EIA procedure should be applied to all actions likely to have a significant environmental effect.

The responsibility for undertaking an EIA can be allocated in several ways, the choice depends upon the particular EIA system in operation, and the nature of the project. The four major alternatives that are commonly used (although others exist) to assign responsibility are: (i) to the authorising agency; (ii) to the developer; (iii) shared responsibility between the authorising agency and the developer; or (iv) an independent specialist body. Similarly, the cost of an EIA study can be allocated (or recovered) in a variety of ways. The impartiality of an EIA may be achieved by, for example:

(a) establish guidelines or minimum standards for the form and content of an EIA;
(b) supervision by a reviewing or controlling body with no vested interest in the project;
(c) mandatory consultation with relevant and competent organizations;
(d) publication and provision for public discussion of the impact statements.

5.4 EIA REPORTS

Many different EIA manuals are available (the majority produced or originally developed in the USA) which provide advice about identifying, predicting and displaying potential impacts or effects. Manuals often contain advice beyond these basic aspects including the structure and scope of the assessment, responsibility for the study, review procedures, implementation, monitoring, etc. Manuals are available which deal with specific types of developments; the EIA and resulting EIS requirements will depend upon the particular type of proposed project. The following discussion relates to general guidelines for performing an EIA study and preparation of the EIS document. It should not be considered to apply to every (or any) particular situation.

(i) Acquisition of Information on a Proposal

The developer should be invited to prepare a *short* brief concerning the proposal(s) including a list of key plant-location criteria, brief details of the proposed development, and the processes to be used when operational. The information available may be limited at this stage and the discussions are usually of a feasibility/evaluation nature.

As the project design and planning proceed, the developer is asked to complete a comprehensive *Project Specification Report* containing the following details:

 (i) physical characteristics of the proposed site;
 (ii) employment requirements;
 (iii) financial data;
 (iv) infrastructure requirements;
 (v) factors of environmental significance;
 (vi) emergency services;
(vii) hazards.

Information can also be obtained by site inspection, consideration of existing planning policies, and consideration of similar proposals.

(ii) Identification and Assessment of Possible Impacts

It is necessary to establish the nature of the existing environmental, social and economic conditions in the area surrounding a proposed development. The possible (or probable) impacts can then be identified and their implications assessed. Information may be required concerning the physical characteristics of the site and its surroundings, local human activity patterns, infrastructure services, social and community services, existing levels of environmental discharges, etc.

For each possible impact it is necessary to undertake an analysis of the scale and significance of the potential change. The environmental consequences of a project will not all be similar in degree or kind — they may tend to cancel each other out or to reinforce each other. The assessment should therefore distinguish between impacts which are:

Permanent/irreversible	or	Temporary/reversible
Short term	or	Long term
Local impact/acute	or	Distant impact/strategic/widespread
Direct	or	Indirect

It may also be necessary to assess not only the consequences of the project under consideration, but also any subsequent actions which become inevitable (or more likely) as a result of the proposal proceeding. Quantitative assessment of the environmental impact should be presented whenever this is possible, even though the available numerical techniques often vary in their precision.

(iii) Presentation of Findings

The conclusions of the EIA study regarding the potential impacts are presented in the EIS report. The EIS should include:

(a) a brief description of the proposed development;
(b) a brief description of the local area;
(c) potential impacts;

(d) mitigating actions;
(e) examination of effects on the area if existing trends were to continue;
(f) consultations and objections.

Some consideration of the probable situation without the proposed development is useful, and should include:

(i) employment prospects;
(ii) dynamic environmental factors;
(iii) changes due to planning policies.

The findings of the draft EIS report can be interpreted and summarized in a *Major Issues Report*, thus providing a shorter and simplified summary of the options. Both reports should be available for public scrutiny and comment, feedback from all interested parties can provide valuable information and should be considered (and included, if appropriate) in the preparation of the final EIS report.

5.5 AUSTRALIA

In Australia, Federal (Commonwealth), State and local governments all have responsibilities and powers in relation to environmental assessment and protection. If proposals are jointly financed by a State Government and the Commonwealth, then both government bodies share the responsibility for the EIA. The Commonwealth Environment Protection (Impact of Proposals) Act (December 1974) provides for the assessment of the environmental effects of proposals so that these factors are included with the consideration of economic, technical, financial, and other factors, before decisions are made. The Act requires environmental examination of all proposals which involve the Commonwealth Government, and which have a significant effect on the environment. The Act contains administrative procedures for the early provision of appropriate information to the Minister responsible for the Act, in order to decide whether an EIA statement is required. State Governments have major responsibilities for the environment. State legislation has been developed to provide support for impact statement requirements, e.g. the Western Australian Environmental Protection Act (1986).

5.6 UNITED KINGDOM

Development projects in the UK are subject to a wide range of controls and consultations under which permission is needed from public authorities prior to commencement. Particular permissions required depend upon the scale of the project, the activity or use for which it is intended, the detailed design, the location, and other factors. Planning control and pollution control, including hazard control, are of primary importance. The assessment of environmental impacts is achieved through existing legislation for planning and discharge control on development projects, including appropriate changes to the existing systems where necessary. New (and specific) legislation has not been implemented, partly due to the flexibility of the existing system.

No significant development of any kind can be undertaken in the UK without the prior permission of a public authority. The local authority normally considers such proposals, although sometimes the decision is made by central government. Public hearings are often held for major development projects and a final decision is made by the appropriate Secretary of State. Discharges and hazards of various kinds are subject to separate controls and all major discharge sources are subject to authorisation procedures, e.g. air, water and solid wastes, and noise nuisances.

5.7 UNITED STATES

The National Environment Policy Act (NEPA) of 1969 shaped US environmental protection efforts. It asserted a collective social responsibility for the quality of the nation's environment, and provided:

(a) a general policy statement of federal environmental responsibility;
(b) formation of the Council on Environmental Quality within the Executive Office of the President;
(c) requirements and guidelines for preparation of environmental impact statements.

All federal agencies are required to produce an EIS for any federal action which significantly affects the quality of the human environment.

Under NEPA, all federal departments and agencies are required to improve, co-ordinate and orientate their planning and development programmes towards:

 (i) minimising the long-term environmental effects of all federal actions;
 (ii) the right of society to a safe, healthy and aesthetic environment;
(iii) multiple use of environmental resources;
(iv) preservation of historic and natural landmarks;
 (v) balanced population growth and resource utilisation;
(vi) recycling of scarce natural resources.

The Environmental Protection Agency (EPA) is an operating line agency responsible for administering and conducting all federal pollution control programmes. It focuses attention on pollution control as a strategy for securing environmental quality as well as preservation of wildlife and natural resources.

Under the guidelines for the preparation of an EIS, the following must be included:

(a) a comprehensive technical description of the proposed action;
(b) an analysis of the probable impact on the overall environment;
(c) a description of any probable adverse environmental effects;
(d) analysis, studies and descriptions of possible alternatives to the recommended course of action, and their environmental effects;
(e) detailed consideration of any irreversible or irretrievable commitments of scarce environmental resources.

Many states have now passed legislation and adopted administrative actions to establish procedures similar to the NEPA.

Federal agencies issue permits for a wide range of activities, e.g. air and water emissions, dock construction, etc., covering most industrial projects. Therefore, all major plant projects and most minor works now require an EIA. The company prepares the EIS (or has it prepared by a consultant) for the appropriate federal agency which then issues the EIS document. The preparation of an EIS report requires detailed knowledge of the plant site and its surrounding territory. It must also include many other aspects, e.g. air and water quality, visual aspects, impacts on local flora and fauna, etc. The EIS may be several hundred pages long (especially for a new plant), so that only a few of the more important points have been mentioned here.

5.8   ISO-14000

The design and subsequent operation of a chemical plant requires consideration and compliance with numerous regulations and standards including Responsible Care, Standard for Exchange of Product Model Data (STEP, or ISO-10303), environmental audits, etc. The most recent is the option of plant certification as prescribed in ISO-14000 (*ISO — International Standards Organisation*, which coordinates the publication of international standards) which is a model for an environmental management system. ISO-14000 consists of a series of standards and guidelines which are divided into two groups, these cover organisation standards (ISO-14001, 14010–14014, 14031) and product standards (ISO-14020 to 14024, 14040–14043). The ISO-14000 series is concerned with the environmental management approach of an organisation, and is intended to establish practices which are internationally acceptable. Certification according to ISO-14000 is site-specific and the group of organisation standards focus on environmental performance and audits, whereas the product standards address environmental labeling and product life-cycle issues. More details of the requirements of ISO-14000 are given by Shah (1996). Discussion on the use of Standards and Codes of Practice in design work is included in Section 9.2.

5.9   LEGISLATION

Students involved in undergraduate design projects are usually more concerned with the feasibility and technical design aspects than in considering the myriad of compliance and regulatory aspects. However, it should be a part of the students' education that they are made aware of these aspects and have some understanding of the issues and regulations which are important in an industry context. Hence, a brief overview of the essential legislation is included. The legislation is of course developed for each country, and specific local requirements need to be considered and adopted for a project.

Relevant regulations are issued in the UK by the UK Health and Safety Executive (HSE) under the Health and Safety at Work Act (1974). Of particular importance for chemical plant operation (but also for plant design) are the *Control of Substances Hazardous to Health (COSHH)* regulations (1988) which cover any hazardous substance in use in any place of work. Control over manufacturing sites and storage installations where there is a potential for a major accident is governed by the *Control of Major Industrial Accident Hazards (CMIAH)* regulations (1984). These regulations followed a series of directives and amendments issued by the European

Economic Commission (EEC). The main UK environmental legislation is the *Environmental Protection Act (EPA, 1991), Water Resources Act (1991)* related to discharges into inland waterways, *Water Industries Act (1991)* and *Trade Effluent Regulations (1989)* concerning discharges into sewers, and the *Town and Country Planning Act (1990)* which covers planning permission. More details concerning the legislation are given in the references by Croner (1991). Legislation in the USA has been reviewed by Davenport (1992). Regulations that have been produced by the EEC are reviewed by Haigh (1990). *Environmental auditing* (see Grayson, 1992) is the operational equivalent of the *Environmental Impact Assessment* discussed in Sections 5.1–5.4, and is a comprehensive assessment of all the effects of plant operation upon the environment. This is now becoming associated with site certification using ISO-14000 (see Section 5.8).

Action: *Prepare an environmental impact analysis for the proposed chemical plant, including aspects of normal operation and major spills.*

Detail the relevant legislation affecting this plant.

References

Detailed reference lists are included in Baasel (2nd edn, 1990; pp. 518–524), and Peters and Timmerhaus (3rd edn, 1980; Chapter 3).

Bahu, R. and Crittenden, B. (eds), *Management of Process Industry Wastes*, IChemE, UK (1996).

Cascio, J., *ISO-14000 Guide: The new international environmental management standards*, McGraw-Hill, New York (1996).

Cavaseno, V. (ed.), *Industrial Air Pollution Engineering*, reprints of papers from *Chem. Eng. (N.Y.)*, McGraw-Hill Publications, New York (1980).

Cheremisinoff, P.N. and Morresi, A.C., *Environmental Assessment and Impact Statement Handbook*, Ann Arbor Publishers, Michigan (1977).

Crittenden, B. and Kolaczkowski, S., *Waste Minimization Guide*, IChemE, UK (1994).

Croner's Environmental Management (1991) and *Croner's Waste Management* (1991), Croner Publications, London.

Davenport, G.B., A guide to environmental law (3 parts), *Chem. Eng. Prog.*, 88(4), 30–33; 88(5), 45–50; 88(9), 30–33 (1992).

Environmental Protection Authority, *Best Practice Environmental Management in Mining*, EPA, Canberra, Australia (1995).

Environmental Protection Authority, *Cleaner Production Case Studies*, EPA, Canberra, Australia (1996).

Grayson, L., *Environmental Auditing*, Technical Communications, UK (1992).

Haigh, N., *ECC Environmental Policy and Britain*, 2nd edn, Longman Publishing, UK (1990).

Hills, J.S., *Cutting Water and Effluent Costs*, 2nd edn, IChemE, UK (1995).

Hobbs, V., *Environmental Auditing: Case Studies of Artificial Waterway Developments in Western Australia*, Murdoch University Press, Western Australia (1990).

Hollick, M., *Report on Environmental Impact Assessment Procedures in Western Australia*, University of Western Australia, Perth, Western Australia (1981).

Hoyle, D., *ISO-9000 Quality Systems Handbook*, 2nd edn, Butterworths, UK (1994).

Lykke, E., *Achieving Environmental Goals: The concept and practice of environmental performance review*, Belhaven Press, UK (1996).

Moilanen, T. and Martin, C., *Financial Evaluation of Environmental Investments*, IChemE, UK (1996).

Muir, D.M. (ed.), *Dust and Fume Control (A User Guide)*, revised 2nd edn, IChemE, UK (1992).

Nemerow, N.L., *Industrial Water Pollution*, Addison-Wesley, Massachusetts (1978).

Nestel, G.K., *Road to ISO-14000: An orientation guide to the environmental management standards*, Irwin Professional Publications, USA (1996).

Newton, D. and Solt, G. (eds), *Water Use and Reuse*, IChemE, UK (1994).

Owen, F. and Maidment, D., *Quality Assurance: A guide to application of ISO-9001 to process plant projects*, 2nd edn, IChemE, UK (1996).

Rothery, B., *ISO-14000 and ISO-9000*, Gower Publishing, UK (1995).

Pratt, M. (ed.), *Remedial Processes for Contaminated Land*, IChemE, UK (1993).

Sanks, R.L., *Water Treatment Plant Design*, Ann Arbor Science Publishers, Michigan (1978).

Shah, G.C., ISO-14000: To be or not to be? *Hydrocarbon Process.*, March, 132D-132E (1996).

Sharratt, P. and Sparshott, M. (eds), *Case Studies in Environmental Technology*, IChemE, UK (1996).

Shillito, D. (ed.), *Implementing Environmental Management*, IChemE, UK (1994).

Strauss, W., *Industrial Gas Cleaning*, 2nd edn, Pergamon Press, Oxford (1975).

Tibor, T., *ISO-14000: A guide to the new environmental management standards*, Irwin Publications, Chicago (1996).

5.10 CASE STUDY — ENVIRONMENTAL CONSIDERATIONS

Summary

The environmental impact due to day-to-day operation and the potential environmental damage that would result from a plant accident or spill have been considered. Potential emissions to the environment from the proposed phthalic anhydride plant have been assessed in three categories: (a) airborne emissions; (b) waterborne emissions; and (c) solid waste. Airborne emissions will contain low concentrations of PAN and other organics but these will be satisfactorily dispersed by a shared stack on an adjoining site. Waterborne emissions will be minimal but available treatment facilities will be utilised where required. Bioremediation facilities are also available from the adjoining utilities plant but will seldom be required as only minimal solid waste will be produced.

The main process hazard which may occur during normal operation of phthalic anhydride plants is *the risk of PAN dust clouds forming from minor process breaches*. PAN dust clouds are both toxic (at concentrations below 1%) and explosive (at concentrations of 1.5–10.5%). Air quality monitoring will be utilised to identify process breaches producing dust clouds before they become hazardous. Appropriate breathing equipment (dust masks, respirators, etc.) will always be available.

Safety will be a priority of the site management team and detailed policies will be developed to ensure safe working practices are cultivated. Employees from all groups and levels will be involved in safety on a day-to-day basis. The use of appropriate protective equipment will be mandatory for both employees and visitors in all process areas. Noise will be controlled through good design and appropriate insulation, and will not exceed recommended levels.

5.10.1 Purpose

A new plant must satisfy the Environment Protection Authority (EPA), or similar government agency, that emissions to the environment will not create any substantial damage, and will be minimised through appropriate design and operating practices. Employee safety must also be guaranteed as far as possible by management policies, provision of protective equipment and precautionary practices. Risk must also be assessed in relation to the possibility of serious equipment failure that could lead to hazardous situations.

5.10.2 Airborne Emissions

The Low Air Ratio process is the cleanest of the processes for phthalic anhydride production. Normal operating conditions will produce only two significant discharges to the environment which are shown in Figure 5.1:

(a) non-condensable reaction by-products that remain with the air as it is rejected from the process to the environment;
(b) heavy residue from the bottoms of the rectification column.

The reaction by-products are mostly light organic vapours. A scrubbing unit will be installed in order to reduce the concentrations of contaminants to less than 25 ppm PAN, 10 ppm maleic anhydride and 3 ppm benzoic acid prior to discharge to the atmosphere. A 50 m stack should disperse these concentrations to acceptable levels. The total amount of PAN vented to the atmosphere will be less than 1 kg/hr. The rectification column bottoms will be transferred to another local chemical manufacturer for reprocessing to recover valuable components and, therefore, will not contribute to daily emissions.

The process will not release any combustion products (CO_2, NO_x and SO_x) as all process heating requirements are met by the heat released in the reaction phase. However, up to 15 T/hr of carbon dioxide will be produced as a by-product of the main synthesis reaction and this gas will need to be vented to the atmosphere. This is a negligible quantity when compared with the discharges from other local industries.

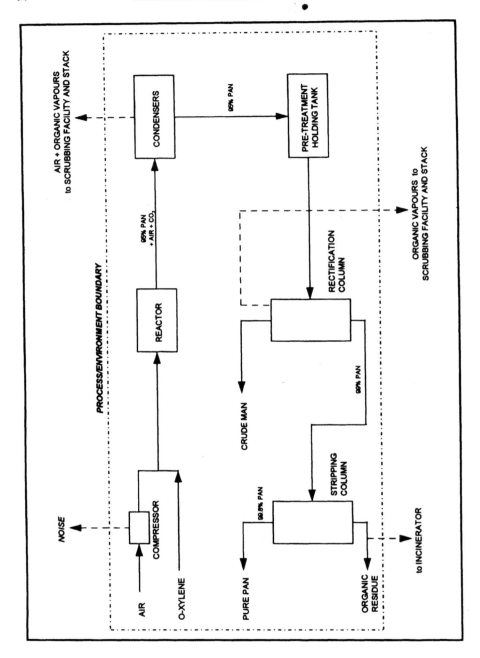

Figure 5.1 Phthalic anhydride process block diagram with emission sources.

5.10.3 Waterborne Emissions

Water is not part of the phthalic anhydride process and will not come into direct contact with any process stream in the system. Air coolers will be used to satisfy most of the process cooling requirements so that cooling water usage will be minimal. However, there will still be a cooling water link with the joint venture utilities plant that will service several local industries, and small amounts of PAN and other organic products could possibly enter the shared cooling-water system. Through economies of scale, the utilities plant is able to use the latest technology (e.g. automated dosing pumps, on-line analysers and advanced control applications) to remove contaminants and minimise emissions to the environment. The cooling-water circuit is a closed system and does not use either sea or river water, and it makes only small discharges of pH-neutral water to the environment. Biological fouling is controlled with phosphate additives rather than the more environmentally hazardous chromate additives.

Steam requirements will be essentially met by the process itself using the heat generated by the PAN reaction. An interconnection with the adjoining utilities plant will be installed but will generally only be used to export steam. Process contaminants that might enter the steam system through exchanger leaks or other process disturbances will be scrubbed at the shared utilities plant, so that condensate can be recycled to reduce energy consumption and chemical treatment costs.

Storm water (run-off from the plant) will be collected in drainage systems and directed to the utilities plant where it will be treated and used for cooling water make-up. Except during the summer months, this will make the PAN plant, and other plants using the same facilities essentially independent of WAWA (Western Australian Water Authority) except for on-site domestic users.

5.10.4 Solid Waste

No solid residue is expected from the phthalic anhydride process. However, bioremediated waste from the adjoining utilities plant which may be partially sourced from the PAN plant wastewater, can be cleanly incinerated in a combustion unit if the heating value is sufficiently high. Heat released by this process will be used for boiler feed-water preheating to minimise energy consumption in the utilities plant.

The requirements for an incinerator in the common utilities plant will have to be assessed with respect to the particular needs of the other industries sharing the facility. Other options such as extended bioremediation of heavy organic residues, may be more economical as the incinerator will require a complex control system to monitor its performance, to regulate the fuel: air ratio, and to safeguard operation against process disturbances that could potentially result in unburned product being emitted directly to the environment. Due to the small volumes of waste to be processed, the incinerator will only be operated on a batch basis, and a small sump will be installed to collect residue during normal operation. This would allow incinerator operation to be scheduled only when wind conditions were suitable (e.g. not calm days). It would also reduce the risk of incinerator failure as it would allow a routine cleaning and maintenance program to be established.

5.10.5 Process Hazards

The most serious process hazard is the potential for phthalic anhydride dust clouds to form following process breaches. The condensers will be the primary point of risk but other equipment, including the storage tanks (if proper heating is not maintained) and the reactor are also possible sources. At low concentrations (less than 1%), PAN dust is a serious health risk. At higher levels (1.5–10.5%), PAN dust is a major explosion hazard [1,3].

A level of 10,000 ppm (1%) PAN in the atmosphere is immediately dangerous to life, but a lower concentration also poses a health risk and is an eye, nose and skin irritant. Respirators will always be available from the control room for use in susceptible areas, i.e. dust and mist respirators at less than 10 ppm, air purifiers at less than 50 ppm, and positive-pressure respirators with face masks for emergency conditions up to 2000 ppm. Health and safety standards require a minimum of 3 ppm PAN and 0.05 ppm MAN for safe working environments before acute or chronic effects are detectable. These standards will be met through appropriate process design and operating precautions.

Phthalic anhydride dust is explosive at concentrations of 1.5–10.5% in air. As there is always the possibility of process leaks (even at very low probabilities), all sources of ignition and non-intrinsically safe equipment will be excluded from the site, except within the main buildings. High-risk zones will be identified and equipped with air quality monitors to detect dust before it reaches a hazardous level. The three areas of highest risk are the condensers, reactor and storage vessels. The areas where these items are located are separated from other items of equipment to prevent an explosion in one area triggering an explosion in another. Major fires in the plant will burn hot and be difficult to extinguish, but will not release large quantities of harmful vapours [1]. A fire station, manned by specially trained process operators, will be located near the process equipment in order to access and contain any fires before they become difficult to manage.

5.10.6 Accidental Spills and Tank Breaches

Significant volumes of reactants, products and intermediates will be held on site in the three product storage tanks, two reactant storage tanks and an intermediate PAN pretreatment tank. Spills from these sources are clearly the most serious due to the potential volume of material that could be lost. Subsequently, all of these vessels require containing walls (or bunds) to be built around them to prevent loss of hazardous materials in the event of a tank breach. All tanks will also be constructed on concrete bases to prevent leaching to the surrounding soil. The containment areas normally drain, via underground pipelines, to the storm-water treatment area in the adjoining utilities plant, but they can be isolated in order to contain a spill. Figure 5.2 shows the modified plant layout. Alterations

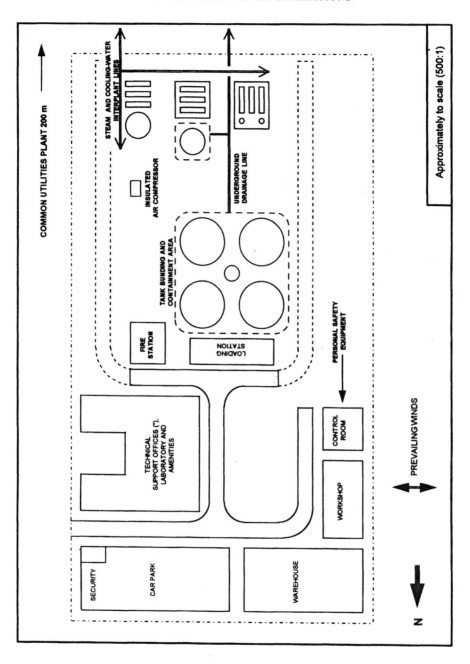

Figure 5.2 Phthalic anhydride plant layout with relevant environmental and safety additions.

made for safety or environmental reasons have been highlighted by bold fonts and thicker lines.

Ortho-xylene is a flammable and moderately toxic liquid at ambient temperatures and represents a serious hazard if a significant volume is spilled. Process operations will be paused during a spill until satisfactory recovery can be completed using temporary storage facilities (isotainers) which will be readily available on the site. Phthalic anhydride is solid up to 131°C, although it will be held as a liquid in the main product tanks and intermediate pretreatment tank. Consequently, any spill will solidify quickly after contact with cool air. This helps to contain spills and aids the recovery process, but increases the risk of dust cloud formation and may block drains. Recovery can be affected with shovels and drums, but adequate protective equipment must be provided for the workers. The recovered product can then be returned to the pretreatment tank to avoid any discharge to the environment. Hot water will be used to clean up any remaining residue [39].

5.10.7 Personnel Safety Precautions and Procedures

Safety will be the primary priority for the site management. A permanent safety officer will be shared with some adjoining plants, and each shift will have a designated safety representative. The 'safety pyramid' principle will be utilised to eliminate near misses and minor incidents (which will all be reported and logged) that have the potential to lead to more serious injuries in the future. The senior operator will have full authority over the process area and will be required to approve any activities, including routine maintenance, undertaken in the process area.

Protective equipment will be made available to all employees and a mandatory policy for the use of safety glasses and hard hats will be implemented. Dust respirators and filters will be available in the control room at all times. Monitoring programs will be established to ensure that the time-weighted average daily exposure of workers to PAN or MAN is below the acceptable safety limits. Similar regulations will apply to visitors.

The primary air compressor (providing air feed to the reactor) is the only loud noise source in the process. Appropriate design modifications will be made to limit noise from the compressor, and it will be housed in an insulated isolation enclosure to further restrict noise emissions. Hearing protection will be mandatory for any employees who are required to work near the compressor while it is operating. Workshop employees will also be required to wear hearing protection while operating or working near heavy equipment.

5.10.8 Conclusions

- The plant layout was modified to protect workers and vital buildings, and to separate potentially hazardous process items.
- The Low Air Ratio process is the cleanest of the commercial processes for phthalic anhydride production.

- Low levels of phthalic anhydride vapour (and some other light organics) will be discharged to the atmosphere, but will be dispersed to acceptable levels by a stack on an adjoining site.
- Water emissions will be minimal but treatment facilities are available and will be utilised where required.
- Bioremediation will be used to dispose of small amounts of solid residue.
- *PAN can form toxic and flammable dust clouds.* Monitoring equipment will be used to detect dust before it reaches hazardous levels.
- All tanks will be built with containing walls or bunds to contain spills, and appropriate cleanup procedures will be established.
- Safety will be a management priority and will involve all employees.
- Protective equipment will be readily available at all times, and its use will be mandatory for employees in the process area.

5.10.9 Recommendations

- Perform more detailed mass and energy balances for the plant items to confirm equipment sizes.
- Determine the cost of the major items of equipment required for a 40 kT/yr PAN plant, and estimate materials transportation costs for a detailed profitability analysis.
- Ensure design modifications are made to minimise process hazards.
- Complete detailed hazard and operability (HAZOP) studies after the P&ID is developed.

6. ECONOMIC EVALUATION

6.1 INTRODUCTORY NOTES

The economic evaluation of a chemical plant covers a wide range of topics and techniques. The scope is much too broad to allow inclusion of detailed descriptions in this short section, indeed entire books have been written on this topic alone. Sections 6.1 to 6.4 contain brief notes covering the main topics to be considered in the introductory stages (feasibility study) of a design project.

Several excellent reference sources are available, those written by and for chemical engineers are especially useful. Several references are included after Section 6.4.5. The most prominent text in this field is by Peters and Timmerhaus, 1991 (see reference in Section IV here). The following chapters, containing over 200 pages, are included in that text:

5. *Cost and Asset Accounting*; presents a survey of accounting procedures for the analysis of costs and profits as used for industrial operations.
6. *Cost Estimation*; provides information regarding the estimation of fixed capital costs, and also recurrent operating expenditure.
7. *Interest and Investment Costs*; discusses the concept and calculation of interest, i.e. payment as compensation for the use of borrowed capital.
8. *Taxes and Insurance*; taxes represent a significant payment from a company's earnings and although insurance rates are only a small fraction of annual expenditure, adequate insurance cover for a plant is essential.
9. *Depreciation*; this is a measure of the decrease in value of an item with respect to time, and can be considered as a loss in value of an asset — incurred for the use of the equipment. Consideration of depreciation often causes students some problems, both in terms of whether it is a cost or a 'revenue'(?), and how it should be included in the economic evaluation. It is perhaps preferable to avoid using the term 'cost' to describe depreciation? This is discussed in more detail in Section 6.3.1.
10. *Profitability, Alternative Investments, and Replacements*; the profitability of an investment is a measure of the amount of profit generated. It is important to assess the profitability accurately, and also the profits that could be obtained from alternative investments.

Other useful chemical engineering design books containing chapters for the economic evaluation of projects are those by Baasel (1990); Backhurst and Harker (1973); and Ulrich (1984). The advantages of the books by Baasel (1990); Peters and Timmerhaus (1991); and Ulrich (1984) are the inclusion of extensive references to the literature and substantial compilations of economic data. Other useful books by Brennan (1990) and Breuer and Brennan (1994) present an Australian perspective. The IChemE (UK) has published the following guides:

Economic Evaluation of Projects, 3rd edn (1990) by D.H. Allen
Project Procedure, 2nd edn (1985) by J.C. Rose, G.L. Wells and B.H. Yeats
A New Guide to Capital Cost Estimating, 3rd edn (1988), Report of a Working Party

The book by Ulrich (1984) contains the following chapters in the section covering economic analysis: Chapter 5 describes capital cost estimation; Chapter 6 discusses procedures for evaluating plant operating costs (and interpretation and preparation of balance sheets); Chapter 7 covers techniques for determining the optimum design from several alternatives; Chapter 8 discusses profitability analysis, discounting to present day values (Discounted Cash Flow, DCF), and determination of the viability of a project. It should be remembered that government taxation policy varies widely between countries (and between successive governments), and it can have a significant effect on project profitability.

The reader can refer to Ulrich (1984) for more detailed descriptions of these topics, especially pages 279–280 containing a review of capital cost estimation procedures. A more detailed (although still necessarily brief) discussion is included here in Sections 6.2 to 6.4, the references quoted provide an excellent in-depth coverage of this subject.

6.2   CAPITAL COST ESTIMATION

The first task in preparing an economic evaluation of a project is usually to obtain a capital cost estimate (usually in the minimum time possible) for the entire plant, remembering that different methods have different errors and accuracy. A pre-design estimate (mentioned in Section 2.2: Preliminary Market Survey/Economic Analysis) based upon capacity corrections and the use of cost indices, when applied to data from an existing plant, will have an accuracy of ± 20–40% (see Section 6.2.1(I) following).

A more accurate (?) cost estimate for the specific plant under consideration may be obtained by costing the individual plant items (see Section 6.2.1(II) following). This may require better estimates of the equipment sizes than was obtained in the preparation of the Equipment Schedule in Sections 3.3, 3.4 and 4.2, depending upon whether more detailed/accurate information has since become available. Obtaining costings of the individual plant items provides a better insight into the design aspects which are going to be important at the detailed equipment design stage. Improved cost estimates can be achieved either by the use of tabulated data and nomographs in the literature (see Ulrich, 1984; pp. 281–323), or from an equipment vendor/supplier who specialises in 'off-the-shelf' equipment, usually common items (e.g. heat exchangers) available in a range of standard pre-determined sizes. The literature data is usually presented graphically as capacity (or size) plotted against purchase price for particular types of equipment. The accuracy of such a cost estimate will probably still only be ± 15–30% but it is sufficient for initial evaluations, and it is not necessary to spend more time than is required to obtain 'ball park' estimates.

More accurate cost estimates (say ± 5–10%) are only required (and possible) after the detailed design work has been completed, including the design and sizing of all equipment, the determination of pipework layouts, and specification of the control and instrumentation schemes. Final accurate costings cannot be obtained, and indeed are not required, at the feasibility stage of a project. This is because detailed mass and energy balances have not been completed, and these calculations will provide the basis for the detailed equipment designs! Final design specifications must be available before an equipment fabricator will spend time preparing detailed construction quotations.

6.2.1 Cost of Equipment (Major Items)

Major items of equipment include reactors, heat exchangers, columns, pressure vessels, storage tanks, etc. Ancillary equipment such as process piping and insulation can be estimated after the total cost of the major items is known.

(1) Cost Correlations

Cost correlations provide a convenient method for *estimating the capital cost of major items* of equipment. Correlations are usually provided graphically as plots (log–log coordinates) of capital cost of a particular item versus capacity (e.g. volume, surface area, throughput, or power rating). These cost correlations do not start from the origin, as even at very low capacities there are some costs (e.g. overheads) associated with the equipment. The cost (C) increases to infinity (i.e. slope of the curve $=1$), at which point it is more economic to use multiple units of the same size. For an intermediate capacity (Q) range, the slope (m) of the curve is approximately constant and the following convenient equation can be used:

$$C = C_B (Q/Q_B)^m$$

where C_B and Q_B are the cost and size of a predetermined 'basic' size. Useful data can be obtained from: *A New Guide to Capital Cost Estimating* (1988; IChemE, UK), and the other sources mentioned in this chapter.

The cost of individual equipment, or of a complete plant, must be up-graded to account for the reduced purchasing power of the dollar (or pound) from a given time datum to the present, i.e. *the effects of inflation.* Various cost indices are published annually (and monthly, see *Chemical Engineering* journal). Those of particular relevance to chemical plant costs are the Nelson Refinery Construction Index, Chemical Engineering *(CE)* Plant Cost Index, and the Marshall and Swift *(M&S,* previously Marshall and Stevens) Equipment Cost Index. These indices apply to complete plants rather than individual items of equipment. The appropriate equation is:

$$C_2 = C_1 (I_2/I_1)$$

where I is the relevant index, and the suffixes represent different time periods.

Values of the *CE* and *M&S* indices are tabulated in Baasel (1990; p. 263, Table 9-1) for the period 1953–1988. The *CE Plant Cost Index* is composed of four components having the following weightings:

Equipment, machinery and supports   0.61
Erection and installation labour   0.22
Building materials and labour   0.07
Engineering supervision manpower   0.10

It is possible to modify the index to represent a particular situation by changing the component weightings, and hence recalculate the values of the modified index for particular years. An alternative index is the Process Engineering *(PE)* Plant Cost Index, values are published monthly in *Process Engineering* journal.

The cost of a complete plant can be estimated using the 'six-tenths rule' which states that: 'the ratio of the costs of two plants producing the same product is proportional to the ratio of their capacities raised to the power of 0.6'. This statement can be written as:

$$(C_1/C_2) = (Q_1/Q_2)^{0.6}$$

where the accuracy of such an estimate is $\pm 30\%$. Using this rule, if the capacity is doubled then the cost increases by 52%. This emphasizes the economic advantages of building very large plants, provided duplicate plant items are not required. However, the effects of a major accident (or failure of an item of equipment) are more serious in a single large plant than for two smaller plants!

(II)   Factored Estimate Method

The complete plant cost can be estimated by extrapolation from the cost of the major items of equipment. An accuracy of ± 15–25% is possible if appropriate factors are chosen. The technique is described by the equation:

$$C_F = C_E \, (1 + \Sigma f_i)(1 + \Sigma f_i')$$

or

$$C_F = C_E F_L$$

where C_F is the fixed investment cost of a complete plant; C_E is the cost of the major items of processing equipment; f_i are factors for the direct costs, e.g. piping, instruments, etc.; f_i' are factors for the indirect costs, e.g. contractor's fees, contingency allowance, etc.

The value of the *Lang factor* (F_L) depends upon the type of process. For most plants it has a value of approximately 3. The following values were suggested by Lang (1948) and have been widely used as a guide:

$F_L = 3.10$ for solids processing plants
$F_L = 3.63$ for a mixed fluids–solids processing plant
$F_L = 4.74$ for fluids processing plants

If possible, the Lang factor should be obtained from the cost files of the particular company. Because of the large difference between the values of F_L (3.10 to 4.74), it appears important to define the type of plant under consideration (although this is sometimes difficult). The validity of the original Lang (1948) factors is discussed in some detail by Baasel (1990; pp. 271–283) and suggestions for obtaining improved values are presented. It was found by Cran (1981) after statistical analysis of the original data used by Lang (which was for plants of comparatively small size by modern standards), that the mean Lang factor was 3.45 with a standard deviation of 0.47, and hence the need for three categories of plants was unnecessary.

Ratio factors for estimating capital-investment items based on delivered-equipment cost are given in Peters and Timmerhaus (1991; p. 183, Table 17) and the values obtained for the three classes of plant listed above are 4.55, 4.87 and 5.69 respectively. Also presented (p. 184, Table 18) are factors for the fixed-capital investment (3.9, 4.1 and 4.8 respectively) and for the total capital investment (4.6, 4.9 and 5.7 respectively). Obviously some care and consideration should be given to the use of appropriate Lang factors for chemical-plant capital-cost estimating. Historical values are likely to be low if used for modern plants as costs of advanced process control and instrumentation need to be included.

The complete cost of a plant can be estimated by using a particular value of F_L, this can be done in the early stages of the design after the preliminary flowsheet and equipment list are available. A better estimate of the Lang factor (F_L) for a specific type of plant can be obtained by calculating the individual cost factors (f_i and f_i') that are compounded into the Lang factor. Lists of direct and indirect costs and typical values of their factors are given in Coulson and Richardson Volume 6 (2nd edn, 1993; pp. 216–218). In the early stages of a design study some items may be overlooked, this results in a low estimate of the capital cost. A contingency allowance of between 10% and 50% (depending upon the completeness of the information) should be included to account for this possibility.

6.2.2 Module Costs

An estimate of the plant capital cost can also be obtained by considering the plant as a set of modules. Each module consists of similar items of equipment, e.g. heat exchangers, steam generators, etc. The standard cost of each process module is calculated (for a particular year), corrections are applied for the use of materials other than carbon steel and for high pressures. Variations in the process piping cost are corrected in the contingency allowance. A modified Lang factor is calculated for each module, including freight costs and sales taxes. Site development, construction and structural work are considered as separate modules. This method is described in more detail by Guthrie (1969).

6.2.3 Auxiliary Services

It is necessary to include the full cost of auxiliary services, e.g. steam, water, electricity supply and distribution, roads, buildings, communications, etc., in the capital cost estimate for a new plant. If a chemical plant is to be constructed on an existing complex, then a proportion of the cost of existing services (based upon the estimated consumption) is usually charged. A very approximate estimate of the cost of auxiliary services can be obtained by taking 20–25% of the total installed (new) plant cost. However, in some situations a figure of 40–70% may be more appropriate. It must be emphasised that a more accurate estimate should be obtained during the latter stages of the feasibility study.

6.3   OPERATING COSTS — FIXED AND VARIABLE

The plant capital cost represents a one-off expenditure, although the capital (and interest) will usually have to be repaid over several years. In order to assess the economic viability of a project it is also necessary to estimate the operating costs which are incurred annually in the production of the chemical. The operating costs should be considered when the alternative process routes are being evaluated, and they can significantly influence the final process choice. Operating costs can be divided into two groups (although in some cases the division may be arbitrary), these are:

(a) *Fixed costs* such as laboratory costs, operating labour, capital repayment charges, insurance, etc. *These costs do not depend upon the production rate*, and they must be paid even if no chemical is produced.
(b) *Variable costs* such as raw materials, utilities (services), shipping, etc. *These costs are dependent upon the amount of chemical produced.*

The plant supervisor/manager has no (or very little) control over fixed operating costs, whereas he/she is held accountable for the variable costs. In addition to those costs incurred due to the construction of the plant and/or its operation, each plant, site or product is usually required to contribute towards the general operating expenses of the parent company. These expenses include general overheads, research and development costs, sales expenses, etc. Each company decides how these costs are apportioned, however as a general indication they may add 20–30% to the direct production costs at site.

The following items represent the more common operating costs, although the list should not be considered complete for any/every plant. Operating costs are usually calculated on an annual basis, and subsequently re-calculated per tonne of product (for example) when determining the necessary selling price of the chemical, and hence the profitability of the plant. The following items, (a)–(i), would normally be considered as variable costs, although in some cases there may be both a fixed and variable component, e.g. the fixed cost of a laboratory and its equipment and the additional variable costs of consumables used.

(a) *Raw materials* are determined from the process flowsheet and from material balances.
(b) *Miscellaneous materials* include items such as safety clothing, chart recorder paper, etc., that are not included as raw materials or maintenance materials. These are usually calculated as 10% of the total maintenance cost, and sometimes referred to as "consumables".

(c) *Utilities (services)* include electricity, water, steam, compressed air, etc. Quantities are determined from the flowsheet and from mass and energy balances; current costs (and anticipated price rises) should be obtained for these items.

(d) *Shipping and packaging costs* depend on factors specific to the process, including the location and type of product. In some cases these costs are negligible, but sometimes they can be significant.

(e) *Maintenance* includes materials and labour costs. This cost is typically between 5–15% of installed capital costs and should be estimated by using data obtained from a similar plant.

(f) *Labour costs* should be estimated from reasonably detailed supervision estimates. The operating labour costs may not decrease if production is reduced, however overtime payments will be required for significant increases in production. Operating labour costs do not normally exceed 15% of the total operating cost (most plants employ few personnel).

(g) *Supervision* includes the management team directly responsible for the overall plant operation and for directing the work of the plant operators (item (f) above). The personnel requirements of the management team should be determined, although an approximate figure of 20% of item (f) can be used to provide an initial estimate.

(h) *Laboratory costs* are incurred for analyses associated with quality control and process monitoring. An approximate estimate can be obtained as 20–30% of the operating labour cost (item (f) above), or 2–4% of the total production cost.

(i) *Plant overheads* include general operating costs such as security, canteen, medical, administration, etc. This item is often estimated as 50–100% of item (f).

The following items (j–m) would normally be considered as fixed costs which must be paid even if no chemical is produced or sold.

(j) *Capital charges* are recovered from the project to repay the initial capital investment and the interest accrued on the debt. The procedure adopted depends upon the accounting practice of the company. Capital is often recovered as a depreciation charge of 10% per annum (for example) based on a plant operating life of 10 years, although the plant is not necessarily replaced after that time! The concept of depreciation can be quite difficult to grasp, therefore it is discussed separately and in more detail in Section 6.3.1. Interest is charged on the capital which is borrowed to finance the plant, and this must be repaid (monthly or annually) until the capital debt is completely repaid. The capital may be obtained from various sources, e.g. a loan from a finance company, share issues, internal company reserves, a fixed-interest government loan, or some combination of these alternatives. If company reserves are used then it should still be repaid with interest — based upon a consideration of alternative investments and their return on capital, i.e. applicable market rates.

(k) *Rates* are payable to the local authority or shire based upon the assessed rateable value of the site. A typical figure is 1–2% of the capital cost.

(l) *Insurance* for the site, the plant and employees is usually about 1–2% of the fixed capital.

(m) *Royalties and licence fees* are payable to the company or individual responsible
for developing the process. This payment may be a lump sum or an annual fee,
and is typically either 1–5% of sales price or 1% of the fixed capital.

More detailed discussion of these items and a worked example can be found in
Coulson and Richardson Volume 6 (2nd edn, 1993; pp. 224–234). The advantage of
identifying fixed and variable costs separately is that it is then easier to determine
the breakeven point for a project, i.e. the level of production necessary to cover
essential costs if short-term market fluctuations mean that no profit can be made.
This type of profitability analysis is discussed in Section 6.4.

6.3.1   Depreciation

Students often find the concept of depreciation difficult to grasp, and even when the
idea of a depreciating asset becomes clear the correct place to include it in the profitability analysis may not be obvious. Hence this section, which attempts to provide
a straightforward explanation. It should be emphasised that this is an engineer's
explanation not a statement of accounting practice or economic theory, those wanting
a more definitive approach should refer to a specialist textbook, or the appropriate
chapter in Peters and Timmerhaus (1991).

Capital (money) is borrowed in order to finance a project, this money is
spent and it then becomes a capital asset. In most cases the (paper, or book) value
of the asset declines as time passes. The parallels with the purchase of a motor car
should be obvious. It is possible that the value of the land will increase, although
this cannot be guaranteed. Regardless of the decline in book value of the plant,
the total capital borrowed will have to be repaid in full (including agreed interest charges). Depreciation of an asset is determined so that the company accounts
can be prepared and the shareholders can identify the current value of the assets
of the company. As an example, for taxation purposes a depreciation charge
of 10% per year may be applied for the first nine years of plant life, at which time
the residual (scrap) value of the plant would be 10% of the inital capital cost
(although in 9 years time this will not have the same dollar purchasing power
due to the ravages of inflation!). After nine years the asset will not be depreciated further.

Depreciation needs to be calculated for taxation purposes. Taxes are paid for a
particular year based upon the revenues received, minus the costs incurred, for that
specific 12 months. Each year the taxation office will allow depreciation of a capital
asset to be claimed (following published rules) as a deduction from profits before
tax is paid. Therefore, when all costs associated with the business operation are
deducted from the total (gross) revenues this yields the gross profit. The agreed
depreciation charge can then be deducted from the gross profit, and tax is only paid
on the remaining net profit. It should be realised that if the company does not make
a profit in a particular tax year, then a depreciation allowance cannot be claimed. If
no profit is earned, then taxes are not paid, and hence no depreciation deduction!
Also if a profit is not made in a particular year then the following year the company

can only claim a depreciation allowance for the current year, the previous year's allowance cannot be carried forward!

The essential points to realise are:

(i) Depreciation allowance is deducted from gross profits, i.e. this money (or tax allowance) stays within the company and is not taxed. As the tax to be paid is calculated on the net profit, so the tax payment is reduced. The company may decide to use the depreciation allowance (i.e. tax saved) to fund new projects which will eventually replace the existing plant.

(ii) If no profit is made in a particular tax year, then no tax is paid, and hence no depreciation allowance can be claimed for that particular year. The allowable depreciation can be claimed for the subsequent years if profits are obtained. If a depreciation allowance is not claimed in a particular year, then it cannot be carried forward and claimed in the next year (it has been lost!).

(iii) Depreciation is sometimes referred to as a capital charge — which can be confusing. It may be preferable not to refer to depreciation as a 'cost' as it is not a direct cost against production, but it is a loss/decline in value of an asset from its original book value. The assets of a company are only the value that could be realised if they were sold at that particular time — regardless of the initial purchase price (or the current depreciated book value)! However if the principles are understood then it should be possible to include depreciation in an initial economic evaluation of a project, even if rigorous accounting practices are not applied!

(iv) A company will usually aim to replace its assets as their value declines, e.g. by embarking on new projects to replace ageing chemical plants. The money saved (or rather not paid as tax) by the depreciation allowance can be seen as an opportunity to replace the declining value of the asset.

6.4 PROFITABILITY ANALYSIS

The first stage in evaluating the profitability of a proposed new project is to compare the total cost of the product (for example, per tonne produced) with the current market price. This is also necessary to estimate future demand for the product, and to determine the trend in the selling price over several years. It is also necessary to estimate the number of years that the product can be sold at a satisfactory profit, this is more useful than estimating the possible operating life of a plant! Backhurst and Harker (1973; p. 47) make the following observation:

'... it is suggested that a large chemical organisation will not invest in a new process unless it is possible to sell the product for less than half the current market price'.

This statement may not be strictly true in the 1990s, but it does serve to emphasise the importance of the detailed economic evaluation of a project! Two simple measures that can be used to evaluate the profitability of a project are the *payback period* and the *return on investment*.

6.4.1 The Payback Period

The *payback period* (or payout time) is the number of years from plant start-up required to recover all expenses involved in a project, if all the pre-tax profits were used for this purpose. Depreciation charges are not included in the operating costs. Expenses not incurred directly in the design and construction of the plant are excluded, the analysis is intended to demonstrate the best means of allocating the present and future resources of the company. A payback period of less than five years is usually required for a project to proceed (preferably closer to 3 years). However, the payback period does not consider the timing of the payments or the profits earned by the plant after the payback period. It is a reasonably quick parameter to calculate but it only provides a very approximate indication of the economics of a project.

6.4.2 Return on Investment (ROI)

The *Return On Investment (ROI)* is the expected profit divided by the total capital invested, expressed as a percentage return. It must be clearly stated whether the profit is based on pre-tax or after-tax earnings. The after-tax ROI is compared with the earnings that could be achieved by an alternative investment, e.g. capital bonds. An after-tax ROI of at least 15–20% is usually expected (or pre-tax ROI of 30–40%), assuming that the project is not particularly risky! Economic indicators such as the ROI provide a useful and quick appraisal of project performance (if calculated correctly), however there are often other considerations and alternatives to be evaluated as discussed in the following sections.

6.4.3 Evaluating Different Scenarios

The aim of an economic evaluation is not merely to provide numbers or data, but to provide an assessment of what they actually mean and to enable a plan of action to be formulated. If numbers alone are provided then the reader has to interpret why they are important, and what to do next. This should be what the graduate engineer is providing — direction and recommendations for a project, not merely facts and figures.

It is necessary to investigate various scenarios and options in order to arrive at the best overall project. The assessment is probably tackled initially by considering normal operation and the expected production rate. It will also be necessary to consider the effects of both increased production (increased variable costs, plus additional equipment?) and any decreases in market demand, the latter is probably more significant! The technical evaluation determines whether it is possible to turn-up or turn-down the plant production, and, if so, what penalties are incurred. The best and worst case profitability scenarios need to be evaluated in terms of changing market size, raw materials price increases, decline in product price, increases in interest rates, utilities costs, etc. For decreases in production (or product selling price), determine the breakeven point where sales (i.e. revenue) can just cover operating costs but there are no profits. This case means that the plant can survive short-term

(3–6 months?) market fluctuations without the need to shut down. A more serious case is where revenues cover only the fixed costs of production and the plant operates at a loss. This would probably only be acceptable for a few weeks, and the debts incurred (e.g. payments for raw materials) would then have to paid out of future profits. Downturns in the market situation will be less serious (i.e. there is a bigger margin of safety and more flexibility) after the capital borrowing (and hence the interest charges) have been repaid.

The above scenarios help to focus the project team on the most critical aspects of plant operation from the economic viewpoint. If the initial evaluation calculations are set up on a spreadsheet program, then changes in production levels, interest rates, etc., can be easily accomodated to provide assessment of alternative scenarios. However, students should be warned against producing large amounts of computer printout merely by changing every possible variable in the program! It is still your job to provide the evaluation, not simply to generate more data. Summarise the results of these changes in tabular form, and identify and discuss the main factors which are important for this project.

6.4.4 Economic Evaluation and Analysis

Sometimes there are political influences which can affect the decision-making process, and hence the evaluation of a project. Government preference for the use coal as a feedstock or the need to reduce unemployment may lead to low-interest government loans, subsidies, or taxation advantages.

When evaluating the economic viability of a project it is necessary to consider the source of capital financing. Whether this is a capital loan, a share issue, use of company capital, or a government-financed joint venture, whatever the source, the capital borrowed will have to be repaid — with interest. The economic evaluation is intended to indicate whether financing for the project is likely to be approved, or if special circumstances need to be considered. In the initial years of operation, the capital repayments and interest charges will be a significant proportion of the fixed costs. After the payback period the profitability of the plant should increase, although predictions of events in say 5 years time are only predictions!

A major consideration in the economic evaluation is the best means of financing the project and reducing the interest repayments, if this is possible. It may be possible to use all profits to repay capital and hence reduce the total interest charges. However, if profits are not made (i.e. they are absorbed as capital charges) then depreciation allowance on the plant cannot be claimed for those years. Clearly it would be necessary to evaluate the advantages and disadvantages of each strategy. Shareholders may be reluctant to wait several years for a (bigger) return on their investment.

The economic evaluation considers not only the expected situation but also ways in which the economics can be improved. Therefore, it is more useful to concentrate on the major costs, e.g. capital and interest repayments, before looking at less significant items. If the variable costs of plant operation are dominated by raw materials costs then it is necessary to investigate the possibility of fixed-price contracts (to ensure stability of operation), and also to consider the option of building (and owning) a plant to supply the raw materials directly, if this is possible.

6.4.5   Evaluating Different Projects: Use of DCF and NPV

It may be useful in the economic evaluation of a project to consider the total returns (profits) over the plant lifetime. However, the effect of inflation will be to reduce the purchasing power of the dollar with time, and cash received sooner rather than later will have more 'value'. Therefore, it is a better indication of the total value of a project to apply concepts such as Net Present Value (NPV) or Discounted Cash Flow (DCF) which attempt to convert the value of future dollars into present day worth. These concepts are defined and explained in more detail in several texts, e.g. Peters and Timmerhaus (1991; pp. 301–308). However, merely to present a total dollar value return for a plant is not very helpful when considering the economic evaluation. This figure needs to be related to the overall capital investment (e.g. return on investment) or to the gross revenues in order to appreciate its significance. Numbers presented on their own do not evaluate a project — it is the analysis and comparison with the alternatives that are required.

When required to make a choice between two projects, then techniques such as DCF and NPV are particularly useful as they take into account factors such as the timing of profits received. Alternative projects may have several differences such as amounts of fixed and variable costs, time to payback the capital borrowed (and hence interest repayments), overall plant life, profits received at various stages of the project. The use of DCF or NPV can help to evaluate different projects on a 'level playing field' by comparing the total value of investments at present day dollar values.

The Engineer's Approach to Economic Evaluation

It should be stressed that in this chapter (and throughout the entire book) the aim is to provide guidance, ideas and suggestions *for undergraduates* considering a feasibility study and design project for the first time. This chapter has tried to provide ideas for ways in which the preliminary profitability of a plant can be assessed, not to present standard accounting practices or to apply the current economic theory. These may be available in other course units or the student may need to enrol in a postgraduate management course. Most companies have their own preferred in-house methods of project evaluation or screening, and even these approaches may not conform to accounting practices required for the preparation of the company accounts, annual reports or taxation returns. Leave the detail to the specialists and concentrate on understanding the 'big picture' — you won't be allowed to borrow or spend money unless an accountant approves it anyway!

The Final Word?

For more detailed discussion of plant economics and profitability analysis, the reader should refer to a specific textbook such as Peters and Timmerhaus (1991). The paper by Sweeting (1997) provides some interesting insights into the difficulties of cost estimating, and some of the problems to be overcome with industrial projects. The following quotation from Backhurst and Harker (1973; p. 47) provides an appropriate summary for this chapter:

'The viability of an investment lies, sad to say, mainly in the hands of the economist and the financial expert, and it is important to realize that an increase of 0.5 per cent in the bank rate has probably far more effect on the profitability of a project than an increase of, say, 10 per cent in the efficiency of a distillation column.'

Action: Perform a detailed economic analysis of the project including:
   Capital cost estimates (by different methods)
   Operating cost estimates
   Profitability analysis

   Re-assess the economic feasibility of the project based upon the complete economic analysis.

References

Detailed reference lists are included in the relevant chapters in Baasel (1990); Peters and Timmerhaus (3rd edn, 1980); and Ulrich (1984).

Anon., Accurately estimate piping costs, *Process Eng. (London)*, 18 January, 17 February (1991).

Bikic, D., Sencar, P. and Glavic, P., Planning for future growth: Economics of oversized plants, *Chem. Eng. (N.Y.)*, December, 104–107 (1991).

Brennan, D.J., *Process Industry Economics: An Australian Perspective*, Longman Publishing, UK (1990).

Brennan, D.J., Relative capital costs of plants built in Australia, *Chem. Eng. Australia*, **16**(1), 4–5 (1991).

Breuer, P.L. and Brennan, D.J., *Capital Cost Estimation of Process Equipment*, The Institution of Engineers, Australia (1994).

Cran, J., Improved factored method gives better preliminary cost estimates, *Chem. Eng. (N.Y.)*, 6 April, **65** (1981).

Edwards, D.W. and Lawrence, D., Assessing the inherent safety of chemical process routes: Is there a relation between plant costs and inherent safety? *Process Safety Environ. Prot.*, **71**(B4), 252–258 (1993).

Elliott, T.R. and Luyben, W.L., Capacity-based economic approach for the quantitative assessment of process controllability during the conceptual design stage, *Ind. Eng. Chem. Res.*, **34**(11), 3907–3915 (1995).

Garnett, D.I., The "dimensions" of economic systems, *Chemtech*, August, 477–481 (1992).

Garrett, D.E., *Chemical Engineering Economics*, Reinhold Publishing, New York (1989).

Guthrie, K.M., Data and techniques for preliminary capital cost estimating, *Chem. Eng. (N.Y.)*, 24 March, 114–142 (1969).

Hall, R.S., Vatavuk, W.M. and Matley, J., Estimating process equipment costs, *Chem. Eng. (N.Y.)*, 21 November, 66–75 (1988).

Holland, F.A., Watson, F.A. and Wilkinson, J.K., Engineering economics for chemical engineers, a 20-part series in *Chem. Eng. (N.Y.)*, from 25 June 1973 (p. 103) to 20 October 1974. This series is available in total as *Chemical Engineering* Reprint no. 215 (1975).

Humphreys, K.K. and Wellman, P., *Basic Cost Engineering*, 2nd edn, Marcel Dekker, New York (1987).

Lang, H.J., Simplified approach to preliminary cost estimates, *Chem. Eng. (N.Y.)*, **55**(6), 112 (1948).

Lindley, N.L. and Floyd, J.C., Piping systems: How installation costs stack up, *Chem. Eng. (N.Y.)*, January, 94–100 (1993).

Linnhoff, B., Use pinch analysis to knock down capital costs and emissions, *Chem. Eng. Prog.*, **90**(8), 32–57 (1994).

Modern Cost Engineering: Methods and Data, Volume I (1979) and Volume II (1984), reprints of papers from *Chemical Engineering*, McGraw-Hill Publications, New York.

Moilanen, T. and Martin, C., *Financial Evaluation of Environmental Investments*, IChemE, UK (1996).

Plavsic, B., Estimate costs of plants worldwide, *Chem. Eng. (N.Y.)*, August, 100–104 (1993).

Ranade, S.M., Jones, D.H. and Suarez, A.Z., Impact of utility costs on pinch designs, *Hydrocarbon Process.*, **68**(7), 39–43 (1989).

Remer, D.S. and Chai, L.H., Estimate costs of scaled-up process plants, *Chem. Eng. (N.Y.)*, April, 138–175 (1990).

Remer, D.S. and Chai, L.H., Design cost factors for scaling-up engineering equipment, *Chem. Eng. Prog.*, August, 79–82 (1990).

Rodriguez, A.L.M. and Coronel, F.J.M., Determining plant costs, *Chem. Eng. (N.Y.)*, March, 124–127 (1992).

Sweeting, J., Stargazing (the difficulties of cost estimating), *The Chemical Engineer (Rugby, Engl.)*, 9 January, 14–15 (1997).

Ulrich, G.D., Calculating utility costs, *Chem. Eng. (N.Y.)*, February, 110–113 (1992).

Various, Techno-economic analysis (special topic issues), *Chem. Eng. Res. Des.*, 71(5), 465–522 (1993); **73**(8), 869–975 (1995).

Ward, T.J., Estimate profitability using net return rate, *Chem. Eng. (N.Y.)*, March, 151–156 (1989).

Ward, T.J., Economic evaluation of projects, *Chem. Eng. (N.Y.)*, January, 102–107 (1994).

6.5 CASE STUDY — ECONOMIC EVALUATION

Summary

The costs of individual items of plant equipment were estimated from preliminary designs. A total equipment cost of A$6.8 million was calculated. The overall installed plant cost was estimated to be A$37 million using a Lang factor, which was determined specifically for the PAN LAR process, applied to the total equipment cost. This value is comparable to cost data for four recently constructed phthalic anhydride plants in the region. The production costs were estimated at A$762/tonne including capital repayments, and A$491 excluding capital repayments. This produces an overall, pre-tax return on investment (ROI) of 42% and a discounted cash flow return (DCFR) of 27%, making the project attractive if sufficient markets (including exports) can be established and if related industries are involved in a co-development. The ortho-xylene price is the major vulnerability, and the DCFR could fall to less than 5% if margins become very tight as they did in 1995.

6.5.1 Background and Objectives

The feasibility study and market assessment (Chapter 2) established that a phthalic anhydride plant would be disadvantaged by being located in Western Australia, and that South-East Asian plants could be expected to dominate the market. However, if production costs could be maintained below those of competitive plants by employing new technology (i.e. the LAR process for phthalic anhydride manufacture) then a small market share could be obtained and the construction of a new plant in WA would possibly be viable. The development of downstream processes in WA and the assurance of government assistance were also identified as being critical to the feasibility of the project.

A preliminary equipment list was developed (Chapter 3) to allow a plant layout to be specified (Chapter 4) and the environmental impact of a new PAN plant in WA to be assessed (Chapter 5). However, these estimates were based only on preliminary mass and energy balances and were not made to a sufficient level of accuracy to allow individual equipment costs to be determined.

A capital cost estimate can be obtained by using data from recently constructed plants, or from nomographs if that is not available. The LAR process for the manufacture of phthalic anhydride is a new development and offers potential size reductions in equipment and, consequently, cost savings. Therefore, a more accurate estimate should be made by sizing and costing individual pieces of equipment. The total capital cost for the plant can then be estimated either by applying a Lang factor or by estimating other direct and indirect costs separately. Once the capital cost is available and raw materials and product costs have been assessed, a preliminary economic analysis can be performed and indicators such as the return on investment and the payback period can be determined.

6.5.2 Equipment Costs

A four-step method was used to estimate the equipment costs for a 40 kT/yr PAN plant:

1. Determine the base-cost from sources such as Peters and Timmerhaus (1991) and Ulrich (1984), or make a personal estimate if a nomograph is not available.
2. Apply factors for specific construction materials, extreme operating conditions and unusual design features, where applicable [30,40].

Table 6.1   Purchased equipment costs for 40 kT/yr phthalic anhydride plant.

Equipment Item	Base Cost (1990 US$ or 1982 US$)	CE Index Relative to Base Cost	Product of all Factors	Current Cost in 1996 US$	Current Cost in 1996 A$
Reactor (R101)	$150 k	1.061	6.3	$1003 k	$1270 k
Vaporiser (E101)	$105 k	1.061	1.0	$111 k	$141 k
Air preheater (E102)	$40 k	1.061	1.0	$42 k	$54 k
Salt cooler (E103A & B)	$70 k ea.	1.061	1.4	$208 k	$263 k
Gas cooler (E104)	$40 k	1.061	1.2	$51 k	$64 k
After-cooler (E105)	$8 k	1.061	3.8	$32 k	$41 k
Switch condensers (E106 & E107)	$10 k ea.	1.061	4.5	$96 k	$121 k
Purification feed preheater (E108)	$2000	1.061	3.2	$7000	$9000
Stripping column reboiler (E109)	$6000	1.061	4.2	$27 k	$34 k
Rectification column reboiler (E110)	$11 k	1.061	4.2	$49 k	$62 k
Stripping column condenser (E111)	$15 k	1.061	3.2	$51 k	$64 k
Rectification column condenser (E112)	$20 k	1.061	3.2	$68 k	$86 k
Vapour sublimer (E113)	$5000	1.061	4.5	$24 k	$30 k
Pretreatment tank (Tk201)	*$13 k*	1.210	1.9	$30 k	$38 k
Stripping column (T101)	$70 k	1.061	1.0	$74 k	$94 k
Rectification column (T102)	$90 k	1.061	1.2	$115 k	$145 k
O-xylene storage tanks (Tk101 & Tk102)	*$150 k ea.*	1.210	1.0	$363 k	$460 k
PAN storage tanks (Tk103 & Tk104)	*$150 k ea.*	1.210	1.9	$690 k	$873 k
MAN storage tank (Tk105)	*$8000*	1.210	1.9	$18 k	$23 k
Misc tanks and drums (D101 to D103)	*$4000 ea.*	1.210	1.5	$22 k	$28 k
Air compressor (C101)	*$140 k*	1.210	1.0	$169 k	$214 k
Pumps (G101 to G114)	$4000 ea.	1.061	1.2	$71 k	$90 k
Air filter (M101)	$5000	1.061	1.0	$530	$7000
Steam ejectors (M102 & M103)	$5000 ea.	1.061	1.0	$11 k	$13 k
Sub-total				*$3337 k*	*$4224 k*
Contingency	30%			$1001 k	$1267
25% of cost of utilities plant	$1000 k	1.0	1.0	$1000 k	$1266 k
Total equipment cost				*$5340 k*	*$6760 k*

3. Apply the Chemical Engineering Index to adjust costs into 1996 dollars [41].
4. Apply the 1996 exchange rate to convert cost estimates to Australian dollars [42].

This information is summarised in Table 6.1. An overall contingency of 30% has been added as the technology is new and sizing was performed without detailed design data.

6.5.3   Installed Plant Cost by Lang Factor

The Lang factor provides a method for estimating total capital costs from the purchased equipment costs [30]. Percentage estimates for all direct and indirect costs were made specifically for a phthalic anhydride plant using the LAR process (Table 6.2). An additional allowance was included for instrumentation and electrical costs to allow plant-wide advanced controls to be installed. The overall calculated

Table 6.2 Determination of Lang factor for phthalic anhydride LAR process.

Cost Component	Factor Estimate
Direct Costs	
Purchased equipment	1.00
Installation	0.48
Instrumentation	0.22
Piping	0.55
Electrical	0.12
Buildings	0.22
Yard improvements	0.10
Service facilities	0.60
Land	0.05
Sub-total	*3.34*
Indirect costs	
Engineering and supervision	0.32
Construction	0.40
Contractor's fee	0.19
Contingency	0.40
Working capital	0.85
Sub-total	*2.16*
Total capital cost Lang factor	*5.50*

cost multiplier (Lang factor) was 5.50, giving an overall installed plant cost of US$29.4 million (A$37.2 million).

6.5.4 Installed Plant Cost from Recent Plant Construction Data

The LAR process is substantially new technology, and ratio estimates or estimates from nomographs are not available. A key incentive for selecting the LAR process is the reduction in equipment sizes (and costs) that this process and its catalyst permits. Therefore, nomograph estimates based on other processes should yield a higher capital cost. Similarly, the cost of recently constructed plants using older technology should be higher than for the proposed new plant.

A nomograph estimate [28] suggests that a 1996 investment of US$35 million (A$44 million) is required. Cost data is available for four phthalic anhydride plants of a similar size to the current proposal which have been built within the region in the last 15 years: a 30 kT/yr plant in Singapore; an 18 kT/yr plant in Thailand; a 27 kT/yr plant in Thailand; and a 60 kT/yr plant in Japan [43]. Using a size exponent of 0.6, an inflation factor based on the CE Plant Cost Index [41], and a current typical exchange rate of A$1.00 = US$0.79 [42], the costs of these plants were determined in 1996 Australian dollars as shown in Table 6.3.

The Lang factor method produced a capital cost estimate of A$37 million. The average of the recently constructed plants is A$45 million, which is consistent with

Table 6.3   Capital costs for four recently constructed phthalic anhydride plants.

Location	Year	Capacity (kT/yr)	Base Cost	Adjusted Cost	Inflation Factor	1996 US$	1996 A$
Singapore	1995	30	US$36 M	US$43 M	0.997	US$43 M	A$54 M
Thailand	1990	18	US$20 M	US$32 M	1.070	US$35 M	A$44 M
Thailand	1989	27	US$20 M	US$25 M	1.061	US$27 M	A$34 M
Japan	1983	60	¥4030 M	US$32 M	1.199	US$39 M	A$49 M

the current estimate based on itemised equipment costs. Therefore, the installed capital cost is hereafter assumed to be A$40 million ±A$10 million (25% uncertainty). A more accurate costing is not possible until detailed mass and energy balances have been completed.

6.5.5   Production Costs

The phthalic anhydride price (molten product) has been relatively constant since January 1994 but the o-xylene price has fluctuated substantially during this period, probably due to shifting demands for p-xylene [6]. Figure 6.1 shows the recent trends. An o-xylene price of A$450 /tonne is assumed as a basis for estimating the production cost for phthalic anhydride.

Operating costs were estimated using specific LAR process data [7] and general phthalic anhydride data [3]. The total working capital employed by the project was estimated to be A$2.0 million, comprising A$0.50 million for catalyst, A$0.80 million for PAN inventory, A$0.63 million for o-xylene inventory, and A$70,000 for MAN inventory. An interest rate of 11% was assumed and repayments were scheduled over five years. The distribution of costs is given in Table 6.4. The total cost of production was determined to be A$762 per tonne of phthalic anhydride. This compares with a current PAN market value of A$975 per tonne.

6.5.6   Profitability Analysis

At a production cost of A$762 per tonne and market value of A$975 per tonne of phthalic anhydride, the project payback period is 4.7 years at full utilisation of capacity. The pre-tax return on investment (ROI, or return on capital employed, ROCE), during this period is 21%, which is significantly better than current interest rates but not overly attractive for the capital investment required. After capital repayments have been completed, the ROI increases to 48%, assuming production can be sustained at maximum capacity. Over the 20 year project life, the pre-tax ROI is 42%. The discounted cash flow return (DCFR) is 27%, assuming a 20 year project life and capital repayments during the first five years only.

The project is viable considering the base case economics, but the sensitivity to changes in raw material costs, product selling price, utilisation of capacity and

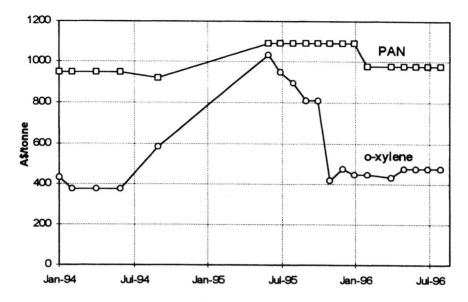

Figure 6.1 Historical phthalic anhydride and *o*-xylene prices.

Table 6.4 Total operating costs for a 40 kT/yr phthalic anhydride plant.

Cost Component	1996 A$/tonne
Raw material	409
By-products	−20
Net feedstock	*389*
Utilities	−55
Catalyst	12
Operating labour	25
Maintenance	19
Laboratory	4
Total direct costs	*5*
Plant overheads	18
License fee	5
Rates & insurance	15
Interest on working capital (5 yrs @ 11% p.a.)	14
Sales, marketing, research	45
Total fixed costs	*97*
Operating cost before capital repayments	*491*
Capital repayments (5 yrs @ 11% p.a.)	271
Total operating cost	*762*

Table 6.5 Profitability scenarios for a 40 kT/yr phthalic anhydride plant.

Profitability Scenario	Pre-Tax ROI	DCFR
Base case	42%	27%
High o-xylene cost	10%	4%
Low PAN value	34%	22%
Low utilisation of capacity	29%	20%
High fixed costs	37%	24%
High interest rates	39%	25%

other factors should be determined to assess the overall viability of the project. Five profitability scenarios were examined:

1. *Ortho*-xylene increases to A$800/tonne,
2. PAN selling price falls to A$900/tonne,
3. Plant capacity utilisation is reduced to 75% due to poor sales,
4. Increase in total operating costs (excluding net raw material cost) of 20%,
5. Interest rates rise to 15%.

Table 6.5 indicates the overall pre-tax ROI and the DCFR for each scenario and suggests that the project can withstand most market threats but is vulnerable to the operating margin between the o-xylene cost and the PAN value. If the cost differential between o-xylene and PAN drops below A$200/tonne (as it did in 1995) the project DCFR falls to around 4%. This confirms the importance of securing a stable and reliable supply of o-xylene before any significant investment is made.

The break-even point can be determined by scheduling the non-variable costs over a lower production rate so that the cost per tonne increases. At a capacity utilisation of 32%, the operating costs reach the estimated product selling price of A$975/tonne and revenue from sales is only enough to meet capital repayments and ongoing fixed expenses. After capital repayments have been completed, the break-even capacity utilisation falls to 17%. Of course, more pessimistic scenarios (e.g. a higher o-xylene cost or lower PAN value) require higher break-even production rates.

The current assessment is that the project is attractive but further studies are required before the construction of a new phthalic anhydride plant can be fully endorsed. The current domestic market is not large enough to independently support a new development, and new export markets must be found. The worldwide consumption of PAN has been steady for several years and, without the co-development of local downstream industries, sufficient markets might not be found to fully utilise plant capacity. Unless definitive contracts can be negotiated before construction then the justification for development remains uncertain. Furthermore, it has already been identified that the sourcing of an appropriate feed supply is a critical consideration so that co-development of a xylene separation plant might also be required.

6.5.7 Conclusions

- The total purchased cost of equipment for a 40 kT/yr plant using the LAR process was estimated at A$6.8 million using basic specifications for all major items of equipment.
- The total capital investment was estimated at A$37 million using a Lang factor of 5.5 (determined specifically for a PAN plant using the LAR process).
- Nomograph and ratio estimates were used to estimate the total capital cost at A$44 million.
- *A value of A$40 ± 10 million was assumed for the total capital investment.*
- *An initial production cost for phthalic anhydride was estimated at A$762 per tonne.* This would drop to A$491 after capital repayments have been met and compares to an average market selling price for PAN of A$975 per tonne.
- The project payback period was estimated at 4.7 years, the overall pre-tax return on investment was estimated at 42%, and the discounted cash flow return was estimated at 27%.
- The breakeven production rate (to cover fixed costs only) before capital repayments have been completed is 13 kT/yr (32% capacity utilisation).
- Five alternative scenarios were examined. The highest sensitivity was shown to be for increases in the *o*-xylene cost.

6.5.8 Recommendations

- Perform detailed mass and energy balances for the plant to allow equipment to be sized more accurately.
- Identify units with large or excessive energy requirements to target areas of potential operating cost savings.

7. MASS AND ENERGY BALANCES

7.1 PREPARATION OF MASS AND ENERGY BALANCES

Mass and energy balances must be completed for each item of equipment before the detailed design of these items can begin. A mass balance is performed for each individual unit shown on the process flowsheet and included in the equipment list (see Section 3.3). Sometimes an energy balance is performed over a group of units (or a section of the plant) where energy is being transferred between various process fluids, e.g. the feed-preheating heat exchangers or the reactor and a subsequent cooler/condenser. After the energy balances have been completed it is possible to consider the efficient utilisation of energy within the plant, and hence prepare an energy conservation scheme (see Section 8.1).

Even at this stage in the design work, it may not be possible to finalise all details of the mass and energy balances. A specification of performance should be available for each unit, but only estimates of certain flowrates and temperatures may be available. These mass and energy balances will need amendment and checking following the detailed and final design of each item of equipment. For example, it may be specified that a liquid-washing absorption tower is to remove at least a certain percentage of a gaseous product stream. It is usually possible to estimate the flowrate of liquid required, but this quantity will not be known accurately until the detailed design of the unit is completed. The design will consider the type and height of packing materials required, and the liquid flowrate necessary for efficient operation of the tower. The final design may be capable of removing more gas than the minimum specified for the original estimated liquid flowrate (however, it is important that the unit designed does not remove less of the gas stream!).

A summary of the mass and energy balances should be presented in a tabular format using a separate sheet of paper for each unit. This information *must* be presented clearly and concisely. The calculations are detailed separately, and often included as an appendix to the design report. The mass and energy balance for each unit should include the following details:

(a) Mass flow of all streams into and out of the equipment per unit time. Molar or volume flowrates must not be used. A convenient and appropriate unit of time is chosen, e.g. per minute, hour, day, etc., so that the numbers being handled can be easily appreciated, e.g. between 0 and 1000, and as far as possible avoiding the use of 10^3 or 10^{-6}, etc. All balances *must* be performed for *the same time period*.

(b) Composition (mass %) of all streams.

(c) Sometimes the molar flowrates (and mole %) of gas streams are also included.
(d) Temperature of each stream; use °C or K consistently (for temperatures below 0°C, absolute temperatures may be preferred).
(e) Pressure of each stream, or state the pressure once if no significant pressure drop occurs.
(f) Enthalpy content of each stream (J, MJ, GJ, etc.).

The tabular format for mass and energy balances should be familiar to students when they undertake a design project, otherwise refer to Himmelblau (1996; p. 147, Figure 3.5; reference at the end of this section), Austin and Jeffreys (1979) or Ray and Johnston (1989). The convention is to include information for all streams entering the unit on the left side of the page, and for all output streams on the right (flow from left to right, as on the P&ID). Each sheet of the mass and energy balances should be numbered consecutively, e.g. sheet 1 of 32 (or 1/32). The reference number for each unit (see Section 3.3) should also be included. If an energy balance is performed over several units, then the combined and summarised balance sheet may be presented. The total mass and total energy into and out of each unit should be given at the bottom of the left-hand and right-hand sides of the page, respectively. Hence, the use of the term 'ledger balance', this provides an immediate indication of whether a balance has actually been obtained.

Data such as flowrates, compositions, temperatures, etc., that is summarised in the mass and energy balances should be transferred to the process flowsheet and the equipment listing (noting any amendments from the original estimates). The process flowsheet should be continually updated as the project proceeds, and the date of the most recent amendments clearly recorded.

The mass and energy balances should not be considered merely as a set of data for use in the detailed design of equipment, they should also be used and analyzed. The mass balances can be used to show those areas of the plant having a high chemical requirement, or by-product production, or process water requirement. Similarly, the energy balances can identify equipment with a high energy requirement or a large surplus of energy to be removed. It is often useful to draw a simple block diagram showing particular sections of the plant, including the major mass flows and the heat output and energy requirements. Such a diagram enables an immediate appreciation of the operational requirements of the plant.

Note to Students

Mass balances often do not balance exactly, some error is allowable (1%, 5%, 10%?). When an exact balance is obtained, this is usually because the flow of one stream has been obtained by difference. Therefore, an exact balance does not mean that the calculations have been performed correctly!

When presenting mass and energy balances, some discretion should be exercised regarding the number of significant figures quoted. An eight-figure calculator display does not imply eight-figure accuracy in the calculations — and such accuracy is not required. Quoting data to three significant figures (or even two figures) is usually sufficient. The inclusion of more figures merely makes the results of the balances more difficult to appreciate, and it becomes difficult to obtain an immediate impression of the process situation. Students often seem unwilling to adopt this suggestion! However, an exception to this general rule is the calculation of small quantities of impurities within the process streams. Low-levels of impurities may be present at different stages, and if ignored they may build-up to unacceptable levels in recycle streams or in the final product stream. It is important to identify and calculate the amounts of impurities at each stage of the process.

Action: *Prepare individual mass and energy balances for each item of chemical engineering equipment, or heat balances over several items of equipment.*

Transfer this information to the process flowsheet and the equipment list.

References

Peters and Timmerhaus (3rd edn, 1980; p. 50) for a reference list.

Austin, D.G. and Jeffreys, G.V., *The Manufacture of Methyl Ethyl Ketone from 2-Butanol (A Worked Solution to a Problem in Chemical Engineering Design)*, The Institution of Chemical Engineers (UK) and George Godwin Ltd, London (1979).

Felder, R.M. and Rousseau, R.W., *Elementary Principles of Chemical Processes*, 2nd edn, John Wiley, New York (1986).

Himmelblau, D.M., *Basic Principles and Calculations in Chemical Engineering*, 6th edn, Prentice-Hall, New Jersey (1996).

Hougen, O.A., Watson, K.M. and Ragatz, R.A., *Chemical Process Principles Part II*, 2nd edn, John Wiley, New York (1959).

Ray, M.S. and Johnston, D.W., *Chemical Engineering Design Project: A Case Study Approach*, Gordon and Breach Publishers, UK (1989).

7.2   PRELIMINARY EQUIPMENT DESIGN

Following completion of the mass and energy balances, preliminary equipment designs can be undertaken. The final equipment designs and specifications are completed after performing a detailed economic evaluation of the project (see Chapter 6) and an energy conservation study (see Section 8.1).

The preliminary design reassess and modifies the data and assumptions made in earlier stages of the project. Any major modifications that become apparent and any changes to the plant or process design aspects are noted, and their effects upon the project feasibility are assessed. The specific type of equipment to be used (as detailed in the equipment list) is confirmed, the appropriate design method is determined, and all necessary data is obtained (hopefully!). Any major differences from the original equipment specifications are investigated. At this stage the design engineer stops short of performing the detailed design calculations.

Action: *Prepare detailed specifications for all major plant items to be designed.*

Obtain all necessary design data.

References

Baasel (1990; Chapter 5); Peters and Timmerhaus (1991; Chapters 14–16); Ulrich
    (1984; Chapter 4); also refer to Aerstin and Street (1978); Backhurst and Harker
    (1973); and Coulson and Richardson Volume 6 (2nd edn, 1993), details given in
    Section IV here.

7.3 COMPUTER-AIDED DESIGN

(Program names in capitals are registered trademarks)
The term *computer-aided design* (*CAD*) is widely used within engineering but is often
interpreted in different ways. It should not be confused with the use of computers to
prepare engineering drawings via draughting packages, this is sometimes referred to
as computer-aided design and draughting (*CADD*). For the chemical engineer
engaged in design work, computers and CAD now have several common applications. These include the preparation of flowsheets and the P&ID, calculation of the
mass and energy balances for the overall process and individual units, costing and
economic analysis, the optimization of process schemes, and the detailed design of
individual chemical engineering equipment, e.g. heat exchangers, distillation columns, etc. Although there is no generally accepted standard terminology, the detailed
design of major units is best described as CAD, whereas the other computer-based
operations listed above can be referred to as *computer-aided process engineering*
(*CAPE*) or simply *computer-aided engineering* (*CAE*).

It is now expected that all graduating chemical engineers will have had at least
some basic experience of CAD in their course, and in many cases the students will
have had extensive hands-on training with a variety of CAD programs and packages.
This experience will vary depending upon the university, the funding and resources
available, and the department philosophy regarding the teaching of design. A book
of this type cannot devote more than a few pages to this subject, and even then
presenting only an overview of the use of computers for design. This discussion must
be limited to a general description of the types of CAD packages available, their
usefulness, and what the student can realistically expect to achieve. Specific details
of particular packages are omitted because of the wide range of CAD programs now
available, the introduction of updated versions of each package, and the variation
between departments in terms of what is available for student use. However, some
of the more-widely used packages are mentioned.

At this stage it is worth considering briefly what experience in CAD a university
course should provide for a student, and what experience the student should expect
to have when leaving university and beginning a career in industry. Hopefully these
two aspects will not be too far different. Most employers of chemical engineering
graduates now expect that the students will have had some hands-on experience of
CAD in their course. How much experience they consider necessary probably
depends upon the use of CAD packages within the company. However, just as
students are expected to be adept at computer programming and to be able to apply
their skills to whatever system is available, it would also be expected that CAD
experience would be such that the student becomes capable and confident of being

able to use new packages as they become available (with some instruction and help). What is really important is that the student appreciates the need for and use of CAD packages for solving chemical engineering design problems, and that the operation and structure of such packages is understood, rather than the need to become fully conversant with the use of a large number of packages. Wherever the graduate engineer is employed, if CAD packages are used then instruction in their use and subsequent advice and assistance should be available. It is necessary for the student to become familiar with the use of such packages, and to be able to adapt to the use of new CAD facilities.

When using a calculator to perform numerical calculations it is necessary to know the sequence of operations and (just as important) to be able to assess the 'correctness' of the answer obtained. The same requirements apply to the use of CAD packages. The programming operations depend upon the particular package, but an assessment of the validity of the results is essential and depends upon the student understanding the fundamental principles of chemical engineering. Students need to be taught (or rather they need to develop the approach) to assess the results produced by CAD packages, and not blindly assume that the results are correct or necessarily apply to a particular situation. The aim of using CAD packages is to reduce the time spent performing manual calculations, and also to enable the engineer to consider a wide range of alternative situations/values/variables in order to optimize the problem and arrive at the 'best' solution. It is neither useful nor practical to suggest that the student (or engineer) should perform many detailed calculations manually to check that the computer results are valid. However, some checks should be performed for appropriate calculations and it should be possible for students to perform the calculations if necessary, and to appreciate the design method or calculation sequence being performed. CAD packages will not replace the need for engineers who understand and can apply the basic principles of chemical engineering, however they can make the engineer's work more productive by reducing the time spent performing laborious or repetitive calculations.

The use of computer programs or packages to carry out mass and energy balances is now commonplace, and probably one of the first encounters the student has with CAD. Unless the equipment or process is fairly simple, hand calculations become difficult and time consuming due to the iterative nature of the method. The design project may be the student's first encounter with the calculation of mass and energy balances for an entire process. Calculation by computer not only saves time but allows many changes to be made to the arrangement of equipment and to the process variables in order to specify the optimum flowsheet and process conditions.

The computer program for the material balance contains several parts. First, a description of each item of equipment in terms of the input and output flows and the stream conditions. Quite complicated mathematical models may be required in order to relate the input and output conditions (i.e. performance) of complex units. It is necessary to specify the order in which the equipment models will be solved, simple equipment such as mixers are dealt with initially. This is followed by the actual solution of equations. The ordering may result in each equation having only one unknown and iteration becomes unnecessary. It may be necessary to solve sets of linear equations, or if the equations are non-linear a suitable algorithm applying some form of numerical iteration is required.

The work involved in preparing a mass balance computer program can be considerable, even for a simple process involving few units. The usefulness of such programs has lead to the development and availability of commercial "Process Flowsheet" programs or packages such as HYSIM, PROII, HYSYS, PROSYS, etc. These programs contain sets of programmed, general-purpose unit-operation subroutines based upon the data input. The flowsheet-type program can also be connected to a physical property database (or databank) so that appropriate data can be obtained automatically for a wide range of components without the need for user input. Further details of the use of flowsheeting programs for material and energy balances are given in Wells and Rose (1986; pp. 285–294).

Depending upon the complexity of the process, it is possible to set up the mass and energy balances using a spreadsheet package such as Microsoft EXCEL or LOTUS 1-2-3. If the student intends to develop their own approach in this way, then it is preferable to have had some prior experience/attempts *before* starting a serious design study in order to avoid time wastage.

Flowsheeting programs can provide much more information than just the material and energy balances for the process. In performing these calculations, each item of equipment has effectively undergone a preliminary design evaluation. The results (or output) can usually include details of equipment heat loads, reflux ratios, utility requirements, and preliminary sizing. If the program is connected to a physical property database then extensive component property data can also be obtained for each process stream. Although a large number of flowsheeting packages are commercially available, many have similar features and program structure. When evaluating packages for use (or purchase) it is desirable to compare the scope of operations included, the range of equipment models, the ease of use (user-friendliness), the size of the database, and the quality of the support documentation.

All design studies require physical property data and the efficient use of CAD methods means that this data must be available in a convenient form for computer use. Data in the form of graphs or tables is of limited use, although some databases contain only measured data. A more useful form of data storage is to correlate the available data and then store the calculated parameters required for use in appropriate empirical equations (stating the limits over which the correlations apply). Some databases contain only the calculated parameters whereas others also store the available raw data. The following are examples of some common physical property databases: ASPEN-PLUS (AspenTech, USA); CHEMCO (Eurecha, Switzerland); DATABANK (Imperial Chemical Industries, UK); FLOWTRAN (Monsanto, USA); PPDS (IChemE, UK).

The most useful databases for chemical engineering design problems are those where missing data can be obtained from appropriate predictive methods, e.g. vapour–liquid equilibria (VLE) data obtained from a group contribution method. Unavailable data may also be obtained by substituting data for a corresponding and similar compound. If the required data does not have a significant effect upon the calculated result then a rough estimate can be used. The user of CAD packages needs to know which predictive methods are being applied, and their accuracy and effect upon the calculated results. This is where an understanding of the structure and application of CAD packages is required, and also a knowledge of fundamental engineering principles. More details concerning the use of physical property databases are given in Wells and Rose (1986; pp. 343–364).

To gain the maximum benefit from the use of a flowsheet program, the operator/designer must be adequately trained. A suitable program will have 20–30 standard units available, numerous equation-solving procedures, control facilities and probably optimisation facilities. The unit-equipment subroutine must adequately represent the process equipment, recycle streams need to be specified, and suitable solution convergence is required. For the effective use of CAD packages, it should be obvious that engineering software will not replace the engineer but a better trained engineer with wider abilities and judgement will be required.

The use of a suitable flowsheet package includes the following four stages: simulation, design, case-study, and optimisation. The initial simulation stage requires definition of the input data and selection of appropriate equipment models. Simple units can replace more complex equipment at this stage, e.g. a 'component splitter' for a distillation column. At this stage repeated substitution usually achieves convergence, and this can then be followed by use of a more suitable accelerated convergence algorithm. The simulation stage provides a simplified model which will converge to an initial solution, although the output and product specification are probably incorrect. This initial stage is followed by the upgrading of the simulation model into a suitable design mode, usually by adding further recycles (as information loops). The design process is iterative and suitable convergence procedures are required. The required result is a model that produces a meaningful solution in a reasonable number of iterations. The case-study stage involves the use of actual plant data and rules-of-thumb for calculating initial design variables. The design model is used to provide a sensitivity approach to indicate which variables are significant and whether it is worthwhile implementing full-scale optimisation procedures. The optimisation stage requires a significant allocation of the designer's time. The appropriate software must be available and the user should possess some experience with these procedures. It will be necessary to optimise sections of the plant separately, and subsequently to make appropriate changes and repeat calculations. A full economic analysis will also be required at this stage.

The application of computers to chemical engineering design, and the potential savings, have lead to the commercial development of a large number of CAD packages. Some of these packages deal with particular aspects of design work, such as detailed heat exchanger design (e.g. HEATEX by Humphreys and Glasgow, UK); flowsheet drawing and draughting packages (e.g. PROCEDE, AUTOSKETCH, PRODESIGN, TURBOCAD), and general drawing packages which are easy to use and can be programmed to include a personal library of flowsheet drawing symbols (and also produce detailed engineering drawings); and databases (e.g. Physical Property Data Service (PPDS) by the IChemE, UK). Many other programs of the flowsheet-type (mentioned earlier) are available for the simulation and design of entire processes, including calculation of the material and energy balances. These include HYSYS which now includes a dynamic modeling section and replaces the original steady-state package HYSIM (by Hyprotech Ltd, Canada), PROTISS (previously PROII, by Simulation Sciences, Inc., USA), ASPEN PLUS (by Aspen Technology, Inc., USA), and SPEED-UP (developed at Imperial College, London).

Early versions of these packages required use of a mainframe computer although they are now available (and most often used) on personal computers or a PC-network. The UNICORN flowsheet program was developed by Eurecha, Switzerland (European Committee for the Use of Computers in Chemical Engineering Education) specifically for teaching purposes, and it can be used with the CHEMCO databank. Details of these two packages are given in Wells and Rose (1986; pp. 649–692). It is important that undergraduate chemical engineers obtain experience in the use of CAD packages, but the actual packages used will depend upon the individual department. It is more important to appreciate the applications of CAD and the skills required by the computer-based designer, than to have superficial hands-on experience of many different systems.

What design work is there left to do now that we have simulation packages?

Students sometimes ask this question, or earlier in the course they ask: "Why do we have to learn all this stuff when its all in the ABC package?" My answer is that they are training to become professional engineers who can perform highly technical detailed design work. In order to do this they must understand the basic theory associated with the unit operations and be able to perform design calculations. Computer packages will help in this work by allowing calculations to be performed quicker, and a wider range of options to be investigated. The computer is a tool to make life easier, if used correctly. However, the design engineer must be able to recognise poor design solutions and incorrect answers and to appreciate that the results from a computer package are not necessarily correct. Students who do not agree with the last sentence should read the paper by Sloley *et al.* (1995) which investigated why a significant number of distillation towers in industry did not perform as well as expected. There were several reasons but surprisingly a large number were due to incorrect designs based on computer simulations, and many of these should have been corrected at the design stage! Computer simulation is no substitute for knowledge and understanding of the basic principles, as demonstrated in the paper by Sadeq *et al.* (1997). *If the simulation package can do the design work, then what are the professional engineers going to do?* We will have been replaced by computer-literate technicians or computer science graduates!

A Word of Caution. The software suppliers accept no responsibility for any damage or loss caused by the use of CAD packages, even if the results obtained from the package are in error. The design engineer chooses to use the program and is responsible for checking and approving the results obtained (see Sadeq *et al.*, 1997).

References

Sadeq, J., Duarte, H.A. and Serth, R.W., Anomalous results from process simulators, *Chem. Eng. Education,* 31(1), 46-51 (1997).

Sloley, A.W., Martin, G.R. and Golden, S.W., Why towers do not work, Paper 30G presented at the *AIChE Spring National Meeting,* Houston Texas, 23 March 1995.

7.4 CASE STUDY — MASS AND ENERGY BALANCES, AND UTILITIES

Summary

The partial oxidation of o-xylene, in the presence of excess air, produces phthalic anhydride and releases large quantities of heat (21 MW). The PAN product is recovered by cooling the reaction gases in a two-stage process which recovers a liquid product and a solid product (by sublimation). Finally, the crude product is purified by distillation. The LAR process requires less air at the reaction stage and hence equipment sizes throughout the plant are smaller and more energy efficient than for the alternative processes. A significant amount of energy (14.6 MJ per kg phthalic anhydride) can be exported from the process as high-pressure steam.

The masses and energy are concentrated in the reaction and recovery sections. The reactor, and its associated heating and cooling equipment, are the critical areas for energy integration as the combination of high flow rate and high temperatures produces the largest enthalpy flows. The purification equipment operates at comparatively mild conditions and with low flow rates, but has the potential for integration between the two distillation columns. Two hot-utilities (steam and diathermic oil) and four cold-utilities (air, cooling water, foundry salt and diathermic oil) are required. Detailed heat integration schemes need to be applied to reach minimum utility targets, and to maximise the export of high grade heat.

7.4.1 Scope and Objectives

The design of equipment for any process must start with detailed mass and energy balances. The results of these calculations determine the overall size of the equipment, in terms of flow rate and heat duty. They also provide the first estimates for the process objectives (target temperatures, purities, etc.). When combined with a P&ID (piping and instrumentation diagram), the basic operations of a plant are defined. The results need not be fixed or final at this stage, as units will be optimised and amended many times before the 'final' engineering drawings and specifications emerge.

The mass and energy balances can be used to identify critical units and variables, and provide the basis for many control and instrumentation decisions. The temperatures and heat flow information contained in the mass and energy balances can be used to target specific areas for energy conservation. Utility demands can be estimated and heat integration problems can be defined from this data so that the final plant design can be as energy efficient as possible.

In practice, process simulation packages can be used to provide thermodynamic data and reduce the calculation time required. However, the results are not always compatible with specific system requirements and may need to be modified or corrected. In this case study, much of the modelling was performed on HYSYS (by Hyprotech) [45] and PRO/II (by SimSci) [38].

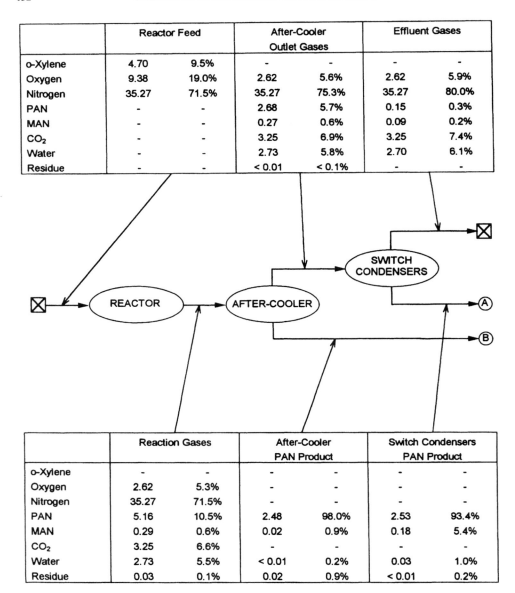

	Reactor Feed		After-Cooler Outlet Gases		Effluent Gases	
o-Xylene	4.70	9.5%	-	-	-	-
Oxygen	9.38	19.0%	2.62	5.6%	2.62	5.9%
Nitrogen	35.27	71.5%	35.27	75.3%	35.27	80.0%
PAN	-	-	2.68	5.7%	0.15	0.3%
MAN	-	-	0.27	0.6%	0.09	0.2%
CO_2	-	-	3.25	6.9%	3.25	7.4%
Water	-	-	2.73	5.8%	2.70	6.1%
Residue	-	-	< 0.01	< 0.1%	-	-

	Reaction Gases		After-Cooler PAN Product		Switch Condensers PAN Product	
o-Xylene	-	-	-	-	-	-
Oxygen	2.62	5.3%	-	-	-	-
Nitrogen	35.27	71.5%	-	-	-	-
PAN	5.16	10.5%	2.48	98.0%	2.53	93.4%
MAN	0.29	0.6%	0.02	0.9%	0.18	5.4%
CO_2	3.25	6.6%	-	-	-	-
Water	2.73	5.5%	< 0.01	0.2%	0.03	1.0%
Residue	0.03	0.1%	0.02	0.9%	< 0.01	0.2%

Figure 7.1   Overall mass balance for 40 kT/yr phthalic anhydride plant (Basis: All flows in T/hr).

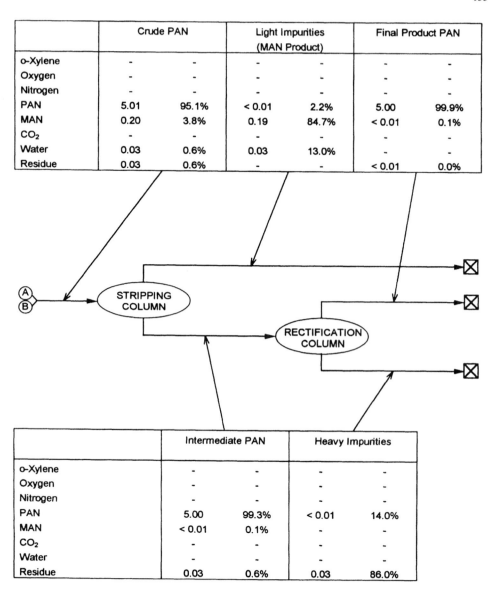

	Crude PAN		Light Impurities (MAN Product)		Final Product PAN	
o-Xylene	-	-	-	-	-	-
Oxygen	-	-	-	-	-	-
Nitrogen	-	-	-	-	-	-
PAN	5.01	95.1%	< 0.01	2.2%	5.00	99.9%
MAN	0.20	3.8%	0.19	84.7%	< 0.01	0.1%
CO_2	-	-	-	-	-	-
Water	0.03	0.6%	0.03	13.0%	-	-
Residue	0.03	0.6%	-	-	< 0.01	0.0%

	Intermediate PAN		Heavy Impurities	
o-Xylene	-	-	-	-
Oxygen	-	-	-	-
Nitrogen	-	-	-	-
PAN	5.00	99.3%	< 0.01	14.0%
MAN	< 0.01	0.1%	-	-
CO_2	-	-	-	-
Water	-	-	-	-
Residue	0.03	0.6%	0.03	86.0%

Figure 7.1 (*Continued*)

7.4.2 Mass Balances

The phthalic anhydride process divides readily into two separate sections which are characterised by very different mass flow rates: first, the reaction section containing the reactor and recovery equipment; and second, the purification section containing the two distillation columns. The after-cooler and switch condensers (in the first section) recover the PAN product from the reaction gases and, consequently, reduce the mass flow rate by about 90–95%. The effluent gases are vented to the atmosphere after further processing. The process mass balance is summarised in Table 7.1 and Figure 7.1. Detailed mass and energy balances are shown for each unit in Figures 7.2–7.10. The key difference from the results of earlier mass balances is the increase in o-xylene feed rate required to account for higher than expected PAN losses in the effluent gases.

At the reaction stage, the hydrocarbons (o-xylene feed or phthalic anhydride product) comprise approximately 10% of the mass flow, while at the purification stage, the hydrocarbon component is at least 98% of the total flow. This difference in relative flows requires equipment in the reaction stages to be generally much larger than equipment in the purification stages. A further consequence of the changing hydrocarbon concentration is the increased explosion risk in the reaction stages. The reaction gas concentrations are just below the upper explosive limit (UEL) for phthalic anhydride. The risk of explosion in the purification stages is greatly reduced as a leak is likely to be detected before the lower explosive limit (LEL, about 2% by weight) is reached.

The effluent gases (non-condensables from the switch condensers) still contain significant quantities of hydrocarbon, particularly maleic anhydride which condenses at a much lower temperature than the PAN. MAN is highly toxic and the limit for human exposure is 0.25 ppm in air. Therefore, the vapour concentration must be reduced by 1000 times, using either scrubbing or incineration. Where scrubbing is used, the maleic anhydride can be recovered for sale. The concentration of PAN in the gases must be reduced by 50 times to reach a safe level. Other components can be safely vented directly to the atmosphere.

Table 7.1 Mass balance summary for a 40 kT/yr phthalic anhydride plant (LAR process).

	O-Xylene	Air	PAN	MAN	Water	Residue	Total
Feed	4.70	44.65	—	—	—	—	49.35
Reaction products	—	41.14	5.16	0.29	2.73	0.03	49.35
After-cooler outlet (liquid only)	—	—	2.48	0.02	0.00	0.02	2.53
Switch condenser outlet (liquid only)	—	—	2.53	0.18	0.03	0.00	2.74
Crude PAN	—	—	5.01	0.20	0.03	0.03	5.27
Rectification column feed	—	—	5.00	0.01	0.00	0.03	5.04
PAN product	—	—	5.00	0.01	0.00	0.00	5.01

Basis: All flows in T/hr.

7.4.3 Energy Balances

Detailed analysis of the process energy balances resulted in several changes to the operating conditions and refinement of the process layout. Originally, a reactor inlet temperature of 300°C was specified but this would require heating with superheated steam or with the reactor product gases. However, a lower reactor inlet temperature is acceptable as the hot circulating salt will provide the necessary heat input to reach the required reaction conditions. Therefore, the preheater is only required to vaporise the o-xylene feed and to provide some heat input to the air stream to prevent cool spots forming in the reactor which could lead to solidification of the heat transfer salt. The reactor product gases can then be used to preheat the BFW to close to its boiling point at 6000 kPa (276°C), so that the HP steam generators are providing latent heat only and can then be operated more efficiently. The boiling point of o-xylene at 250 kPa (the process pressure at the vaporiser) is 194°C so that a reactor inlet temperature of 205°C is adequate. The effect on the reactor is to reduce the heat liberated by 1.47 MW, but this is balanced by an equal reduction in the heating duty of the o-xylene vaporiser and air preheater. The more moderate reactor inlet temperature allows MP steam, (2000 kPa) to be used as the heating medium.

Heat flows within the process are highest around the reactor where high temperatures and high flow rates coincide. The heat of reaction liberates 20.93 MW of heat which is mostly transferred directly to the high temperature coolant (a eutectic mixture of sodium nitrate and potassium nitrite, commonly called foundry salt). Heat not transferred to the cooling salt, about 2.61 MW, remains with the process and causes a temperature rise of approximately 155°C. Heat is recovered from the cooling salt in high pressure process boilers, creating approximately 24 T/hr of superheated HP steam which can be exported and/or used to produce electricity. Control of the reactor temperatures is important to maintain an overall high energy efficiency and to maximise steam production. The temperature rise across the reactor should be minimised without sacrificing conversion or selectivity. The characteristic mass flows, temperatures and energy flows (i.e. the process energy balance) are summarised in Table 7.2.

The reaction gases leave the reactor at approximately 370°C and a significant amount of heat is still available for direct recovery. This heat could be used either to preheat the reactor feed (both air and o-xylene), or to preheat water for steam production in the salt coolers, or to generate more steam directly. Detailed consideration of these options resulted in the selection of the second option as it minimised overall equipment requirements. The latent heat of condensation is still available for recovery, and this is achieved with the production of more MP steam in the after-cooler and in the diathermic oil exchangers associated with the switch condensers. The effluent gases leave the system at about 70°C, taking about 1.9 MW of heat. It is inefficient to recover more of this energy, via scrubbing or incineration, prior to further processing. A scrubbing system offers the advantage of maleic anhydride recovery, but an incineration system will have a much lower capital cost and could be combined with a waste heat boiler to produce some low pressure steam.

The heat transfer processes in the purification section are less significant as the flows are much smaller. Only slight improvements in efficiency in the reaction stages would be sufficient to satisfy all the heating requirements in the pretreatment and distillation equipment. The overall process-heat demand in the purification area is only approximately 760 kW.

The net heating and cooling loads in the plant (excluding heat transfer between utilities) were calculated to be 4.33 and 24.31 MW, respectively, distributed according to Table 7.3. The net production of energy using the LAR process

Table 7.2 Energy balance summary for a 40 kT/yr phthalic anhydride plant (LAR process).

	Flow (T/hr)	Temperature (°C)	Enthalpy Flow (MW)
Feed	49.35	25	0 (ref. point)
Reactor inlet	49.35	205	3.20
Reactor outlet	49.35	370	7.40
HP Steam from reactor			18.32
After-cooler outlet	46.82 (vap) 2.53 (liq)	135	3.22 (vap) 0.06 (liq)
Switch condensers outlet	44.08 (vap) 2.74 (liq)	70	1.90 (vap) 0.03 (liq)
Crude PAN	5.27	140	0.15
Rectification column feed	5.06	205	0.25
PAN product	5.01	170	0.20

Table 7.3 Heating and cooling duties in a 40 kT/yr PAN plant.

	Heating Duty (MW)	Cooling Duty (MW)
Air preheater	2.29	—
O-xylene vaporiser	0.91	—
Reactor	—	18.32
Gas cooler	—	2.59
After-cooler	—	1.29
Switch condensers	—	1.21
Crude PAN pretreatment (tank and preheater)	0.28	—
Stripping column preheater	0.09	—
Stripping column reboiler	0.11	—
Stripping column condenser	—	0.24
Rectification column reboiler	0.65	—
Rectification column condenser	—	0.66
Totals	4.33	24.31

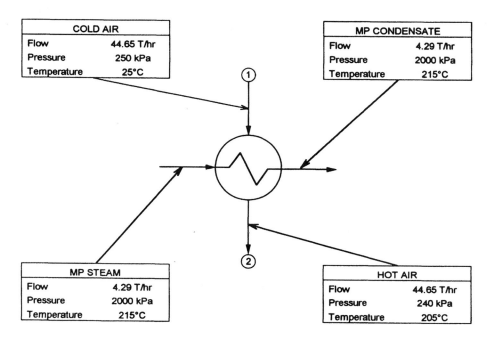

Figure 7.2   Air preheater (E101).

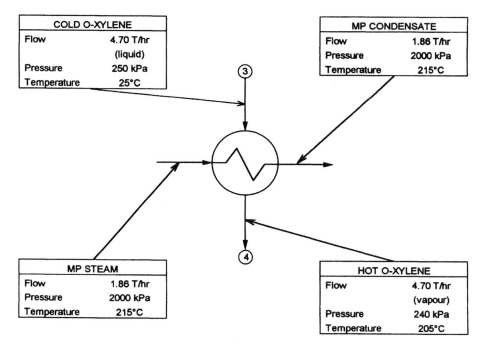

Figure 7.3   *Ortho*-xylene vaporiser (E102).

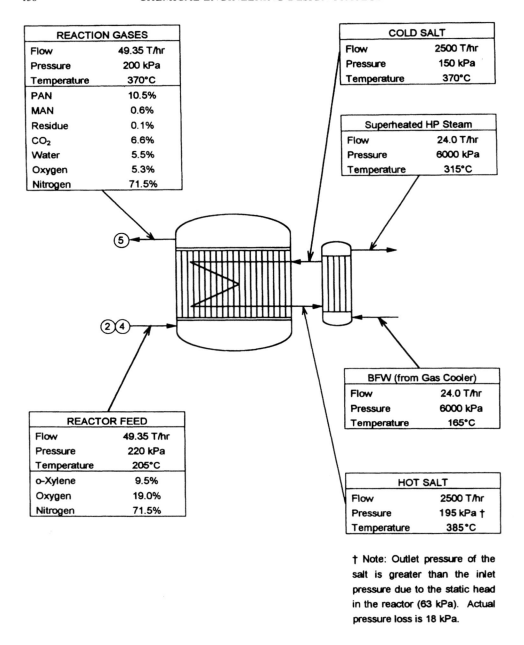

REACTION GASES	
Flow	49.35 T/hr
Pressure	200 kPa
Temperature	370°C
PAN	10.5%
MAN	0.6%
Residue	0.1%
CO_2	6.6%
Water	5.5%
Oxygen	5.3%
Nitrogen	71.5%

COLD SALT	
Flow	2500 T/hr
Pressure	150 kPa
Temperature	370°C

Superheated HP Steam	
Flow	24.0 T/hr
Pressure	6000 kPa
Temperature	315°C

REACTOR FEED	
Flow	49.35 T/hr
Pressure	220 kPa
Temperature	205°C
o-Xylene	9.5%
Oxygen	19.0%
Nitrogen	71.5%

BFW (from Gas Cooler)	
Flow	24.0 T/hr
Pressure	6000 kPa
Temperature	165°C

HOT SALT	
Flow	2500 T/hr
Pressure	195 kPa †
Temperature	385°C

† Note: Outlet pressure of the salt is greater than the inlet pressure due to the static head in the reactor (63 kPa). Actual pressure loss is 18 kPa.

Figure 7.4 Reactor (R101) and salt cooler (E103).

is, therefore, up to 19.98 MW for a 40 kT/yr plant. Verde and Neri (1984) [7] report that an energy export of 3.081 MMkcal (12900 MJ) is possible per tonne of phthalic anhydride produced. On the basis of 8000 operating hours, this is equivalent to 17.9 MW. As 2.1 MW of energy will be lost in the product and effluent,

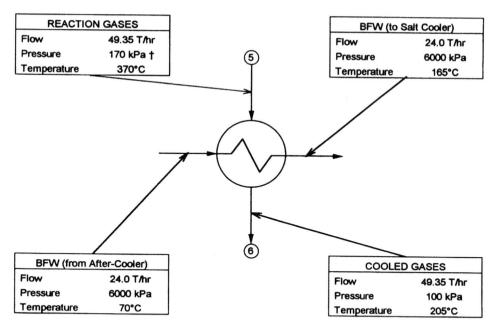

Figure 7.5 Gas cooler (E104).

the overall energy balance is shown to be accurate to within 1%. The error may be even less because much of the 0.2 MW (or 140 kJ per kg of PAN produced) is likely to be accounted for by equipment inefficiencies, pumping duties and other energy losses.

Some of the duties shown in Table 7.3 can be 'matched-up' to reduce the overall utility demand. As already discussed, the gas cooler can be used to pre-heat the BFW which then passes to the salt coolers. The after-cooler can also be used in this service and is best employed in series with (and upstream of) the gas cooler. The water can be heated to approximately 165°C using the available heat without condensing PAN in the gas cooler, or reducing the purity of the PAN liquid from the after-cooler. This effectively increases the HP steam production by 26%. The reboiler and condenser heat duties are similar (for both columns) and could be matched by using heat pumps, but this is unlikely to be economically attractive.

7.4.4 Optimisation of Mass and Energy Balances

The main difference between the LAR process and other comparable technologies for the production of phthalic anhydride is the hydrocarbon loading in the reactor. A new catalyst allows the air requirements to be halved which significantly reduces

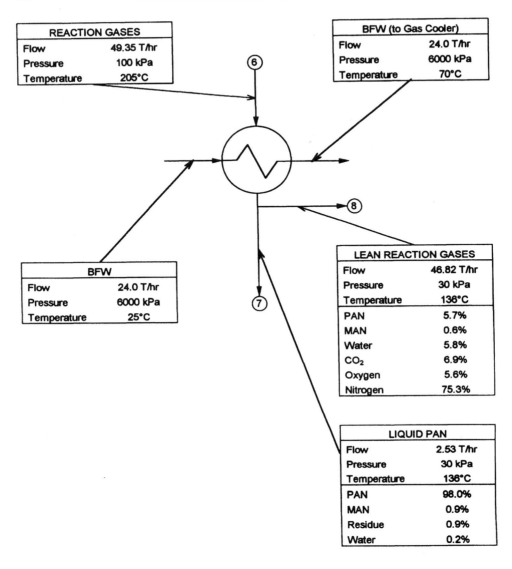

REACTION GASES	
Flow	49.35 T/hr
Pressure	100 kPa
Temperature	205°C

BFW (to Gas Cooler)	
Flow	24.0 T/hr
Pressure	6000 kPa
Temperature	70°C

BFW	
Flow	24.0 T/hr
Pressure	6000 kPa
Temperature	25°C

LEAN REACTION GASES	
Flow	46.82 T/hr
Pressure	30 kPa
Temperature	136°C
PAN	5.7%
MAN	0.6%
Water	5.8%
CO_2	6.9%
Oxygen	5.6%
Nitrogen	75.3%

LIQUID PAN	
Flow	2.53 T/hr
Pressure	30 kPa
Temperature	136°C
PAN	98.0%
MAN	0.9%
Residue	0.9%
Water	0.2%

Figure 7.6 After-cooler (E105).

the heat losses, the preheating energy requirements and the blower power. Operating the reactor at close to the maximum hydrocarbon loading maximises the benefits of the LAR catalyst. Similarly, maximising the selectivity for phthalic anhydride by optimal control of the temperature profile, reduces heat losses and the blower duty. The expected increase in recoverable energy is 70% compared with other processes, while the expected decrease in the compressor power requirement is 60% [7].

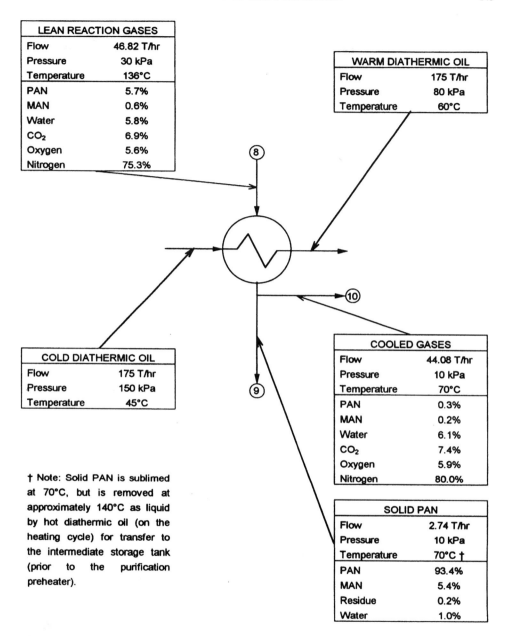

Figure 7.7 Switch condensers (E106 & E107; condensing cycle).

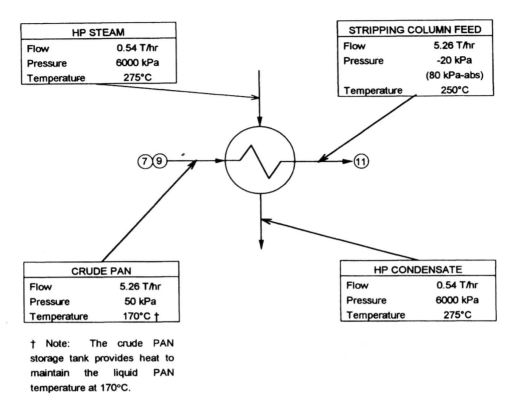

Figure 7.8 Purification preheater (E108).

The high concentration of hydrocarbon in the reactor feed produces an additional saving in the product recovery section. The higher concentration of phthalic anhydride in the reaction gases allows about 50% of the PAN vapour to be condensed above its sublimation point. Consequently, the duty of the switch condensers is halved with large capital savings. The electricity generation capacity is also improved by the LAR process. High-pressure steam from the gas cooler (6000 kPa) is superheated by contact with the cooling salt and then expanded through a multistage back-pressure expander down to 1700 kPa. The final product cost (before tax) is nearly 10% less than from other processes [7]. This saving is comprised of 60% from reduced capital costs and 40% by energy efficiency improvements that allow a greater steam (or electricity) export.

Although distillation columns are traditionally very large users of heat, they are not critical in a phthalic anhydride plant. The small flows in the purification section of the plant shift the focus to the reaction stage. The two most important process areas contain the reactor and associated cooling equipment, and the product recovery equipment (after-cooler and switch condensers). These very large sources of heat must be handled efficiently to ensure high recoveries of energy.

The reactor produces 19.47 MW of heat that passes directly to the circulating foundry salt. This heat is then recovered by the production of high pressure steam.

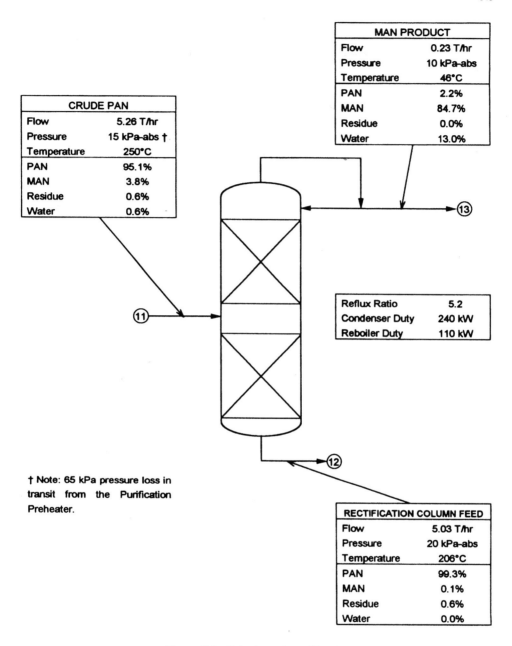

Figure 7.9 Stripping column (T101).

The rapid release of energy due to the reaction requires a number of protection systems to be installed to prevent explosions. However, if these back-up systems (vents, etc.) operate too frequently then large amounts of energy will be lost. Therefore, good operating techniques and careful monitoring will be required at all times.

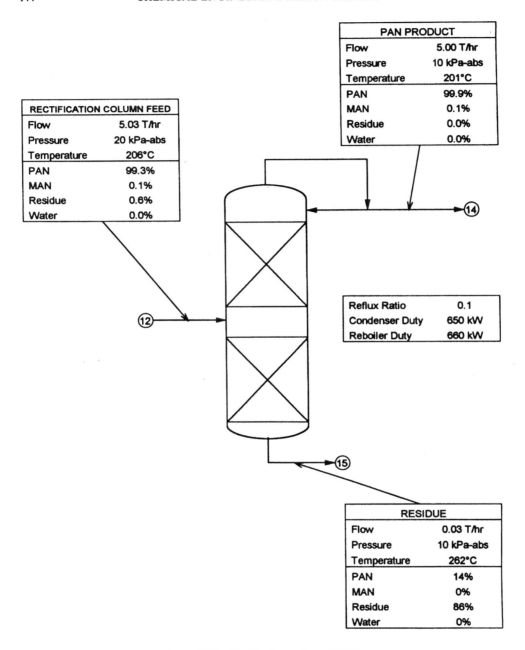

Figure 7.10   Rectification column (T102).

The heat flow to the salt, and therefore the heat recovery, is dependent on the temperature rise in the reactor. The temperature rise also affects the selectivity for the phthalic anhydride reaction, and, therefore, the temperature profile in the reactor is critical and must be well controlled and continually optimised. A low yield of PAN is wasteful of raw materials but increases the amount of heat produced as the complete oxidation reaction (to CO_2) is more exothermic than the partial oxidation (to PAN). If the temperature rise is too great, a runaway reaction is possible. In this scenario, little or no phthalic anhydride will be produced and the highly exothermic complete oxidation will take precedence, which will subsequently load the cooling salt excessively. In less severe cases, the energy losses will increase as more heat passes to the effluent gases.

As discussed previously, the LAR process provides two means of PAN recovery; via the after-cooler and the switch condensers. The operation of these units is intimately linked so that a balance between recovery in the after-cooler and recovery in the switch condensers must be found to maximise the *overall* PAN recovery. Effectively, this requires the after-cooler temperature to be optimised. The liquid product is of significantly higher purity and requires less energy to produce than the solid product. However, the temperature in the after-cooler must be maintained high enough to avoid solidification of phthalic anhydride crystals (the sublimation point is 131°C). Verde and Neri [7] have suggested that 49% of the product should be collected in the after-cooler, corresponding to a temperature of 136°C (from HYSYS). With improved control methods, a lower temperature may be possible. Table 7.4 indicates the changes in recovery and purity with changing outlet temperatures from the after-cooler, for a constant switch condenser outlet temperature of 70°C.

The remaining phthalic anhydride in the reaction gases is recovered in the switch condensers. Here, the exit temperature must be optimised to recover as much product as possible without condensing maleic anhydride and other contaminants in the gases. Verde and Neri [7] suggest an exit temperature of 70°C (10°C higher than in conventional processes). If a scrubbing system (rather than an incinerator) were to be used to clean the effluent gases, it may be possible to use a higher exit temperature and recover more PAN from the liquor in addition to the MAN (refer to Cavaseno, [46]). For the conditions shown in the unit balances (Figures 7.2–7.10), the crude PAN from the switch condensers is only about 93% pure (compared with 98% from the after-cooler). Table 7.5 indicates the changes in recovery and purity for different outlet temperatures from the switch condenser, for a constant after-cooler outlet temperature of 136°C.

The division of cooling duty between the gas cooler and the after-cooler must also be optimised. In this case, the primary consideration will be energy integration. A temperature should be selected based on the utility demands as determined from pinch calculations.

Table 7.4 Effect of after-cooler outlet temperature on PAN recovery and purity.

Outlet Temperature (°C)	After-Cooler Recovery (%PAN)	After-Cooler Product Purity (%PAN)	Overall Crude Purity (%PAN)
134	53.3	97.9	95.7
136	49.3	98.0	95.6
138	45.0	98.0	95.5
140	40.4	98.0	95.4

Table 7.5 Effect of switch condenser outlet temperature on PAN recovery and purity.

Outlet Temperature (°C)	Total Recovery (%PAN)	Switch Condenser Product Purity (%PAN)	Overall Crude Purity (%PAN)
60	98.4	92.0	94.8
65	97.9	92.7	95.2
70	97.2	93.4	95.6
75	96.2	94.1	96.0

Crude phthalic anhydride is pretreated before distillation. The temperature is raised to convert phthalic acid (formed from any water remaining in the product) to phthalic anhydride. The reaction is favoured by higher temperatures, but if the temperature is too high there will be excessive vapourisation (boiling point is 280°C). In the vapour state, the product attracts impurities again. At 250°C, only 0.25% PAN is vapourised. If the temperature is raised a further 10°C, the vapour fraction increases to 13% (results from HYSYS [45]).

7.4.5 Utilities

Five utilities are required by the process. The principal cold-utilities are air and water which are both used in process coolers (heat exchangers). Steam is used as the principal hot-utility. Foundry salt is used as a cold-utility in the reactor due to the very high temperatures (400–500°C). Diathermic oil is used as both a cold- and a hot-utility in the switch condensers. Although the temperatures in the switch condensers are moderate (approximately 150°C when heating and 50°C when cooling), the same utility must be used for both heating and cooling to simplify the mechanical design. High pressure water (e.g. 1000 kPa) would suffice but diathermic oil is more convenient as it can be used at low pressure and is less fouling and less corrosive. The costs and operating range for each of the utilities are summarised in Table 7.6.

There are significant transfers of energy between these utilities and only air and water cross the system boundary to remove energy (no external energy is added to the process). About 18 MW of surplus steam can be exported to the adjacent Utilities Plant for use by other local industries, or used to generate electricity (which may then be sold to Western Power).

Three pressures of steam are required: low pressure (400 kPa), medium pressure (2000 kPa) and high pressure (6000 kPa). The HP steam, which will be produced by the reactor, will also be superheated to drive turbines for power generation. The exhaust steam from power generation will be used to meet high temperature process heating duties. All condensate will be recycled to save energy and reduce the steam system chemical costs (e.g. utility plant deionisation costs and additives to prevent fouling). Cooling water will only be available from the adjacent Utilities Plant. Therefore, its use will be minimised to reduce dependence

Table 7.6 Utilities specifications.

Utility	Hot or Cold	Process Temperatures	Estimated Cost
Air	C	35–80°C	—
Cooling water	C	25–60°C	$40/ML
LP steam	H	25–120°C	$10/tonne
MP steam	H	100–160°C	$14/tonne
HP steam	H	150–250°C	$18/tonne
Foundry salt	C	350–550°C	
Diathermic oil	H/C	50–150°C	

on other facilities. Air coolers will be used wherever possible (i.e. if the cold stream outlet temperature is above 35°C).

Foundry salt will be used in a closed-loop system. It is costly, highly corrosive and will contain significant quantities of energy when hot. Therefore, extreme care in design and operation will be required to prevent leaks. If leaks occur, the elements of the process equipment which are made from carbon steel will be at risk. Salt losses will also have a large detrimental effect on the energy recovery.

7.4.6 Conclusions

- *Process flow rates in the reaction section are approximately ten times higher than in the purification section.*
- *Energy is concentrated in the reaction section of the plant.*
- Heat requirements in the purification stages can be met easily by heat generated from the reaction and the cooling of the reaction products. About 18 MW can be exported from the process as steam or electricity.
- Five utilities (three cold, one hot and one dual-purpose) are required.
- Although the distillation columns in the purification section are energy intensive, the critical units are the reactor and the PAN recovery equipment. The heat flows in these units are much larger because of higher temperatures and greater flow rates.
- *The energy available for export from the LAR process is approximately 35% higher than from competitive PAN processes.*

7.4.7 Recommendations

- Perform laboratory test work to obtain physical property data, where required, in order to complete the mass and energy balances to an acceptable accuracy.

8. ADDITIONAL DESIGN CONSIDERATIONS

8.1 ENERGY INTEGRATION AND CONSERVATION

Chemical plants have always been designed to operate efficiently and economically due mainly to product competition. However before 1970, the objective of building a low first-cost plant was generally considered more important than low operating cost. This concept changed due to the 1973 oil crisis and the actions of the USA Environmental Protection Agency in promoting the use of non- or low-polluting fuels to replace heavy fuel oils and coal. Since the first oil (energy?) crisis, particular attention has been paid to such topics as energy conservation schemes, process (heat) integration, heat exchanger network design, cogeneration, etc. This attention is evidenced by the large number of books and journal articles published on these topics in recent years, a few useful references are included at the end of this chapter.

The design engineer must consider appropriate energy conservation/integration schemes that are designed to:

(a) utilise as much of the energy available within the plant;
(b) minimise the energy (and utility) requirements for the plant.

The energy balances performed for the plant items (or plant sections) provide the initial key to identifying areas of high energy availability or demand. An attempt can then be made to utilise excess energy in those areas where energy must be provided. However, this is not always possible because:

(a) a high energy load (from the energy balance) may constitute a large volume of liquid at a relatively low temperature, exchanging this energy may require large and expensive equipment;
(b) the energy source may be distant from the sink, and piping and insulation costs may make utilisation uneconomic; sometimes a re-arrangement of the plant layout (see Section 4.2) is required;
(c) the energy source may be corrosive.

Any energy conservation scheme must also consider the costs involved in removing or transferring the excess energy, i.e. capital cost of heat exchangers, piping, valves, pumps, insulation, and operating costs of pumps and maintenance. Energy conservation is only worthwhile if the reduction in energy costs exceeds the cost of implementation. A scheme may be devised for a plant and then held over until energy prices make the proposal attractive. This type of forward planning requires that the plant layout adopted can be easily modified.

149

Energy conservation can be achieved at three 'levels', these are:

(a) correct plant operation and maintenance (good 'housekeeping');
(b) major changes to existing plants and/or processes (retrofits and revamps);
(c) new plants and/or new processes.

The time required to implement energy conservation measures, the capital cost required, and the potential savings, all increase from level (a) to (c) above. The cost of downtime for level (b) can be significant, and level (c) offers the greatest long-term potential for energy conservation. This latter objective can be achieved either by designing new, energy-efficient plants for established process routes, or adopting new and less energy-intensive process routes. The areas immediately obvious for consideration of energy utilisation are the feed preheating section of the plant (probably utilising high-temperature reaction products), and the distillation section by attempting to optimise condenser and reboiler duties on various columns. The following examples illustrate some applications of basic engineering principles to the design of equipment for improved energy efficiency.

(i)   Plant operation
Energy savings can be achieved by good engineering practice and the application of established principles. These measures may be termed 'good housekeeping' and include correct plant operation and regular maintenance. The overall energy savings are usually small ($<10\%$) and may not be easy to achieve, significant time may be required for regular maintenance and checking. However, such measures do help to establish the commitment of a company to a policy of energy conservation.

(ii)   Heat recovery
Heat recovery (or reuse) is an important and fundamental method of energy conservation. The main limitations of this method are:

(a) inadequate scope for using recovered waste heat because it is too low-grade for existing heat requirements, and/or because the quantity of waste heat available exceeds existing requirements for low-grade heat;
(b) inadequate heat-transfer equipment.

Developments and improvements are continuing in the design and operation of different types of heat exchangers including the use of extended (finned) heat-transfer surfaces, optimising heat-exchanger networks, heat recovery from waste fuels, heat-exchanger fouling, and the use of heat pumps.

(iii)   Combined heat and power systems
Significant energy conservation is achieved by the well-established method of combined heat and power generation (cogeneration). This is often referred to as CHP or COGEN. The heat is usually in the form of intermediate or low-pressure steam, and the power as direct mechanical drives or as electricity generated with turbo-alternators. The choice of systems is usually between back-pressure steam turbines, or gas turbines with waste-heat boilers for the process steam. The amount of power generated is usually determined by the demand for heat.

It is not usually possible to balance exactly the heat and power loads in a system. The best method of achieving this aim is to generate excess electricity for subsequent sale, other balancing methods tend to be less efficient. Therefore, it is important to forecast the heat to power ratio accurately at the design stage to avoid large imbalances and reduced system efficiency.

(iv) Power-recovery systems

A power-recovery (expansion) turbine can recover heat from an exhaust gas, and then use this heat to provide part of the energy required to drive the shaft of a motor-driven process-air compressor. Other examples are the use of a steam-turbine drive and a two-stage expansion turbine with reheating between the stages.

A hydraulic turbine can be incorporated on the same shaft as a steam turbine (or motor drive). This arrangement can be used to provide about 50% of the energy needed to recompress the spent liquor in a high-pressure absorption/low-pressure stripping system.

Power generation using steam or gas turbines is now well established, however power recovery by the pressure reduction of process fluids is more difficult and less common. In general, the equipment is not considered to be particularly reliable. Rankine cycle heat engines have been developed/adapted to use relatively low-grade waste-heat sources (particularly from organic fluids) to generate power in the form of electricity or direct drives. They tend to be used when the heat source would otherwise be completely wasted, the low efficiencies (10–25%) do not then represent a significant disadvantage.

(v) Furnace efficiency

Incorporating an air heater (to utilise low-level heat) on a steam boiler can be more economic than using a hot-oil system which is designed for high-level heat only.

(vi) Air coolers versus water coolers

Air coolers have a higher installed cost but lower operating costs than water coolers.

(vii) Low-pressure steam

Energy savings can be achieved by the efficient use of low-pressure steam, e.g. use of 60–80 kPa gauge (10–12 psig) steam in absorption-refrigeration units (steam condensate is returned to the boiler).

(viii) Distillation

All separation processes use energy but distillation is the most significant energy-consuming unit operation, and has therefore received considerable attention. Other (less versatile?) processes are also being developed as alternative separation methods. The other significant energy-consuming processes are drying and evaporation. The three main ways of reducing the energy requirements of distillation are discussed below.

Improved control and operation of existing columns includes using a lower reflux ratio (or better control of reflux ratio), reducing (if possible) the overhead and bottom product purity specifications, lower operating pressures, and correct feed plate location. Energy can be saved by designing tray-distillation columns to include more trays and hence operate at a lower reflux ratio. A reflux ratio of between 1.10 and 1.20 times the minimum value is used for many installations, except for low-temperature (cryogenic) columns because of their high utilities costs (use a reflux ratio of 1.05–1.10 times the minimum value) and in systems using low-level heat or recovered heat (use 1.20–1.50 times minimum).

The second method involves recovery of the heat content of overhead vapours to be used for generating low-pressure steam (if there is a suitable use for it). The overheads can be used in the reboiler of another column. The overhead vapours can also be used in the same column in an open-cycle heat-pump system known as vapour-recompression distillation. The overhead vapours are compressed and then condensed in the reboiler. Heat of compression raises the temperature of the vapour and provides the necessary ΔT for the reboiler. Condensed vapour is collected in a pressure-controlled reflux drum (no reflux pump), and is returned to the top of the column. This technique is useful if the relative volatilities of the components are 'enhanced' at reduced pressures, and therefore fewer trays and/or a lower reflux ratio are required.

Thirdly, energy savings can often be achieved in multicomponent distillation systems by appropriate sequencing of columns.

(ix)   Drying operations

Dryers may be classified as direct (convection) dryers, radiant-heat dryers, and indirect (conduction or contact) dryers. Direct-contact dryers are inherently less efficient than the other types due to the difficulty of economically recovering heat from the exit gases. Energy conservation could be increased by improving the thermal efficiency (typically <30%). Thermal drying is an energy-intensive process and consideration should be given to the use of alternative drying methods or appropriate pre-treatment, e.g. mechanical separations such as filters and centrifuges for water removal. General methods of reducing the energy requirement of drying operations include:

(a) improved operation and insulation of existing dryers;
(b) reduced moisture content in the feed;
(c) optimisation of the exhaust air rate, humidity and temperature;
(d) heat recovery from the exhaust gas.

Energy conservation in the design of complete processes may be achieved in four ways:

(A) major modifications to existing plant;
(B) new plant using an existing process route;
(C) new process routes and/or alternative raw materials;
(D) new processes for new (substitute) products that are less energy intensive.

Items (A) to (C) represent short-term and medium-term energy conservation measures. Item (D) requiring the use of new products or processes is more appropriate for long-term energy efficiency planning, due to the long lead times for the application of new technology in the chemical industry. Although energy conservation is an obvious objective of all equipment manufacturers and plant designers, more attention is necessary in relation to education, training, and the application of new and existing technology to ensure significant medium-term and long-term savings.

Energy conservation must be considered at various stages of the project, e.g. feasibility study, process selection, plant layout, energy balances (identifying major areas), and in conjunction with the detailed equipment design. If energy utilisation is only an afterthought, either unnecessary and costly modifications may be required to the design work, or the plant may not be as economically feasible as it originally appeared.

8.2 PROCESS CONTROL, INSTRUMENTATION AND ALARMS

Process control and instrumentation have largely been ignored thus far in the notes for Part I, certainly not because these topics are unimportant or because they can be safely left until late in the design work. The opposite applies in both cases. However, this subject is broad and is changing so rapidly that, like the economic analysis, to provide a complete coverage would require a separate volume to be written. The design of process control systems and the specification of plant instrumentation are subjects that apply to specific plants and processes, any general description would be inadequate in most situations, however some typical guidelines and ideas can be presented.

Process control and the associated instrumentation were discussed briefly in relation to the process selection (see Section 3.1.10). To reiterate those comments, these aspects of plant design must be considered during the process selection stage and at other subsequent stages in the design of the plant. Do not assume that any item of equipment or an entire plant that can be designed can also be operated and controlled easily or efficiently. It should be established early in the design that the control and instrumentation aspects are not only feasible but that the cost is within acceptable limits. Ignoring the control and instrumentation requirements may mean that (at best) the capital cost is higher than necessary, and that plant operation and control are difficult.

Process control is such an important topic in the design of increasingly complex chemical plants that the student should have been thoroughly exposed to the basic principles, and to advances and applications in chemical engineering situations. The design project provides an opportunity to apply this knowledge to design problems. A detailed control strategy for an entire plant is beyond the scope of a student project (and the time available), however at least one unit or a section of the plant should be considered in detail.

Before embarking on the detailed design of process control schemes for the plant equipment, the student should 'step back' and view the overall plant and ask: 'What is the overall process control strategy?' It is surprising how difficult it can be for the student to answer this question (or even to understand what it means) despite having taken (and passed) detailed control courses. The heart of most chemical processes is the reactor, and it would be expected that one of the main aims of the plant control would be to ensure that the reactor functions correctly. This means efficient heating or cooling, and control of reactant and product flows to enable maximum conversions to be obtained. After the reactor, the remaining downstream items tend to deal with removal of unreacted products (and their recycle) and purification of products to achieve their specifications. Depending upon the type of process and the product specification, the next major control initiatives may be in achieving and maintaining product quality, or the ability to vary take-offs of different product streams to respond to changes in market demand. An overall view of the plant and its objectives should help to focus the process control strategy on particularly sensitive areas.

Instruments which can monitor the important process variables during plant operation must be specified. These instruments must be capable of measuring the variables and should have an acceptable accuracy and repeatability of measurement, usually the latter attribute is more important than the former for chemical plant measurements. The instruments may be used for manual measurements or included in automatic control loops. Automatic alarms may also be required to indicate deviations outside acceptable limits. If possible, direct measurement of the process variable should be made, however, it is often easier to measure a dependent variable, e.g. temperature measured as an indication of composition for distillation column top product.

The specification of a control scheme and the associated instrumentation for a chemical plant should satisfy several main objectives. First, the plant should operate at all times in a safe manner. Dangerous situations should be detected as early as possible and appropriate action initiated, also the process variables should be maintained within safe operating limits. Second, the plant should operate at the lowest cost of production. Finally, the production rate and the product quality must be maintained within specified operating limits. These objectives may be conflicting and the final control scheme to be adopted is based upon a realistic and acceptable compromise between the various factors. The main conflict is between the need to design and operate as safe a plant as possible and the desire to produce the chemical at the lowest cost. Safe plant operation can be expensive, both in terms of the capital cost of instrumentation and control schemes, and the annual operating costs, e.g. maintenance.

Experienced process control engineers are usually responsible for the design and specification of automatic control schemes on large chemical plants. The book by Shinskey (4th edn, 1996) provides details of the practical aspects of process control-system design. The P&ID is used to specify the preliminary instrumentation requirements and the control actions. In this way, control is linked to and recorded on the P&ID. Control loops are identified for level, flow, pressure and temperature controls to ensure steady plant operation. The first steps in the design of a good control system are to define which variables need to be measured and which need to be controlled. To specify pressure, temperature, flow and level controllers on all units at different locations, and to assume that the measurements will provide the basis for adequate control is unrealistic. It is necessary to specify:

(a) the variables that *need* to be measured;
(b) the location of these measurements;
(c) the variables that *need* to be controlled;
(d) how the desired control can be achieved.

A *control strategy* can then hopefully be developed. It is also necessary to determine the ancillary instruments required for process monitoring, and for plant commissioning and start-up. It is worthwhile at the design stage including connections for instruments which may be required in the future. The types of recorders and their location (local or a distant control room) must be specified, and the alarms and interlocks that are required. Control does not occur independently in different units, and an overall and integrated approach to the control of the plant is required.

The control and instrumentation required are determined partly by an understanding of the equipment function, which comes from the detailed design, and also from the results of the detailed *Hazard* and *Operability (HAZOP) study* for the plant. The preliminary HAZOP was mentioned in Section 3.5 and it is also discussed in more detail in the following section concerning plant safety. The HAZOP is a rigorous evaluation of the operational aspects of the plant equipment, and one of the outcomes of this study is a re-evaluation of the proposed process control scheme. The aim in designing control systems should be to select the most appropriate control scheme which will do the job adequately and in the simplest acceptable way. The use of advanced and expensive control schemes must be justified for the particular application, but in many situations cheaper and less complicated control schemes are often more appropriate. Cost is a major consideration, but safety and 'user-friendliness' for the process operators are also important. The process operators must be able to interpret and initiate control actions in order to correct process disturbances, this requirement should be considered in the specification of an appropriate control scheme.

A second outcome from the HAZOP study is the specification of alarms, and the development of an alarm system for the particular item and for the entire process operation. The aim with an alarm system should be to structure a system of responses such that the original alarm(s) can be identified, and appropriate corrective actions can be initiated. As with process control, the simplest system and minimum amount of hardware/instrumentation that can do the job effectively should be used. More is not necessarily better — neither more alarms, nor a more complex system, nor more cost! An effective alarm system is not merely composed of a set of individual alarms, it must be an integrated system such that the sequence in which alarms are activated can be determined by the process operator. Often an action which sets off an alarm also activates other (later) alarms and corrective action may only require attention to the original deviation, this is the aim of an effective integrated alarm system.

8.3 SAFETY, HEALTH AND THE ENVIRONMENT

Factors relating specifically to the safety of a chemical plant have so far largely been ignored, although there has been some discussion of process control (see Sections 3.1.10 and 8.2) and preliminary HAZOP studies (see Section 3.5). This is mainly because it is impossible to discuss all the aspects of plant design initially in a set of notes. The designer will obviously design items of equipment, and perform other tasks in process design, in order to ensure safety during operation — no one is going to deliberately design a plant to operate in an unsafe manner. However, the designer cannot envisage all possible malfunctions that can occur, whether due to human error or mechanical failure. Also, the occurrence of minor operating changes in one unit can have serious consequences on equipment performance in another section of the plant.

When the first edition of this case study book was published in 1989, this section was called *Safety, Loss Prevention and HAZOP*. In the intervening years the field of *Process Safety Analysis* has developed and expended its horizons. The HAZOP study is still the accepted and most important tool used by the design engineer, but there are now other methods and techniques which should also be considered, e.g. Failure Mode and Effective Analysis (FMEA), Quantitative Risk Assessment (QRA) including analysis of logic trees, and Consquence Modelling. Safety assessment requires answers to the following questions, and use of appropriate techniques.

(i) What can happen?   Use of HAZOP and FMEA.
(ii) How often can it happen?   Use of QRA.
(iii) What are the consequences? Use of HAZOP and QRA.
(iv) Is the risk acceptable?   The answer here is complex and depends upon technical, social and political considerations.

Design engineers have both a legal and a moral duty with regard to safe design and operation. This duty is in relation to avoiding human injury, for protection of property, and prevention of damage to the environment. The general term 'loss prevention' covers these aspects as well as economic losses and loss of goodwill. The latest thinking is to integrate the topics of safety, health and environmental protection, hence the revised title of this section. This philosophy is discussed in more detail by Turney (1990) and Pitblado and Turney (1996) in relation to the design of process plants. Current philosophy on process safety analysis is described very clearly by Skelton (1997).

All process design activities, e.g. process selection, plant layout, etc., and the equipment designs must be carried out with the design objective of ensuring the safe operation of the plant (within acceptable risk levels — here risk may be considered either as the frequency of an occurrence or as a probability). To achieve this aim, safety and plant operation must be considered at several stages in the preliminary design work. After the detailed plant design is completed, it is necessary to perform a detailed and complete safety analysis. This study should identify aspects of the design that may cause operational or safety problems, and also any modifications necessary to the equipment design (or the control scheme) which will minimise the effects of changes in the plant operating conditions. The following activities are carried out in order to achieve these aims.

(a) *A safety review of all equipment designs*, e.g. design to appropriate standards and observation of design code recommendations. This is really checking that a 'good' design has been performed. However, mistakes do occur and a second opinionand review by a group of experts can help to uncover some errors. For further reading on this subject refer to the books by Kletz (1988, 1991); Scott and Crawley (1992); and Wallace (1995).

(b) *Loss Prevention Studies.* Loss prevention is the general name given to the activities that help provide anticipatory safety measures for the *prevention* of accidents. Some of the techniques that are used are described below.

(i) *Hazard and Operability (HAZOP) Studies*, sometimes simply referred to as Operability Studies, provide a systematic and critical examination of the operability of a process. They indicate potential hazards due to deviations from the intended design conditions. The techniques can be applied to existing plants, and should be standard procedure at the process design stage of new plants; more details and design case studies/examples can be found in the books by Kletz (1992), Lees (1996), and Skelton (1997).

(ii) *Failure Analysis* is described by Leach (1972) and provides a crude criterion of acceptance. The overall reliability of a chemical plant can be calculated from the reliability of individual components, additional consideration is then given to parts of low reliability, e.g. duplication of equipment. Powers and Tompkins (1973) describe this procedure as applied to a chemical plant, other useful reference sources are BS 5760, Part 5 (1991) and Davidson (1994).

(iii) *Hazard Analysis (HAZAN)* is described by Lawley (1973) and by Kletz (1992), and is a technique for the quantitative assessment of a hazard, *after* it has been identified by an operability study, or similar technique. HAZAN is used to compare risks to life per hour of exposure. Attention is concentrated on those risks which exceed a given acceptable value, e.g. 3 deaths per 10^5 people per year.

(c) *Quantitative Safety Analysis Techniques* need to be applied to particular and appropriate situations, refer to Skelton (1997).

(i) *Quantitative Risk Assessment (QRA)* investigates the risk involved in an incident and the consequences, decisions are then made based on comparisons with various acceptance criteria. All QRA is based on the construction of logic diagrams (either fault trees or event trees) showing how the various causes of an incident are related. Failure data is used to quantify the logic trees. For further explanation refer to BS 5760, Part 7 (1991).

(ii) *Consequence Modelling* evaluates the consequences of an incident on people, property and land (i.e. the environment). The problem that is specific to a chemical plant is that an incident may result in either the release of a toxic or flammable material, or an uncontrollable release of energy. The effects may be both long-term and distant from the plant site. Four broad areas which are discussed by Skelton (Chapter 8; 1997) are gas dispersion, toxicity, explosions, and fires. This approach is quite complex and it is essential to appreciate the limitations of the methods, models, and data that are being used.

(iii) *Evaluation of Human Factors*: Human error can never be totally eliminated and therefore available models and methods must be used to evaluate the probability of human error and to assess the consequences. The applicability of the techniques to the particular situation, and the reliability of the human-error data, must be carefully assessed; refer to Kletz (1991A) and Mill (1992).

A HAZOP study is a detailed and formal evaluation of a process and is performed section-by-section (or line-by-line of the flowsheet), in order to consider operational deviations and their possible effects. The study is based upon a set of 'guide words' (identifying possible deviations), examples are: 'NONE', 'MORE OF', 'LESS OF', 'PART OF', etc. A list of guide words and their precise interpretation is included in the Chemical Industries Association booklet (1979), and in Coulson and Richardson

Volume 6 (2nd edn, 1993; pp. 336–338). Each guide word is applied systematically to the 'property words' (e.g. flow, pressure, temperature, etc.) which are chosen as being relevant to the design conditions and the design intention. A typical sequence of events comprising a detailed operability study is presented in Coulson and Richardson Volume 6 (2nd edn, 1993; p. 339, Figure 9.4). The results of the HAZOP study are usually presented in tabular form with the following headings:

PROPERTY WORD   GUIDE WORD   CAUSE   CONSEQUENCE   ACTION

It is important to identify not only what may happen (the event and consequences) and why (causes), but also what should be done (actions) to prevent serious accidents. Examples of operability studies in particular situations are included in Coulson and Richardson Volume 6 (2nd edn, 1993; pp. 334–336, 340–342). Modifications initiated in one section of the plant can have effects in other sections, it is therefore necessary to consider the effects of any changes that are made on the overall operation of the plant. The existence of recycle lines requires careful consideration in the study. A HAZOP study needs to include *ALL* process and utility lines associated with a piece of equipment, including pressure-relief lines. No line can be omitted from the HAZOP evaluation because 'it is unimportant' or 'has no effect'! A second operability study will need to be performed after all the actions identified by the first HAZOP study have been incorporated into the design and included on the flowsheet. Subsequent HAZOPs will be required as modifications are made, and when commissioning trials yield operational data — the typical iterative design activity.

Note: In student design projects, the aim is to produce a competent design and also to communicate to the supervisor what has been done. It is therefore essential to include a *summary* of the major findings obtained by performing the HAZOP study. (This is usually only performed on one item of equipment or over a section of the plant for student projects, a complete study would require too much time and is not necessary for teaching purposes.) The student should not assume that the marker will read, analyse and digest several pages (tables) of an operability study in detail. Therefore, identify in a summary the most important consequences and actions arising from the study. Ironically the requirement for a student to summarise the most *important* findings is contrary to the actual intention of a HAZOP study! The HAZOP is required to consider *all* effects (however insignificant they may appear) because they may have serious consequences when combined with other events.

Safety checklists can provide a useful guide to the main actions to be considered during the process design stage. It must be remembered that no list is ever complete and other considerations and techniques should also be applied. Safety checklists can be found in Coulson and Richardson Volume 6 (2nd edn, 1993; pp. 345–347); IChemE Flowsheeting for Safety (1976); Wells (1980); and Balemans (1974).

The preceding discussion may give the impression that loss prevention, operability and safety are only considered during the design phases of a project. This is certainly not the case and the safe operation of the plant must also be evaluated during the construction phase, commissioning, and during normal operation. Plant start-up, shutdown, and performing modifications to the plant, are particularly high risk activities with respect to safe operation. These departures from normal operation must be carefully assessed. The following summary describes the relevant studies during the various stages of a project.

Project Stage	Details of Study
1. Feasibility Evaluation	Identify major hazards and inherent process risks (toxic, explosive).
2. Process Design	Systematic quantitative analysis of hazards and definition of measures to reduce them, refer to the P&ID.
3. Engineering Design	Formal examination of engineering drawings (and the P&ID) with quantitative hazard analysis to ensure the above measures have been implemented.
4. Construction	Systematic check that hardware (as constructed) and operating instructions are as intended in the design.
5. Operation	Post start-up review of all departures from design affecting safety, operating difficulties, equipment failures, etc., with associated hazard implications. Plant modifications and changes in operating procedures or conditions are checked for their effect on plant safety.

Much has been written about Loss Prevention, and HAZOP studies in particular, the volume of publications has grown considerably since the Flixbororugh disaster in 1974. Standards and Codes of Practice (see Section 9.2), and Acts of Parliament provide essential information, and the references included here provide a good overview of the subject, and a basis for more detailed study.

Action: Prepare the preliminary designs for the chemical engineering units in the plant. Consider energy conservation measures and the process control and instrumentation required as the designs are performed. Prepare a design specification sheet for each unit. Detail the specific energy conservation schemes considered and adopted.

Discuss the process control strategy for the overall process and the instrumentation required.

Prepare a complete P&ID for the process.

Prepare a HAZOP study for a particular item (or section) of the chemical plant.

Identify major hazards and necessary actions.

Summarize the HAZOP conclusions in relation to the design of the plant.

References

Energy Conservation

Boland, D. and Linnhoff, B., Preliminary design of networks for heat exchange by systematic methods, *The Chemical Engineer (Rugby, Engl.)*, April, 222–228 (1979).

Broughton, J. (ed.), *Process Utility Systems: Introduction to design, operation and maintenance*, IChemE, UK (1994).

Cavaseno, V. (ed.), *Process Heat Exchange*, reprints of papers from *Chemical Engineering*, McGraw-Hill Publications, New York (1980).

Grant, C.D., *Energy Conservation in the Chemical and Process Industries*, IChemE, UK (1979).

Greene, R. (ed.), *Process Energy Conservation: Methods and Technology*, reprints from *Chemical Engineering*, McGraw-Hill Publications, New York (1982).

Linnhoff, B. *et al.*, *User Guide on Process Integration for the Efficient Use of Energy*, 2nd edn, IChemE, UK (1994).

Linnhoff, B. and Turner, J.A., Simple concepts in process synthesis, *The Chemical Engineer* (*Rugby, Engl.*), December, 742–746 (1980).

Linnhoff, B. and Senior, P., Energy targets for better heat integration, *Process Engineering*, March, 29, 31, 33 (1983).

Linnhoff, B., Choosing a route to energy efficiency, *Processing*, March, 39, 41 (1986).

Linnhoff, B. and Polley, G., Stepping beyond the pinch, *The Chemical Engineer* (*Rugby, Engl.*), February, 25–32 (1988).

Samdani, G.S. (ed.), *Heat Transfer Technologies and Practices for Effective Energy Management*, reprints of papers from *Chemical Engineering*, McGraw-Hill Publications, New York (1996).

Smith, G. and Patel, A., Step-by-Step through the pinch, *The Chemical Engineer* (*Rugby, Engl.*), November, 26–31 (1987).

Smith, R., *Chemical Process Design*, McGraw-Hill, New York (1995).

Taffe, P., Pinch technology for batch processes, *Processing*, December, 20–21 (1987).

Process Control

Chopey, N.P. (ed.), *Instrumentation and Process Control*, reprints of papers from *Chemical Engineering*, McGraw-Hill Publications, New York (1996).

Considine, D.M., *Process/Industrial Instruments and Controls Handbook*, 4th edn, McGraw-Hill, New York (1993).

Luyben, W.L., *Process Modeling, Simulation, and Control*, 2nd edn, McGraw-Hill, New York (1990).

Miller, R.W., *Flow Measurement Engineering Handbook*, 3rd edn, McGraw-Hill, New York (1996).

Sawyer, P., *Computer-Controlled Batch Processing*, IChemE, UK (1993).

Seborg, D.E., Edgar, T.F. and Mellichamp, D.A., *Process Dynamics and Control*, John Wiley, New York (1989).

Shinskey, F.G., *Process Control Systems: Application, Design, and Tuning*, 4th edn, McGraw-Hill, New York (1996).

Stephanopoulos, G., *Chemical Process Control: An Introduction to Theory and Practice*, Prentice-Hall, New Jersey (1984).

Seborg *et al.* (1989) and Stephanopoulos (1984) are student texts which provide comprehensive coverage of basic principles and advanced topics, and also contain extensive referencing to books and technical literature for each chapter.

Safety, Health and the Environment (Including Loss Prevention and HAZOP)

Balemans, A.W.M., a paper in *Loss Prevention and Safety Promotion in the Process Industries*, Elsevier Publishing Co., The Netherlands (1974).

Barton, J. and Rogers, R., *Chemical Reaction Hazards*, 2nd edn, IChemE, UK (1996).

BS 5760, Part 5: *Reliability of Systems, Equipment and Components*, British Standards Institute, London (1991).

BS 5760, Part 7: *Guide to Fault Tree Analysis*, British Standards Institute, London (1991).

Chemical Industries Association, *A Guide to Hazard and Operability Studies*, C.I.A., London (1979).

Davidson, J. (ed.), *Reliability of Mechanical Systems*, 2nd edn, IMechE, London (1994).

Davis, L., *Quality Assurance: ISO-9000 as a management tool*, Copenhagen Business School Press, Denmark (1996).

Hoyle, D., *ISO-9000 Quality Systems Handbook*, 2nd edn, Butterworths, UK (1994).

IChemE User Guides include:

Explosions in the Process Industries (1994).
Flowsheeting for Safety (1976).
Loss Prevention (1979).
Prevention of Fires and Explosions in Dryers, 2nd edn (1990).

Jones, D., *Nomenclature for Hazard and Risk Assessment*, IChemE, UK (1992).

Kenney, W.F., *Process Risk Management Systems*, VCH Publishers, New York (1993).

Kletz, T.A., *Learning from Accidents in Industry*, Butterworths, UK (1988). This author has published widely in the technical literature on the subject of loss prevention and safety, and has also produced several books for the IChemE, see below.

Kletz, T.A., *Plant Design for Safety: A User-Friendly Approach*, 2nd edn, Taylor and Francis, London (1991).

Kletz, T.A., *An Engineer's View of Human Error*, IChemE, UK (1991A).

Kletz, T.A., *HAZOP and HAZAN: Identifying and Assessing Process Industry Hazards*, 3rd edn, IChemE, UK (1992).

Kletz, T.A., *Lessons from Disaster*, IChemE, UK (1993).

Kletz, T.A., Chung, P., Broomfield, E. and Shen-Orr, C., *Computer Control and Human Error*, IChemE, UK (1995).

Lawley, H.G., Operability studies and hazard analysis, *Chemical Engineering Progress Loss Prevention Symposium*, Vol. 8, 105–110 (1973).

Leech, D.J., *Management of Engineering Design*, John Wiley, New York (1972).

Lees, F.P., *Loss Prevention in the Process Industries*, 2nd edn, 3 volumes, 3500 pages, Butterworth, London (1996).

Mill, R.C. (ed.), *Human Factors in Process Operations*, IChemE, UK (1992).

Owen, F. and Maidment, D., *Quality Assurance: A guide to application of ISO-9001 to process plant projects*, 2nd edn, IChemE, UK (1996).

Petersen, D., *Safety Management: A human approach*, 2nd edn, Aloray Inc., New York (1988).

Petersen, D., *Techniques of Safety Management: A systems approach*, 3rd edn, Aloray Inc., New York (1989).

Pitblado, R. and Turney, R. (eds), *Risk Assessment in the Process Industries*, 2nd edn, IChemE, UK (1996).

Powers, G.J. and Tompkins, F.C., Synthesis strategy for fault trees in chemical processing systems, *Chemical Engineering Progress Loss Prevention Symposium*, Vol. 8, 91–98 (1973).

Rothery, B., *ISO-14000 and ISO-9000*, Gower Publishing, UK (1995).

Scott, D. and Crawley, F., *Process Plant Design and Operation: Guidance to Safe Practice*, IChemE, UK (1992).

Shaw, J., *BS EN ISO 9000 made simple*, Management Books 2000, Oxford, UK (1995).

Skelton, B., *Process Safety Analysis: An Introduction*, IChemE, UK (1997).

Turney, R.D., Designing plants for 1990 and beyond: Procedures for the control of safety, health and environmental hazards in the design of chemical plant, *Process Safety and Environmental Protection*, 68(1), 12–16 (1990).

Wallace, I.G., *Developing Effective Safety Systems*, IChemE, UK (1995).

Wells, G.L., *Safety in Process Plant Design*, IChemE, UK (1980).

Wells, G.L., *Hazard Identification and Risk Assessment*, IChemE, UK (1996).

8.4 CASE STUDY — ENERGY INTEGRATION, PIPING SPECIFICATIONS, PROCESS CONTROL AND THE P&ID

Summary

Good operating practices and control systems are essential for the reactor, after-cooler and switch condensers to ensure high product purity and energy efficiency, and to achieve the design production rates. The reactor control system is required to regulate normal operations and provide an adequate protection system against a runaway reaction caused by high operating temperatures. The control systems on the recovery units are required to maintain the outlet temperatures within a narrow operating range for optimum product recovery. Other units in the plant need less sophisticated control schemes but must guarantee safety and basic operational stability.

   Piping throughout the plant was sized from correlations for the economic pipe diameter based on typical flow rates for the different sections of the plant. A colour coding system has been specified for safety and ease of operation. All key items of equipment and instrumentation are shown on the P&ID for the proposed 40 kT/yr plant.

Note: The following case study section does not include a HAZOP study. However, examples are included at the end of Section 10.16 as part of the reactor design, and in Section 11.14 for the after-cooler. Performing a HAZOP is a very time consuming task, and for a book of this type it was decided that the two examples would be sufficient to illustrate the work involved.

8.4.1   Energy Management and Integration

The high value of energy should be acknowledged in plant operation by treating it as a product with monetary value that can be sold or traded, just like a chemical product. This should be the basis for operational policies concerned with energy management or energy conservation. A phthalic anhydride plant is too small to appoint an energy manager but these duties could easily be incorporated into those of the process engineer. EAM&T (energy auditing, monitoring and targeting) is a means to efficient operation in this area, but there must be a commitment from all operational and managerial personnel to the importance of these tasks if they are to be successful.

The reaction and product-recovery areas have been identified as critical units from an energy perspective. Detailed monitoring and targeting should be established in these areas. Variables that should be recorded regularly (hourly or per shift where appropriate, and daily where laboratory analysis is required) for the reactor include the feed and product flows and temperatures, yields of PAN and MAN, cooling salt flow and temperatures, boiler feed-water flow, steam pressure and the reactor temperature profile. A similar combination applies to the switch condensers and the after-cooler. Targets should be introduced and updated monthly or biannually. These targets, if constructed correctly, allow performance to be measured easily and accurately and provide an incentive for operational staff to maintain or improve efficiency.

Energy integration using pinch technology should be employed in the design of all new plants. However, the phthalic anhydride process is comparatively simple and there is limited scope for the application of the Linnhoff techniques. The two main sections of the plant, namely reaction and purification, will be separated by at least 50 m. Therefore, piping costs will prohibit integration between these sections and each area should be considered separately. Excluding utilities, there are only two streams in the reaction section, namely feed and product. Clearly, the heat in the product can be utilised for feed pre-heating or boiler feed-water pre-heating, but otherwise there are few energy-saving opportunities in the network design. There is slightly greater scope in the purification section, but the rewards from an energy efficient design are comparatively small.

The steam circuit will be extensive and should be designed with careful attention to the energy requirements. A boiler feed-water storage tank with appropriate chemical dosing facilities will be required. Steam lines should be lagged and designed to optimise the pressure losses and the installation cost. Power generation equipment, if installed, should be located relatively close to the high-pressure steam generators.

Reductions in performance of any of the heat exchange equipment associated with the reactor will also affect operation. For example, if the boiler feed-water pump fails, the salt will not be cooled properly. The salt temperature will subsequently increase and the reactor outlet temperature will rise, risking a reaction runaway. Disturbances in the steam production will also have the potential to interrupt the operation of other units (e.g. the purification columns).

The product recovery units are also important with respect to energy conservation. Large amounts of energy are involved and the effluent gases are the single largest source of heat loss from the system. The switching system for the condensers, in particular, should be well controlled. A sufficient period for cooling and sublimation should be allowed to prevent any loss of hydrocarbon and excessive heat loss in the effluent gases.

8.4.2 Plant Piping Specifications

Piping is required for a wide range of streams, each of which is associated with different hazards and require different precautions. For example, molten phthalic anhydride has the colour and viscosity of water yet is significantly more dangerous to handle. Colour coding should be used to distinguish between the different types of streams according to the classifications shown in Table 8.1. Although predominantly a safety measure, the use of a colour coding system will also help operators and engineers to correctly identify the service use of various pipes throughout the plant.

The materials specification for each line should be determined for the specific operating conditions at that point. However, a general materials specification can be made based on the type of fluid being handled (see Table 8.1). Similarly, each line will need to be sized for the expected flow at that point, but generally, the flow rates of particular components do not change significantly throughout the process. Therefore, each stream is associated with a typical pipe diameter (shown in Table 8.1) which was estimated from a correlation for the optimum economic pipe diameter [44]. Schedule 40 pipe should be used throughout the plant due to the relatively modest pressures encountered.

8.4.3 Control and Instrumentation

Three levels of control and instrumentation are required: (a) an alarm and warning system to ensure safety of all process units, but especially those where there is a significant risk (e.g. reaction runaway in the reactor); (b) a regulatory control

Table 8.1 PAN plant piping specifications.

Service	Colour	Typical Diameter	Typical Material
Process streams— liquid	white	50 mm	Plain carbon steel for o-xylene and PAN at ambient temperature or stainless steel type 317 for hot PAN liquid
Process streams— vapour	white	up to 900 mm	Plain carbon steel for air or stainless steel type 316 for PAN vapours
Cooling water	green	25–100 mm	Plain carbon steel
Steam	green/ white	25–100 mm	Plain carbon steel
Fire water	red	100 mm	Plain carbon steel
Heat transfer salt (HTS)	yellow	600 mm	High chromium stainless steel
Other	purple	various	Various

system for all process units which is used to maintain steady operation of the process; and (c) an optimising control system which utilises advanced control techniques to maximise the profitability of key units (e.g. the reactor, after-cooler, switch condenser and purification columns).

Protection systems need to be installed to handle emergency situations. A pressure release valve could be installed on the reactor and would be capable of quickly reducing the reactor temperature and pressure in the event of a runaway. However, this would allow large quantities of hydrocarbon to enter the atmosphere, creating a secondary, and possibly more serious, hazard. Therefore, a vent could only be used as a final emergency action and other control systems must be incorporated into the design. The principal emergency control for the reactor should be to incorporate selective control to monitor the reactor temperature at 12 or more locations to detect hot spots before they become dangerous. The selective control unit should initially open the coolant flow control valve, and then trigger a salt deluge system if the temperatures continue to rise. This would require an over-capacity of coolant to be stored in a secondary, gravity-feed vessel near the reactor. This vessel could also be used to store the salt during plant shutdowns.

During normal operation, the temperature profile in the reactor should be continually monitored and optimised in order to control the composition of the products and regulate the energy flow. High temperatures reduce the selectivity for the partial oxidation reaction (forming phthalic anhydride) and increase the fraction of feed that is fully oxidised (to carbon dioxide and water). The complete oxidation reaction releases about four times as much heat as the phthalic anhydride synthesis reaction and, therefore, significantly increases the cooling load and may cause further increases in the reactor temperature (eventually leading to runaway). The cooling process must also be controlled effectively to ensure that a high level of process energy efficiency (a key to process profitability) is maintained.

The large reactor volume suggests that some type of feedforward control would be effective. The feed rate, temperature and composition are all critical variables that will need to be incorporated into the control system. An adequate over-capacity must be installed in both the salt pumps and control valves to provide good controllability and emergency response. The control system should also be linked to the steam circuit to ensure that the salt temperature is stabilised.

The outlet temperature from the recovery equipment (after-cooler and switch condensers) is critical to the overall plant performance. A compromise must be found between quantity (the amount of product condensed) and quality (the amount of impurities condensed with the product). The after-cooler product is liquid PAN with a higher purity than the switch condenser product which is a solid, so that there is an incentive to maximise recovery in the after-cooler. However, if the after-cooler outlet temperature is reduced too far (in order to increase recovery of the liquid product) then solid will begin to form and block the pipes and valves. Consequently, there is a very narrow range of operating temperatures, and an effective control system for the after-cooler is essential. Similarly, in the switch condensers, if the temperature is too low, impurities co-condense with the product PAN and quality is reduced. However, if the outlet temperature is too high then an excessive amount of PAN will escape with the effluent gases. Both recovery units need tight temperature control which may be enhanced by using appropriate advanced control strategies.

The switch condensers present a further problem. The switching mechanism between hot and cold oil flows must be sufficiently regulated to require minimal attention from the operators during routine operation switches (which might occur as frequently as every eight hours). The switching mechanism should be optimised to keep the sublimation rate high as this could easily become the bottleneck of the entire process if poorly operated. If the cycle is too short, capacity is lost; if the cycle is too long then the thickness of solid increases to an extent that prevents further sublimation.

The rectification and stripping columns should be capable of producing product that easily exceeds the purity specifications. Although distillation columns are traditionally very large users of heat, the reduced flows in the purification section shifts this emphasis to the reaction and recovery units. Therefore, the rewards for efficient operation of the recovery columns are small and a more simplified control system can be installed. However, if the control system is too simple, the risk of producing off-specification product (which would require reprocessing through the purification columns) will increase. One-point temperature control should be sufficient and requires only basic regulatory control instrumentation.

Product composition should be monitored regularly throughout the process. One or more process analysers could reduce the level of sampling and testing required and provide continuous process data. However, significant and specialist mainte-nance will be required to ensure that the analysers are calibrated and operating accurately. Otherwise, they are of little or no value to the operators. Daily samples should be taken from the reactor inlet (to check on variation in feed composition which might require process adjustment), reactor outlet (to monitor reactor oper-ation), the after-cooler outlet and switch condenser products, the crude product storage vessel, stripping column bottoms and final product (rectification column distillate). The laboratory turnaround should be typically less than four hours so that appropriate operating adjustments can be made in a timely manner.

Adequate control and instrumentation and good operating techniques are essen-tial to the overall profitability of the process. Safety should always be the first objec-tive of the control system, followed by operability and profitability (via optimisation).

8.4.4 The Piping and Instrumentation Diagram (P&ID)

The LAR process for the manufacture of phthalic anhydride from o-xylene is a sequential process with no recycling. Feed enters the process as o-xylene and air, which are both fed to the reactor via filters and a carburettor which doses the o-xylene into the air stream at a controlled rate. Material leaves the process as PAN product, MAN product, waste gases or residue.

Essentially all o-xylene fed to the process reacts to form either PAN, MAN or CO_2. A gas cooler is used to reduce the temperature of the hot reactor effluent gases (due to heat of reaction). Phthalic anhydride is recovered either as a liquid from the after-cooler (approximately 50%) or as a solid from the switch condensers (the remaining PAN). A two-stage recovery is required due to the tendency of PAN to remain in the vapour phase and sublime before condensing. Crude PAN from the after-cooler and switch condensers is combined in a pre-treatment tank where it is held at elevated temperature to prevent decomposition before being purified in two consecutive distillation columns.

The 'final' layout of the process equipment essentially follows the configuration shown in Figure 3.1 (PFD for the LAR process). However, the consideration of the overall plant-wide energy balance (Section 7.4.3) prompted some modifications. Most significantly, the hot reaction product gases will be used to preheat water prior to the production of HP steam in the salt-cooler, and MP steam will be used to vaporise the o-xylene feed and preheat the air. The P&ID in Figure 8.1 includes some items of equipment that were not shown on the PFD, particularly around the switch condensers and pertaining to the flow of diathermic oil and other utilities. These items were found to be necessary during a more detailed consideration of the process operability which occurred while developing the P&ID. The new items have been numbered sequentially, following the notation developed for the PFD.

The P&ID, Figure 8.1, for the LAR process for phthalic anhydride production shows the complete piping layout and all the basic and essential instrumentation. Equipment numbers are given in Table 8.2 which follows the P&ID.

8.4.5 Conclusions

• An effective energy policy, incorporating detailed monitoring and targeting, should be developed to emphasise the importance of energy as a product with substantial value.
• *The design and location of the steam circuits is important as the heat source and target areas are well separated.*
• *The control and instrumentation for the heat transfer equipment associated with the reactor and product recovery units will be critical because decreases in performance will adversely affect the operation of the primary items of equipment.*
• Liquid-phase process piping is generally 50 mm diameter while gas-phase process piping may be up to 900 mm diameter. Utility piping varies from 25 mm to 600 mm diameter.

8.4.6 Recommendations

• Devise a layout for the hot (steam) and cold (cooling water and air) utility systems that optimises the energy integration in order to minimise the overall cost (capital and energy).

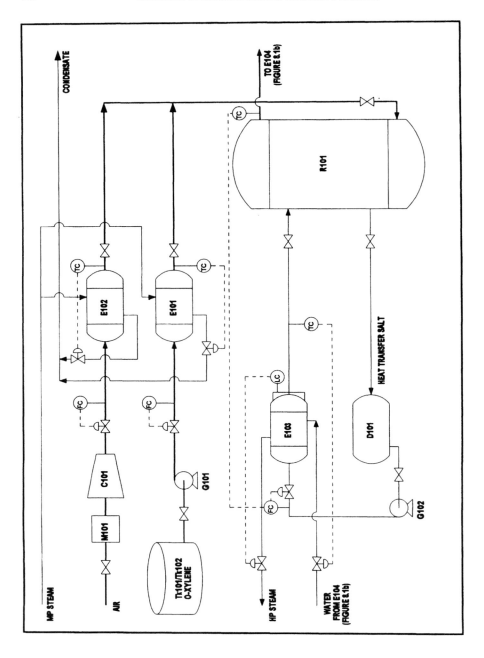

Figure 8.1a P&ID for the feed system and PAN reactor.

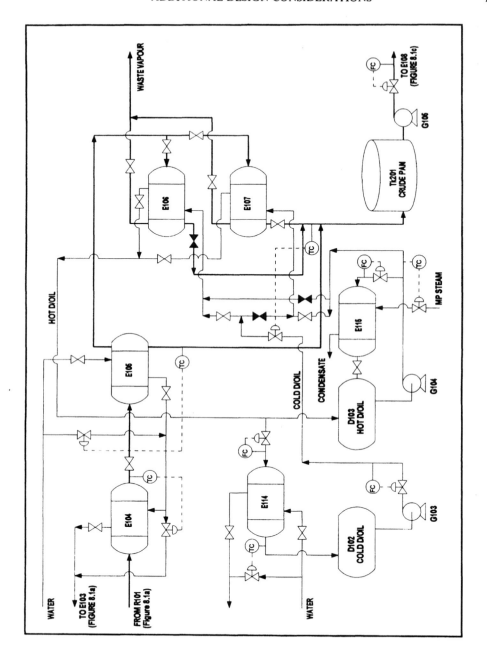

Figure 8.1b P&ID for the PAN recovery system.

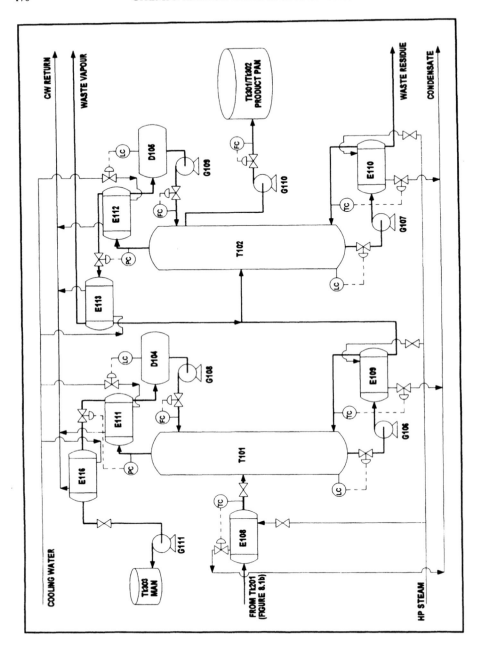

Figure 8.1c　P&ID for the PAN purification system.

Table 8.2 Key for the P&ID of the LAR process for phthalic anhydride production.

Equipment Number	Description
C101	Air compressor
D101	Heat transfer salt drum
D102	Cold diathermic oil receiver
D103	Hot diathermic oil receiver
D104	Stripping column reflux accumulator
D105	Rectification column reflux accumulator
E101	O-xylene vaporiser
E102	Air preheater
E103	Salt cooler
E104	Gas cooler
E105	After-cooler
E106 & E107	Switch condensers
E108	Stripping column preheater
E109	Stripping column reboiler
E110	Rectification column reboiler
E111	Stripping column condenser
E112	Rectification column condenser
E113	Vapour sublimer
E114	Diathermic oil-cooler
E115	Diathermic oil-heater
E116	MAN condenser
G101	O-xylene feed pump
G102	Salt pump
G103	Cold diathermic oil pump
G104	Hot diathermic oil pump
G105	Crude PAN transfer pump
G106	Stripping column bottoms pump
G107	Rectification column bottoms pump
G108	Stripping Column reflux pump
G109	Rectification column reflux pump
G110	PAN product pump
G111	MAN product pump
M101	Air filter
R101	PAN reactor
T101	Stripping column
T102	Rectification column
Tk101/102	O-xylene tanks
Tk201	Crude PAN tank
Tk301/302	Product PAN tank
Tk303	Product MAN tank

REFERENCES FOR CASE STUDY SECTIONS IN CHAPTER 1 TO 8

1. Phthalic Acids, in (Kirk–Othmer) *Encyclopedia of Chemical Technology* (4th ed.), Vol. 18, Wiley, New York (1996).
2. Phthalic Acid and Derivatives, in *Ullman's Encyclopedia of Industrial Chemistry* (5th ed.), Vol. A20, VCH, Germany (1992).

3. Phthalic Anhydride, in (McKetta) *Encyclopedia of Chemical Processes and Design*, Vol. 34, Marcel Dekker, New York (1990).

4. Graham, J.J., The Fluidized Bed Phthalic Anhydride Process, *Chem. Eng. Prog.*, 66(9), 54–58 (1970).

5. Nikolov, V., Klissurski, D. and Anastasov, A., Phthalic Anhydride from *o*-Xylene Catalysis: Science and Engineering, *Catal. Rev. Sci. Eng.*, 33(3/4), 319–374 (1991).

6. Commodity Prices, *Chem. Marketing Reporter*, various issues (1994–1996).

7. Verde, L. and Neri, A., Make Phthalic Anhydride with Low Air Ratio Process, *Hydrocarbon Process.*, 63(11), 83–85 (1984).

8. Ockerbloom, N.E., Xylenes and Higher Aromatics — Part 3: Phthalic Anhydride, *Hydrocarbon Process.*, 50(9), 162–166 (1971).

9. Zimmer, J.C., New Phthalic Anhydride Process, *Hydrocarbon Process.*, 53(2) 111–114 (1974).

10. Anon., Phthalic Anhydride (LAR Process), *Hydrocarbon Process.*, 64(11), 156 (1985).

11. Anon., Phthalic Anhydride (Von Heyden), *Hydrocarbon Process.*, 64(11), 157 (1985).

12. Anon., Phthalic Anhydride (Scientific Design Co.), *Hydrocarbon Process.*, 54(11), 173 (1975).

13. Anon., Phthalic Anhydride (Badger), *Hydrocarbon Process.*, 50(11), 189 (1971).

14. Anon., Phthalic Anhydride, *Chem. Week*, 6 December, 64 (1995).

15. Westervelt, R., Phthalic Anhydride — Tight Supplies Stir Thoughts of Expansion, *Chem. Week*, 21 December, 8 (1994).

16. Brand, A., Phthalic Anhydride Makers Quietly Step Up Capacity, *Chem. Marketing Reporter*, 3 June, 5 (1996).

17. *1991 Worldwide Petrochemical Directory* (29th ed.), Pennwell (1990).

18. *1988 Japan Chemical Annual*, Japan Chem. Week (1988).

19. *Etonwood Guide to Chemical Industries of Australia 1989*, Etonwood Consultants (1989).

20. Chemical Advisory Service, Directory of Australian and New Zealand Chemical Manufacturers and Wholesalers, Vermont, Australia (1994).

21. Petrochemical Future Outlook, Demand, Supply, Pricing, in (McKetta) *Encyclopedia of Chemical Processes and Design*, Marcel Dekker, New York (1990).

22. Anon., Chemical Profile, *Chem. Marketing Reporter*, 24 July (1989).

23. Anon., Phthalic Anhydride, *Japan Chem. Week*, 21 September (1989).

24. Australian Bureau of Statistics, *Foreign Trade Statistics, Imports* (*Phthalic Anhydride, Ortho*-Xylene), June 1989.

25. Australian Bureau of Statistics, *Foreign Trade Statistics, Exports* (*Phthalic Anhydride*), June 1989.

26. Australian Bureau of Statistics, *Merchandise Imports* (*Phthalic Anhydride*), June 1993.

27. Australian Bureau of Statistics, *Merchandise Exports* (*Phthalic Anhydride*), June 1992.

28. Anon., Process Plant Capital Costs, *Process Economics Int.*, V(I) 2–6 (1984).

29. Holland, F.A., Process Economics (Chapter 5), in *Perry's Chemical Engineers' Handbook* (6th ed.), McGraw-Hill, New York (1984).

30. Peters, M.S. and Timmerhaus, K.D., *Plant Design and Economics for Chemical Engineers* (4th ed.), McGraw-Hill, New York (1991).

31. Guthrie, K.M., Capital and Operating Costs For 54 Chemical Processes, *Chem. Eng. (N.Y.)*, 15 June, 140–156 (1970).

32. Nakanishi, Y. and Haruna, Y., Recycling Vent Gas Improves Phthalic Anhydride Process, *Hydrocarbon Process.*, 62(10), 107–110 (1983).

33. Xylenes and Ethylbenzene, in (Kirk–Othmer) *Encyclopedia of Chemical Technology* (3rd ed.), Wiley, New York (1978–84).

34. de Virgiliis, A. and Gerunda, A., Optimize Energy Usage in Phthalic Anhydride Units, *Hydrocarbon Process.*, 61(5), 173–175 (1982).

35. Anon., Phthalic Anhydride. Von Heyden Process. *Brit. Chem. Eng.*, 14(9), 1168A–1168C (1969).

36. Gurevich, D., Manufacture of Phthalic Anhydride. *Int. Chem. Eng.*, 20(8), 277–284 (1968).

37. Krieth, F. and Bohn, M.S., *Principles of Heat Transfer* (4th ed.), Harper & Row, New York (1986).

38. Simulation Sciences, *PRO/II Keyword Input Manual*, Brea, California (1994).

39. ICI, *Safety Data Sheet - Phthalic Anhydride* (1996).

40. Ulrich, G.D., *A Guide to Chemical Engineering Process Design and Economics*, Wiley, New York (1984).

41. CE Plant Cost Index. *Chem. Eng. (N.Y.)*, various issues (1986–1996).

42. Current Exchange Rates, *The West Australian*, various issues (1996).

43. HPI Construction Boxscore, *Hydrocarbon Process.*, various issues (1983–1996).

44. Harker, J.H., Finding an Economic Pipe Diameter, *Hydrocarbon Process.*, 57(3), 74–76 (1978).

45. Hyprotech, *HYSIM User Manual*, Calgary, Canada (1990).

46. Casaveno, V., Belgians Tap PA Wastes for Maleic Anhydride, in *Process Technology and Flowsheets*, McGraw-Hill, New York (1979).

COMMENTS

The reader should by now have made two important observations about plant and process design studies. First, decisions are usually based upon a compromise between conflicting factors, the final design is not necessarily the 'best' in all aspects but it must be technically and economically acceptable for the task. Second, a design study does not follow the usual format for solution of a traditional undergraduate tutorial problem. There is a preferred sequence of activities to be followed, but this does not lead to a set of completed individual stages and a final single correct solution. The design project requires a continual re-evaluation of previous decisions, changes are made as new developments occur, or as additional information becomes available. Several aspects of the design need to be considered at different stages of the work. A preliminary economic feasibility study is performed (see Chapter 6), and a final economic evaluation after the detailed design is completed. The equipment design is considered during the feasibility study, the preparation of the equipment list, the mass and energy balance calculations, and during the preliminary and final design stages. The same work is not repeated each time, but new information is assessed, previous data and decisions are re-assessed, and the design is continually refined and developed. This is the nature of the design project — it is the ultimate iterative, trial-and-error type problem.

The student undertaking a design for the first time often finds this unnerving. There is often a reluctance to carryout the early stages of the project because the detail required for final decision-making is incomplete. However, it should be apparent that to wait until all information is available, and then make the decision and carry out the calculations in 'one-hit' is not the best way to proceed.

Action: Adopt a new way of thinking and approaching engineering problems — be prepared to re-assess decisions at every stage of the design, and incorporate necessary changes. Do not assume every problem has a particular method of solution and a single correct answer.

References

Ray, M.S., *Elements of Engineering Design: An Integrated Approach*, Prentice-Hall International, UK (1985). A general introductory textbook on engineering design, including discussion of creativity, innovation, problem solving and decision making, and other aspects of engineering design such as economics, ergonomics, legal factors, etc.

Sears, J.T., Woods, D.R. and Noble, R.D. (eds), *Problem Solving*, AIChemE Symposium Series, Volume 79, No. 228, New York (1983).

Woods, D.R., *Process Design and Engineering Practice*, Prentice-Hall Inc., New Jersey (1995).

Woods, D.R., *Data for Process Design and Engineering Practice*, Prentice-Hall Inc., New Jersey (1995).

PART II

DETAILED EQUIPMENT DESIGN

Note: For Part II, references are included at the end of each chapter. General references are given in Section IV, Section 2.5 and after Section 8.3.

9. THE DETAILED DESIGN STAGE

9.1 DETAILED EQUIPMENT DESIGN

After all the preliminary work (the technical and economic feasibility study) has been completed, the detailed design work can begin. The equipment can be designed in its final (?) form and full specification sheets prepared for each item. At this stage, the process flowsheet/P&ID and equipment list should be checked and amended. The cost estimates should also be revised to account for any significant changes from the preliminary design specifications. Space precludes the inclusion of design details for a wide range of items of equipment in this book, and only general discussion of equipment design is included with some appropriate references.

Many different process operations are performed in a chemical plant, e.g. distillation, heat transfer, absorption, etc., and for each of these operations a wide range of equipment has been developed to suit particular applications and situations. These include different types of heat exchangers, e.g. plate, shell and tube, carbon block, etc.; packed and plate distillation columns; CSTR or tubular reactors, etc., etc. Many books have been published describing particular process operations, a few of the more prominent volumes are listed in the references at the end of this chapter. By the time the student undertakes a design project, these books should be very familiar from use in other parts of the course. Some books provide only an introduction to the subject, aimed mainly at undergraduate courses and emphasising the basic theory. Other books concentrate upon the application of basic principles to the design of equipment, and include details of relevant design methods and useful design data. It is design handbooks which are needed for design work, some examples are also listed at the end of this chapter.

The reader requiring an overview (or a review) of the different types of equipment available and their relevant design features, should refer to the chapters on equipment design in Baasel (1990); Peters and Timmerhaus (1991); Ulrich (1984); Coulson and Richardson, Volume 6 (1993); and Walas (1988). These books provide extensive references to the literature, for different types of equipment and for details of new developments in the design of equipment. Some detailed design information can be found in Perry's Handbook (1997). Sometimes it is not sufficient to base a design on a method presented in a book published 10 years ago (or even 5 years ago), the design engineer must be familiar with new methods and improvements as they occur. In some areas, chemical engineering design does not change or advance that quickly, also some older reference sources remain very useful, e.g. Kern (1950)! However, it is necessary to be aware of changes that are occuring and to apply recent knowledge in order to produce cost effective designs. Papers published in *Chemical*

Engineering (N.Y.) (including Feature Articles describing the state-of-the-art in particular fields), *Hydrocarbon Processing* and *Chemical Engineering Progress* are good sources of recent information. Other journals and reference sources are listed in Section IV, and Sections 2.3 and 2.5.

The design of an item of equipment should conform to (most of) the following criteria:

(a) It must operate safely, minimising the risks of explosion and fire and the danger to operating personnel.
(b) It must perform its intended function, i.e. fulfil the design brief.
(c) It must be a viable/economic design.
(d) It should operate with minimum maintenance and repair, and with minimum operating costs and utilities requirements.
(e) It should be designed for minimum cost at the specified design rate, however, it is usually required to operate 'satisfactorily' under conditions of both increased and reduced capacity where penalty costs are incurred.
(f) It must operate for the expected life of the plant.
(g) It must be feasible (i.e. capable of construction and operation).

The design must also be acceptable (?) to all concerned parties, and will probably need to meet several other requirements.

When the equipment is designed, a full technical design report must be prepared giving details of all relevant design calculations and the data used, and assumptions/ restrictions inherent in the design method. Full design specification sheets and detailed (scaled) engineering drawings are prepared. Sufficient details must be available so that the item can be costed accurately — and constructed (if approved).

9.1.1 Equipment Design — HELP!!!

Not all the students in a chemical engineering course have mastered all the basic principles that have been taught, many students feel uncertain about their ability to transform basic theory into an engineering design. Some students find it difficult to adapt to open-ended (and often ill-defined) design problems after a 'diet' of predominantly precise textbook-type exercises. It is necessary to realise that the first stage of the design problem is to provide some order to the available (and sometimes contradictory) information that constitutes the basis of the equipment design. These notes are intended to provide some immediate help and guidance to those students who feel apprehensive about tackling an equipment design problem. Reference to other books which detail general approaches to design, e.g. Ray (1985); Peters and Timmerhaus (1991); Walas (1988); Ulrich (1984), should prove helpful. It is also necessary to read journal articles and design case-studies (e.g. Ray and Johnston, 1989; Austin and Jeffreys, 1979) for a particular application. Sometimes the better academic students find difficulty with the required design approach, especially if their good grades have been obtained, in part, by good memory and a grasp of textbook-type problems (which tend to appear on examination papers!). Good design work is not based on memory, but it is based on a sound knowledge and understanding of engineering principles. The better students often possess these attributes but they have yet to realise their importance! These notes should also be useful as the design proceeds, to ascertain that important aspects have not been overlooked. The following brief notes do not represent a detailed analysis of the design process but rather a few general points to indicate the general direction.

How to Begin to Design an Item of Equipment

1. Whatever you have to design there is always some information available. In the absence of 'hard' data, agree with the supervisor what are the necessary assumptions/estimates that will enable the design to proceed. Write down all the available information which might be applicable to the design of this item of equipment. Itemise the information that is required, but not available, for the design.

    DO NOT flounder and delay at this stage! Decide to begin (and begin to decide!), and make the list of available (and required) information.

    DO NOT expect the data to be absolute, this is not a textbook problem, some data will be contradictory and some decisions will be based upon estimates. Some students become paralyzed, unwilling to proceed, because the problem is not clearly defined and all necessary information is not available. Make a start, check the validity of any estimates as the design proceeds.

2. *Formulate a statement of the design criteria for the unit.* What are the general design criteria (or perhaps only one criterion)? What is the driving force to make heat transfer (ΔT) or mass transfer (Δc) occur? What is the basic design parameter that needs to be determined (e.g. heat-transfer area, volume of packing, residence time)? What is the design (or the unit) intended to achieve? For example, a reactor is usually required to achieve a particular conversion of reactants to products, a distillation column must provide a particular separation, an absorption column provides removal of a specific level of impurity, etc.

3. *Identify the design (stream) data* that will be required to complete the design, e.g. temperatures, pressures, flowrates, compositions, etc. Complete a *preliminary design specification* (i.e. performance specification) by listing the known information, and providing estimates (clearly stated as such) for any unknown requirements.

4. Consider general ways in which the design criteria can be achieved, specification of alternative types of equipment, manufacturer's data, economic considerations, materials and fabrication requirements for the design, etc. Consider the traditional (standard or accepted) means of performing the required operation, and then consider all the alternatives, i.e. different types of equipment that could be used, alternative methods of operation, varying the particular requirements, changing stream conditions, etc.

5. *Decision time* — select the particular type of equipment to be used. Justify this choice — list the advantages *and* disadvantages compared with alternative equipment or process schemes. Never consider a design decision to be the only possible correct choice, or to be irreversible. As the detailed design proceeds, more appropriate choices may become apparent. It may be necessary to make major changes, this is unfortunate in relation to the design time already spent but not as serious as the consequences of producing an unacceptable design!

6. *Revise the design criteria*, and the preliminary design specification, *if necessary*.
7. *How is this type of equipment designed?* You need a *design method* to be implemented. Refer to design books, handbooks, journal articles, etc. Remember to do some background reading, determine the basic principles of operation and the theory related to the design of the equipment. Never proceed with a design method unless you are completely satisfied with its theoretical basis, and with your knowledge of the associated chemical engineering principles. If you are not satisfied with your understanding of the justification of the proposed design method — obtain reliable advice.
8. Ensure that the proposed design method is actually applicable to the type of equipment to be designed. This appears obvious, but many times students have applied design methods unsuitable for a particular application!
9. Study the design method in detail. Are all the necessary data available, including physical and empirical constants, etc.?
10. List all the assumptions/limitations inherent in the design method. State all the estimates that must be made (both in the method itself and in the design specification parameters) in order to apply the method.
11. Apply the design method to your situation.
12. *Evaluate the results*, i.e. the design specification, that is obtained using this method. Does the design appear to be sensible? Is the design acceptable? Do not assume that the results obtained by applying an established design method, and by using a simulation package, are necessarily acceptable or sensible. The design is not finished at this stage. Establish the acceptability of the results by comparison with other sources, e.g. published information in the literature, comparison with the performance or size of actual operating units, consideration of operational specifications (e.g. fluid velocities, pressure drop, etc.), and ultimately a common-sense (or engineering-sense?) evaluation of the design results. If the design is not acceptable in all (major) respects, then either corrections or modifications must be made to the design method employed, or a completely new approach must be adopted (return to point 2 above).
13. A complete design includes consideration of all aspects of the equipment specification such that the item could be built and operated. These aspects include the mechanical design, materials selection, fabrication requirement, operational details, safety features, versatility (e.g. turn-down), economic considerations, etc. The scope (i.e. depth) of the design should be clearly specified by the supervisor.

Summary

The most relevant advice when faced with the task of designing an item of equipment is to *get started*! Survey the appropriate literature for useful background information and details of available design methods. Make necessary assumptions/estimates in order to proceed with the design. It is extremely unlikely that a design will be completely original with no precedent. Do not expect the design method to be completely rigorous, many designs are based upon empirical correlations and experience (i.e. 'rules of thumb').

9.2   STANDARDS AND CODES

The terms *Standard* and *Code* are often used interchangeably and the difference
between them may not be apparent to a student commencing a design study. However, it is hoped that at this stage the student knows about the existence of these
documents, and has had some preliminary exposure to their use and application
in basic design studies. A *Standard* usually provides details of standard (or nominal) or preferred sizes (for pipes and flanges, etc.), compositions and testing procedures (of common engineering materials), etc. A *Code* should be reserved for
a *Code of Practice* which covers a recommended design method or an operating
procedure.

Most developed countries have their own national standards organisations, in the
UK this body is the *British Standards Institution (BSI)* and their standards carry
the prefix BS. In Australia, the Standards Association of Australia (SAA) issues
Australian Standards with the prefix AS, e.g. AS1210: Unfired Pressure Vessels. In
the USA, standards are issued by Federal, State and various commercial organisations, and the coordination of information is undertaken by the *National Bureau of
Standards*. Chemical engineers are probably most interested in the standards issued
by the American Society for Testing and Materials (ASTM, for materials specifications), the American Society of Mechanical Engineers (ASME, for pressure vessel
design), the American Petroleum Institute (API), and the American National
Standards Institute (ANSI). The International Standards Organisation (ISO) coordinates the publication of international standards.

9.3   ADDITIONAL DESIGN CONSIDERATIONS

The detailed design stage described so far, describes fairly accurately the way in
which many student projects are carried out. However, several important aspects
have been ignored which should be considered as the detailed equipment designs are
being performed. In many student projects, these other factors are often only
considered after the equipment design is completed and are usually adapted to fit
the final design specification. After reading these notes, it should be obvious that this
procedure is unacceptable and several additional factors *must* be considered *concomi-
tantly* with the detailed equipment design. The most obvious aspects have been discussed in Chapter 8, namely Energy Integration and Conservation; Process Control
and Instrumentation; and Safety, Health, and the Environment (including Loss
Prevention and HAZOPs).

References

Some General Textbooks

Coulson, J.M. and Richardson, J.F., *Chemical Engineering Volume 1: Fluid Flow,
Heat Transfer and Mass Transfer*, 5th edn (1996); *Volume 2: Particle Technology
and Separation Processes*, 4th edn (1991); *Volume 3: Chemical and Biochemical*

Reactors and Process Control, 3rd edn (1994); *Volume 6: An Introduction to Chemical Engineering Design*, 2nd edn (1993), Pergamon Press, Oxford.

Henley, E.J. and Seader, J.D., *Equilibrium-Stage Separation Operations in Chemical Engineering*, John Wiley, New York (1981).

Himmelblau, D.M., *Basic Principles and Calculations in Chemical Engineering*, 6th edn, Prentice-Hall, New Jersey (1996).

Holman, J.P., *Heat Transfer*, 7th edn, McGraw-Hill, New York (1990).

King, C.J., *Separation Processes*, 2nd edn, McGraw-Hill, New York (1980).

Kreith, F. and Bohn, M.S., *Principles of Heat Transfer*, 5th edn, Gordon and Breach Publishers, New York (1993).

Levenspiel, O., *Chemical Reaction Engineering: An Introduction to the Design of Chemical Reactors*, 2nd edn, John Wiley, New York (1972).

Levenspiel, O., *Engineering Flow and Heat Exchange*, Plenum Press, New York (1984).

Levenspiel, O. and Kunii, D., *Fluidization Engineering*, 2nd edn, Butterworth-Heinemann, Massachusetts, USA (1991).

Lock, G.S.H., *Latent Heat Transfer: An Introduction to Fundamentals*, Oxford University Press, UK (1996).

McCabe, W.L., Smith, J.C. and Harriott, P., *Unit Operations of Chemical Engineering*, 5th edn, McGraw-Hill, New York (1993).

McKetta, J.J. (ed.), *Chemical Processing Handbook*, Marcel Dekker, New York (1990).

Morris, G.A. and Jackson, J., *Absorption Towers*, Butterworth Scientific Publishers, London (1953).

Norman, W.S., *Absorption, Distillation and Cooling Towers*, Longman Publishing, London (1961).

Sherwood, T.K., Pigford, R.L. and Wilke, C.R., *Mass Transfer*, McGraw-Hill, New York (1975).

Smith, B.D., *Design of Equilibrium Stage Processes*, McGraw-Hill, New York (1963).

Treybal, R.E., *Mass Transfer Operations*, 3rd edn, McGraw-Hill, New York (1979).

Wankat, P.C., *Equilibrium-Staged Separations*, Elsevier, New York (1988).

Wankat, P.C., *Rate-Controlled Separations*, Elsevier, New York (1990).

Whalley, P.B., *Boiling, Condensation and Gas–Liquid Flow*, Oxford University Press, UK (1987).

Whalley, P.B., *Two-Phase Flow and Heat Transfer*, Oxford University Press, UK (1996).

Zarzycki, R. and Chacuk, A., *Absorption: Fundamentals and Applications*, Pergamon Press, UK (1993).

Design Books

Fraas, A.P., *Heat Exchanger Design*, 2nd edn, John Wiley, New York (1989).

Hewitt, G.F., *Handbook of Heat Exchanger Design*, Begell House Publishing, UK (1989).

Hewitt, G.F. and Whalley, P.B., *Handbook of Heat Exchanger Calculations*, CRC Press, Florida (1990).

Kakac, S., *Boilers, Evaporators and Condensers*, John Wiley, New York (1991).

Kern, D.Q., *Process Heat Transfer*, McGraw-Hill, New York (1950).

Kister, H.Z., *Distillation Operation*, McGraw-Hill, New York (1990).

Kister, H.Z., *Distillation Design*, McGraw-Hill, New York (1992).

Kohl, A.L. and Riesenfeld, F.C., *Gas Purification*, 5th edn, Gulf Publishing, Texas (1997).

McAdams, W.A., *Heat Transmission*, 3rd edn, McGraw-Hill, New York (1954).

Perry, R.H. and Green, D.W. (eds), *Perry's Chemical Engineers' Handbook*, 7th edn, McGraw-Hill, New York (1997).

Reid, R.C., Prausnitz, J.M. and Poling, B.E., *The Properties of Gases and Liquids*, 4th edn, McGraw-Hill, New York (1987).

Ricci, L. (ed.), *Separation Techniques; Volume 1: Liquid–Liquid Systems; Volume 2; Gas/Liquid/Solid Systems*; reprints of papers from *Chemical Engineering*, McGraw-Hill Publications, New York (1980).

Rohsenow, W.M. and Hartnett, J.P., *Handbook of Heat Transfer Fundamentals*, 2nd edn (1984) and *Handbook of Heat Transfer Applications*, 2nd edn (1985), McGraw-Hill, New York.

Rosaler, R.C. and Rice, J.O., *Standard Handbook of Plant Engineering*, McGraw-Hill, New York (1983).

Rousseau, R.W. (ed.), *Handbook of Separation Process Technology*, John Wiley, New York (1987).

Sandler, H.J. and Luckiewicz, E.T., *Practical Process Engineering: A working approach to plant design*, McGraw-Hill, New York (1987).

Soumerai, H., *Practical Thermodynamic Tools for Heat Exchanger Design Engineers*, John Wiley, New York (1987).

Schweitzer, P.A., *Handbook of Separation Techniques for Chemical Engineers*, 2nd edn, McGraw-Hill, New York (1988).

Walas, S.M., *Chemical Process Equipment: Selection and Design*, Butterworth, USA (1988).

Woods, D.R., *Process Design and Engineering Practice*, Prentice-Hall Inc., New Jersey (1995).

Woods, D.R., *Data for Process Design and Engineering Practice*, Prentice-Hall Inc., New Jersey (1995).

10. CASE STUDY — PHTHALIC ANHYDRIDE REACTOR DESIGN

OVERALL SUMMARY

A catalytic, fixed-bed, tubular reactor has been designed to produce 40 kT/year of phthalic anhydride from o-xylene using the Low Air Ratio (LAR) process. The highly exothermic PAN synthesis reaction will be cooled by circulating 'Hitec' (a heat transfer salt) through the shell-side of the reactor.

The chemical engineering design was based on the results of a computer model that can simulate the operation of the reactor over a range of conditions. The catalyst will be supported on 6 mm ceramic spheres in 13,500 reaction tubes (25 mm $\emptyset \times 3500$ mm). An external salt cooler, producing high pressure steam, will remove heat from the salt. The mechanical design of the reactor conforms to the specifications in AS1210 (Unfired Pressure Vessels). The vessel is to be constructed from low-alloy chrome steel and is to be insulated with mineral fibre. A cylindrical skirt and concrete footing will be used to support the reactor.

A hazard and operability (HAZOP) study of the reactor was completed and the results were incorporated into the design of the control and instrumentation system. Recommendations are made for the safe operation, commissioning, shut-down and start-up of the unit. A regular maintenance program is also discussed. The reactor is estimated to cost US$1.26 million, excluding the cost of catalyst and coolant charges, the salt cooler, the control system and an inventory of spares.

PART I: CHEMICAL ENGINEERING DESIGN

10.1 OVERALL DESIGN STRATEGY

The design of process equipment is normally performed by a team of experienced and skilful professionals (and para-professionals) from a variety of fields. The final design must be safe, operable and profitable. The time and resources allocated to achieve this result are usually extensive whereas the time and resources available to the student are much less. Therefore, a design strategy must be formulated so that the required tasks can be performed immediately. Clearly, the final design will not be optimal. However, it should be possible to produce a design that satisfies most of the basic constraints.

The phthalic anhydride reactor is a complex item of equipment which must satisfy both mass transfer and heat transfer requirements. The available information is

Updated Material and Energy Balance for the Phthalic Anhydride Reactor (R101)

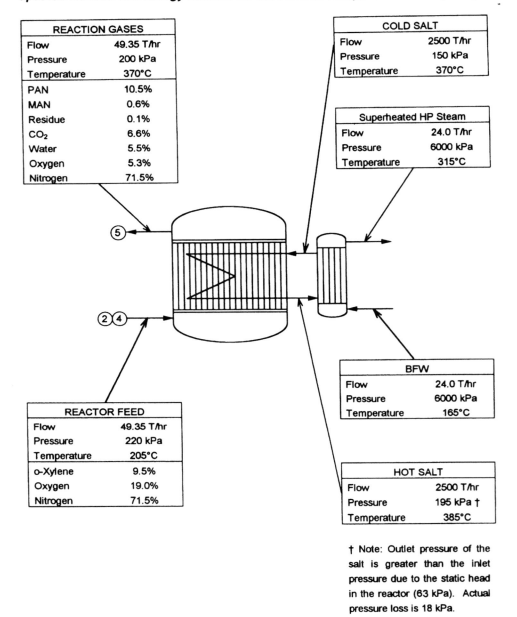

REACTION GASES	
Flow	49.35 T/hr
Pressure	200 kPa
Temperature	370°C
PAN	10.5%
MAN	0.6%
Residue	0.1%
CO_2	6.6%
Water	5.5%
Oxygen	5.3%
Nitrogen	71.5%

COLD SALT	
Flow	2500 T/hr
Pressure	150 kPa
Temperature	370°C

Superheated HP Steam	
Flow	24.0 T/hr
Pressure	6000 kPa
Temperature	315°C

REACTOR FEED	
Flow	49.35 T/hr
Pressure	220 kPa
Temperature	205°C
o-Xylene	9.5%
Oxygen	19.0%
Nitrogen	71.5%

BFW	
Flow	24.0 T/hr
Pressure	6000 kPa
Temperature	165°C

HOT SALT	
Flow	2500 T/hr
Pressure	195 kPa †
Temperature	385°C

† Note: Outlet pressure of the salt is greater than the inlet pressure due to the static head in the reactor (63 kPa). Actual pressure loss is 18 kPa.

restricted because the process is comparatively new and only licensed to industrial users. Therefore, a combination of knowledge from related processes and the application of chemical engineering 'first principles' must be used to satisfactorily complete the design. Some design decisions must be based on 'why not?' logic

Engineering Specification Sheet for the Reactor (R101)

Vessel	Catalytic, tubular, fixed-bed reactor		
Operation	Continuous	Design Code	AS1210 (1989)
	Shell	Tubes	
Orientation	-	Vertical	
Fluid	Hitec (heat transfer salt)	Hydrocarbon vapour and air	
Flow Rate (T/hr)	2500	49.35	
Composition	Pure	10.5% Phthalic anhydride	
Number	1	13,500	
Dimensions	5500 mm Ø i.d.	3.5 m x 25 mm Ø i.d.	
Inlet Temperature	371°C	205°C	
Outlet Temperature	385°C	370°C	
Mechanical Design		Heat Transfer	
Material	Low-alloy chromium steel	Heat Duty	22.5 MW
Construction	Cylindrical walls	Heat Transfer Area	3770 m^2
	Ellipsoidal ends		
Support	Cylindrical skirt	Overall HT Coefficient	195 $W/m^2/°C$
Foundation	Concrete footing		
Notes			
1. Catalyst: vanadium pentoxide			
2. Yield: 1.10 kg PAN / kg o-xylene			

where necessary information is not available and previous experiences of other designers have converged to a common solution. Other design decisions must be based on totally new ideas and, as such, may not be fully mature. The final design is not necessarily the best that is possible but is, hopefully, the best that could be achieved under the circumstances.

This report considers five stages of the design process. Each stage is the answer to a particular question:

1. What must the design be capable of? (*Design Basis*)
2. What design decisions must be made? (*Design Parameters*)
3. How are those decisions to be made? (*Design Criteria*)

Schematic Drawing of the Reactor (R101)

Overall height 9.5 m

Outside diameter 5.65 m

89 mm

63 mm

GAS OUTLET
(1000 mm i.d.)

SALT INLET
(600 mm i.d.)

13,500 TUBES
(3500 x 25 mm)

SALT OUTLET
(600 mm i.d.)

TUBE-PLATE (100 mm)
and SUPPORT (150 mm)

GAS INLET
(1000 mm i.d.)

MANHOLE
(450 x 400 mm)

4. What is the physical, chemical or mechanical basis for those decisions? (*Design Methods*)
5. What are the results of those decisions? (*Detailed Design and Specification*)

After the design has been completed, two other questions must also be answered:

6. How is it to be operated and controlled? (*HAZOP, Control and Instrumentation and Operating Considerations*)
7. How much does the design cost? (*Costing*)

10.2 DESIGN BASIS

A single, packed-bed, catalytic tube-type reactor using a licensed catalyst for the LAR (Low Air Ratio) process is preferred. The reactor must be capable of producing 40 kT/year of phthalic anhydride from pure *o*-xylene and air. The conversion of *o*-xylene should be essentially 100% with approximately 79% selectivity for PAN. The composition of the products must be controlled to maximise the concentration of phthalic anhydride and minimise the formation of impurities.

The feed is available at 205°C from preheating equipment and should leave the reactor at not more than 400°C. The surplus heat should be removed with an appropriate coolant. Any hot-spots in the reactor must be controlled to protect the catalyst, coolant and products from potential damage.

The vessel should be able to withstand an internal explosion of *o*-xylene or phthalic anhydride caused by auto-ignition (or otherwise) which could cause a pressure rise of 1800 kPa. It must also be possible to access the tubes in order to replace the catalyst when it becomes deactivated.

10.3 DESIGN PARAMETERS

The final design must specify all the dimensions to enable the reactor to be constructed. Initially, this requires that the normal operating conditions should be fully specified in terms of the flows, temperatures, pressures and compositions. Essentially, this is the chemical engineering design. The design parameters which require evaluation at this stage are:

a) tubes (diameter, length, fill height, layout)
b) catalyst (type, support, holding method)
c) shell (area, baffles, divisions)
d) pressure drop
e) composition of reaction products
f) type and flow of coolant
g) temperature profile along tubes
h) composition profile along tubes
i) heat transfer rate
j) velocity of reactants, products and coolant
k) method of controlling coolant temperature

10.4   DESIGN CRITERIA

The LAR process selectively oxidises o-xylene to phthalic anhydride over a proprietary catalyst (based on vanadium pentoxide) in a tubular fixed-bed reactor. Although little specific information is available that is directly applicable to the LAR process, several studies have been performed on the traditional process which uses the same type of reactor. Many design considerations are the same for both reactors.

The advantage of the new catalyst (the LAR proprietary catalyst) is that it is able to operate at a significantly higher o-xylene concentration without either reducing the selectivity for the partial oxidation reaction which produces phthalic anhydride or increasing the formation of by-products. The LAR catalyst can oxidise feeds with o-xylene concentrations up to $134 \, g/Nm^3$ while traditional catalysts operate on only $40-70 \, g/Nm^3$. Thus, substantially less air (which is mostly inert) is required and much of the downstream equipment can be reduced in size and duty. However, the heat released by the reaction is also concentrated in a small part of the reactor and difficulties arise in adequately dissipating the heat to prevent excessive temperatures in the reactor.

Heat transfer considerations predominate in many of the design decisions which must be made. The heat transfer medium was selected on the basis of temperature limitations, the heat transfer coefficient, the cost, availability, the hazards of use and corrosivity. The operating temperature and heat transfer coefficient were identified as being particularly important.

A FORTRAN model of the reactor was written to allow its operation to be simulated for a range of operating conditions. The program considers four variables: the conversion of o-xylene; the gas temperature in the tubes; the tube wall temperature; and the coolant temperature. Each variable is calculated at 5 cm intervals along the length of a tube. All the significant reaction conditions can be altered to assess the effect on the conversion and the temperature profile. The model was tested on experimental data from Nikolov and Anastasov (1989) and Calderbank (1974). Once a satisfactory accuracy could be established, the parameters were varied to simulate operation with the LAR catalyst.

An iterative technique was used to determine the optimum design that satisfied all necessary constraints on the conversion and temperature profile. Design parameters that were determined from the computer model included the length of tube required for complete reaction (with both new and used catalyst), the tube diameter necessary to control temperatures adequately, the flow of coolant required to reduce the temperature difference to an acceptable level, and the overall heat transfer coefficient required to dissipate the heat and avoid hot-spots in the reactor.

After a target heat transfer coefficient had been established, various configurations were investigated to determine whether the target could be reached and, if so, the optimal arrangement required. Detailed calculation methods were used to determine the heat transfer coefficients from the gas to the tube, through the tube wall, and from the wall to the coolant. Thus, the limiting resistance (either tube or shell-side) could be found and a strategy could be formulated to reduce this resistance if the target for the overall coefficient could not be met.

Results from the computer model and detailed heat transfer calculations completed the basic design. However, several other chemical engineering design parameters were still to be determined, including the pressure drops on both the shell and tube-sides, the catalyst holding method and the mechanism for removing heat from the coolant. Pressure drops were calculated from literature correlations for packed columns and shell and tube exchangers. A catalyst holding method was selected from those used in traditional phthalic anhydride reactors.

The reaction heat absorbed by the coolant must be dissipated externally to maintain steady-state operation. This heat can be used to produce high-pressure steam which is required for power generation. The heat removed from the gas cooler (downstream) can be used to preheat the boiler feed-water. The basic design requires one of three standard types of power plant evaporator to be selected, and a calculation of the surface area required based upon the LMTD and heat transfer coefficient.

Key Results

- Since 1980, a tubular fixed-bed catalytic reactor has generally been selected to produce phthalic anhydride from *o*-xylene.
- The new LAR catalyst, compared with traditional catalysts, is capable of operating at significantly higher hydrocarbon loadings without deactivation.
- The heat transfer medium was selected to dissipate the heat of reaction. Temperature limitations and the heat transfer coefficient were considered the most important parameters.
- A computer model of the reactor was written in FORTRAN in order to determine the conversion and temperature profiles. The model was tested on openly available experimental data and then modified to simulate the LAR catalyst.
- The final design was selected from the results of many simulations which were performed using different operating conditions and parameters.
- Various reactor configurations were tested in order to find a set of operating conditions that were able to dissipate the heat of reaction effectively.
- Design parameters that were not fixed by the model were calculated from literature correlations or estimated from available details of traditional reactors and economic and safety considerations.
- The cooling system for the reactor coolant will produce HP steam for power generation. The required heat transfer area for the salt-cooler was calculated by using the LMTD available and an estimate of the overall heat transfer coefficient.

10.5 CHEMICAL ENGINEERING DESIGN METHODS

10.5.1 Catalyst Properties

The catalyst is a combination of vanadium pentoxide, titanium dioxide and small amounts of certain proprietary compounds which improve the range of operating conditions. It is prepared by heating a suspension containing the compounds, and spray coating the suspension onto a support which can be either 6 mm porcelain spheres, or similar sized steatite rings. The surface area available for reaction is about 10–12 m²/g. A selectivity for phthalic anhydride of 78.8% is claimed, producing a yield of 110% by weight of o-xylene feed. The maximum safe loading of hydrocarbon is 0.192 kg/m²/s.

Key Result

- The catalyst properties are documented in proprietary data books and the patent literature (Blechschmitt *et al.*, 1977). The operating conditions and method of use are specified in those references but more detailed information is withheld.

10.5.2 Kinetics

The vapour-phase oxidation of o-xylene over a vanadium pentoxide catalyst is essentially first order with respect to the o-xylene concentration. However, above a maximum o-xylene concentration (which depends only on the specific type of catalyst and is independent of temperature while the catalyst remains activated) the catalyst deactivates irreversibly and the reaction rate decreases rapidly. Deactivation occurs when the valence-state of vanadium changes. New developments have focussed on finding additives to prevent this from happening. The LAR catalyst can sustain the reaction up to a concentration of 2.77 mole% o-xylene, compared with only 1.0–1.4% in other catalysts.

There are at least three significant reaction mechanisms which occur at the temperatures encountered within the reactor. Below 370°C, the reaction is first order with respect to the oxygen concentration and the rate constant is low. The selectivity for phthalic anhydride is high but conversion of the feed may not be complete and the product is likely to undergo further combustion. Between 370°C and 440°C, the reaction is first order with respect to the hydrocarbon. The rate constant is significantly higher, the selectivity remains high and the o-xylene conversion is essentially complete. Above 440°C (and below 550°C), the reaction is also first order with respect to hydrocarbon concentration but conforms to a third reaction mechanism. The activation energy is substantially less, the rate constant is higher and, because the activation energy is low, the reaction rate is almost independent of temperature. The selectivity for phthalic anhydride begins to decrease at higher temperatures as side reactions and complete oxidation become more likely. The catalyst is irreversibly deactivated if the surface temperature exceeds 500°C. There will be a temperature drop from the gas to the catalyst surface of 10–50°C. Thus, the maximum allowable gas temperature is 510–550°C. The gas temperature in the reactor should, therefore, be controlled so that the bulk of the reaction occurs from 440–510°C.

Key Results

- The LAR catalyst, compared with traditional catalysts, can operate at significantly higher hydrocarbon concentrations without deactivating.
- At reaction temperatures of 440–510°C, the reaction rate is first order with respect to hydrocarbon concentration and essentially independent of temperature.

- At lower temperatures, the reaction is much slower and dependent on the partial pressure of oxygen, and the PAN yield is reduced.
- At higher temperatures, the catalyst can deactivate irreversibly.

10.5.3 Reactor Simulation

The FORTRAN program was written in order to calculate the temperature profiles for the gas, the tube wall and the coolant and the conversion profile for o-xylene. Thus, four independent equations are required. The first equation relates the conversion to the tube length by assuming the rate constant is independent of temperature or position along the reactor (this introduces little error if the gas temperature is between 440°C and 510°C at the site of the reaction). Calderbank (1974) derived this equation analytically (equation (10.1), following) in "Kinetics and Yields in the Catalytic Oxidation of o-Xylene to Phthalic Anhydride with V_2O_5 Catalysts". In the same work, he derived the relation between the gas temperature and the wall temperature which is given by equation (10.2), below.

$$x = 1 - \exp(-\alpha L) \tag{10.1}$$

$$T - T_w = \beta/(\gamma - \alpha) \times (\exp(-\alpha L) - \exp(-\gamma L)) \tag{10.2}$$

where x = the conversion of o-xylene; T = gas temperature; T_w = wall temperature; L = length; α, β, γ are variable groups which are defined in Appendix B.
   A third equation can be obtained by performing an energy balance over an element of the reactor. A fourth equation is derived from an energy balance over the coolant:

$$(x_2 - x_1) \times m \times \Delta H_r = m \times c_p \times (T_2 - T_1) + (U \times A \times \Delta T) \tag{10.3}$$

$$U \times A \times \Delta T = m_s \times c_{ps} \times (T_{2s} - T_{1s}) \tag{10.4}$$

where x = conversion of o-xylene; m = mass flow of reactants; ΔH_r = heat of reaction per kg of reactants; c_p = average specific heat of reactants; U = overall heat transfer coefficient; A = heat transfer area; ΔT = temperature driving force for heat transfer; m_s = mass flow of salt; c_{ps} = average specific heat of salt.
   The program has provision to change any of the operating conditions or parameters, and to simulate a range of catalysts providing the reaction kinetics are similar. This is a valid assumption for the LAR catalyst as the principal development has been the increase in maximum hydrocarbon concentration allowable before deactivation commences. Results were obtained at 5 cm intervals along the length of the reactor tubes.

Key Results

- The conversion and the wall temperature were determined from equations derived by Calderbank (1974).
- The gas temperature and the coolant temperature were determined from energy balances over finite elements of the reactor.

- The conversion and temperature profiles can be determined for any operating conditions, or a combination of parameters.

10.5.4 Heat Transfer

The overall heat transfer coefficient was calculated in three parts: convection from the gas to the tube wall; conduction through the tube wall; and convection from the tube wall to the bulk coolant. The tube-side coefficient was calculated from a correlation for packed columns (see Kreith and Bohn, 1986) which is given by:

$$Nu = 0.203 \times (Re \times Pr)^{0.33} + 0.220 \times Re^{0.8} \times Pr^{0.4} \tag{10.5}$$

where Nu = Nusselt number; Re = Reynolds number; Pr = Prandtl number.

The physical properties of the gas were estimated using HYSYS and the bed properties (i.e. catalyst support) were estimated from Coulson and Richardson, Volume 2 (1991). The rate of conduction through the wall was calculated by assuming a wall thickness of 3.2 mm and using the properties for chrome steel, and using the method of Kreith and Bohn (1986) which is given by:

$$1/h_k = r \times \ln(d_o/d_i)/k \tag{10.6}$$

where h_k = conductive heat transfer coefficient; r = average tube radius; d_o = outside diameter; d_i = inside diameter.

The shell-side heat transfer coefficient (convection to the coolant) was calculated using the method of Kern (1950) for baffled shell and tube exchangers. The Colburn factor, j_H, is determined graphically as a function of Reynolds number and then the Nusselt number is calculated from j_H. The tube count (and shell inside diameter) were determined by assuming that the tubes were packed without boundary constraints (i.e. infinite reactor diameter). This assumption is acceptable because of the large number of tubes present (13,500). The physical and thermal properties of the coolant were taken at the average coolant temperature from data compiled by Kreith and Bohn (1986).

Key Results

- Correlations for convective heat transfer were taken from Kreith and Bohn (1986), in order to calculate the tube-side coefficient, and from Kern (1950) to calculate the shell-side coefficient.
- Physical properties for the gas were obtained from HYSYS and from Kreith and Bohn (1986) for the coolant.

10.5.5 Pressure Drop

The tube-side pressure drop was calculated from the Darcy equation which is given by equation (10.7), below. The physical properties of the gas which were required for the calculation were estimated from HYSYS.

$$\frac{\Delta P}{L} = \left(\frac{150(1-\varepsilon)\cdot\mu}{D_p} + 1.75G\right)\frac{1-\varepsilon}{\varepsilon}\cdot\frac{G}{D_p\cdot\rho\cdot g} \tag{10.7}$$

where ΔP = pressure drop; L = length of tube; ε = porosity of packing; μ = viscosity; D_p = particle diameter; G = mass flow rate of gas; ρ = average density of gas; g = acceleration due to gravity.

A second estimate was obtained by nomograph (McCabe et al., 1985) for packed columns with no liquid flow. A pressure drop correlation for liquid flowing on the shell-side of a baffled shell and tube exchanger was taken from Kern (1950), and is given by equation (10.8), below. The calculation was supported by physical properties obtained from Kreith and Bohn (1986).

$$\Delta P = (f \times G^2 \times D \times (N+1))/(2 \times g \times \rho \times D_e) \tag{10.8}$$

where ΔP = pressure drop; f = friction factor; G = mass flow rate of coolant; D = inside diameter of shell; N = number of shell-side passes; ρ = average density of coolant; g = acceleration due to gravity; D_e = effective diameter of tubes.

Key Result

- Literature correlations were used to calculate the pressure drop on both the shell and tube sides of the reactor using available physical property data.

10.6 DETAILED DESIGN

10.6.1 Reactor Configuration

A catalytic, tubular, fixed-bed type reactor will be used. It will be operated continuously for 8000 hours/year. The hydrocarbon feed rate is limited by the catalyst activity to 340 g/hr/tube. Thus, the number of tubes which are required to meet the design production requirements can be calculated from:

$$N = \frac{40,000\,\text{kg/yr/8000 hr/yr}}{0.340\,\text{kg/hr/tube}} = 13,500 \text{ tubes} \tag{10.9}$$

10.6.2 Coolant

The preferred coolant is 'Hitec' which is a triple eutectic mixture of 40% sodium nitrite, 7% sodium nitrate and 53% potassium nitrate. Hitec can be used at temperatures up to 500°C, which is well above the normal operating temperatures for the reactor. It has the highest heat transfer rate of any of the commercially available high-temperature heat transfer media. Furthermore, it is readily available and much cheaper than other high temperature coolants,such as mercury, sodium or sodium-potassium alloys. The principal disadvantages are that it is a mild irritant, and that it supports combustion. Care will be required to ensure that it does not come into contact with any hydrocarbon. Shut-downs and start-ups will require special attention as the salt is a solid at ambient temperatures. Hitec will react with some metals (magnesium and aluminium) but has a low corrosivity for common stainless steels and chrome steels.

Key Results

• An industrial heat transfer medium, Hitec, was selected as the reactor coolant.
• The normal operating temperature of the reactor is below the maximum allowable temperature for Hitec, and the heat transfer rate is high.
• The main hazard associated with using Hitec is the potential for contact with hydrocarbon as the salt is strongly oxidising.
• Shut-down and start-up will be difficult as Hitec solidifies at 143°C.

10.6.3 Computer Model

The reactor tube temperature profile and conversion profile were determined for a range of operating conditions, using both the traditional PAN catalyst and the LAR catalysts. A listing of the FORTRAN program is included in Appendix C at the end of this chapter.

The accuracy of the model can be assessed by simulating operation with traditional catalyst (which processes a lower hydrocarbon concentration) and comparing the results with those obtained experimentally (Nikolov and Anastasov,

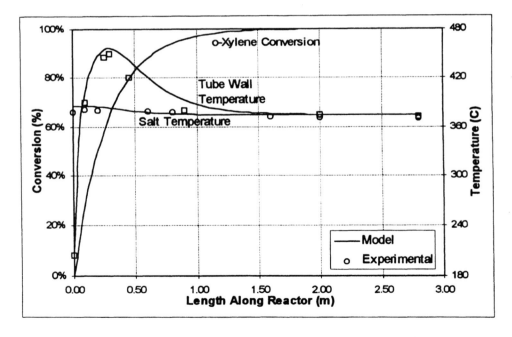

Figure 10.1 Reactor operation with conventional PAN catalyst; comparison of computer model predictions with experimental results.

1989). Figure 10.1 displays the results graphically and indicates that the model is satisfactory and, therefore, suitable for modelling the LAR catalyst.

The principal difficulty in attaining adequate operation of the reactor has already been identified as the need to provide sufficient heat dissipation to prevent hot-spots from forming. Thus, the worst case is when the catalyst is new and heat effects are concentrated near the entrance. As the catalyst ages, the location of the peak temperature will move along the reactor tubes and decrease in intensity. Therefore, models were based on conditions for a new catalyst charge.

Operating limitations were identified for the reaction gas temperature and the maximum tube wall temperature. To maintain favourable reaction kinetics, the gas temperature should be between 440°C and 510°C during the majority of the reaction. Additionally, the catalyst can deactivate above 500°C. There is a temperature gradient of 50–100°C between the bulk gas and the catalyst surface. Thus, the maximum gas temperature should be lower than 550°C. The cooling salt is a strong oxidant at high temperatures and can react with the metal tube walls above 500°C.

The model was run with many sets of operating conditions. The results were correlated to determine the significant variables that controlled the maximum gas and wall temperatures, and the temperature drop in the coolant (which is important in maintaining steady reaction temperatures). It was found that the peak temperatures depended primarily on the overall heat transfer coefficient and that the temperature drop depended on both the flow of coolant and the heat transfer coefficient. Table 10.1 indicates the effect of heat transfer coefficient on the peak gas and wall temperatures. In each case, the peaks occurred within the first 200 mm inside the reactor (less at high values of an overall heat transfer coefficient). At an overall heat transfer coefficient of 500 W/m²/°C, the gas temperature had dropped to less than 440°C within 350 mm of the inlet at a conversion of 83%.

Clearly, none of the conditions presented in Table 10.1, which are also shown in Figure 10.2, are satisfactory. Higher heat transfer coefficients are not physically possible. Another parameter that could be modified is the reaction tube diameter. However, it was found that for larger tubes, the gas temperature could not be controlled adequately and that for smaller diameter tubes, the wall temperature was difficult to maintain below 500°C. Thus, a diameter of 25 mm provides the correct balance between heat passing to the gas further up the tube and heat passing to the wall and coolant. Reversing the direction of flow of the coolant (from counter-current to co-current) had little effect on the reaction temperatures.

Table 10.1 Effect of heat transfer coefficient on maximum operating temperatures.

Heat Transfer Coefficient (W/m²/°C)	Maximum Gas Temperature (°C)	Maximum Wall Temperature (°C)
150	724	612
200	700	574
250	672	552
300	645	535
350	630	519
400	615	508
500	587	496

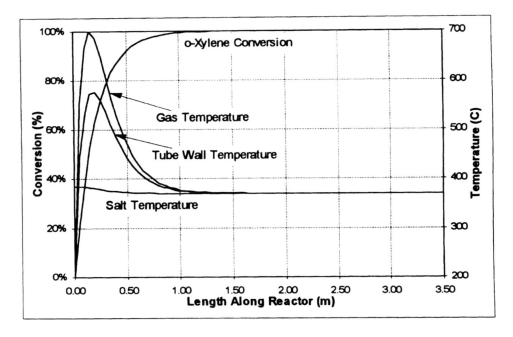

Figure 10.2   Reactor operation with LAR catalyst (no activity gradient).

The solution proposed in order to provide adequate control of reaction temperature is to dilute the catalyst in the initial section of the tubes. The method of dilution recommended is to replace some of the catalyst particles with inert supports that are not coated with the catalyst. Thus, bed characteristics will not be affected. After an initial operating period, the catalyst will become used and assume characteristics similar to the diluted catalyst. It may be economically sound to utilise used catalyst when it becomes available. Care will be required to monitor the effectiveness of the catalyst throughout the catalyst life cycle.

A range of dilutions were investigated. The results are most easily modelled by replacing the length with an effective length in each equation (where the effective length is the length containing an equivalent amount of undiluted catalyst), and by increasing the heat transfer area in the diluted zones as the same heat is being produced in a larger volume.

The preferred combination is to include 80% of inert particles in the first 500 mm and 60% in the following 500 mm. These conditions, with a salt flow of about 2500 T/hr, produce a relatively smooth conversion profile and keep maximum temperatures below dangerous levels, and allow the bulk of the reaction to occur between 440°C and 510°C. This is illustrated in Figures 10.3 and 10.4. The reaction temperature can be controlled by adjusting the inlet temperature of the coolant, and a temperature of 385°C was found to be the most effective. Increasing the flow of HTS improves the controllability only slightly and is not justified from an economic perspective.

A side effect of diluting the catalyst is that the cycle length is shortened (less catalyst is present in the reactor). This can be overcome by extending the length

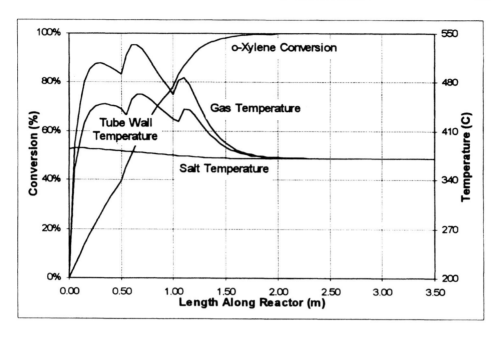

Figure 10.3 Reactor operation with LAR catalyst (using graduated catalyst activity).

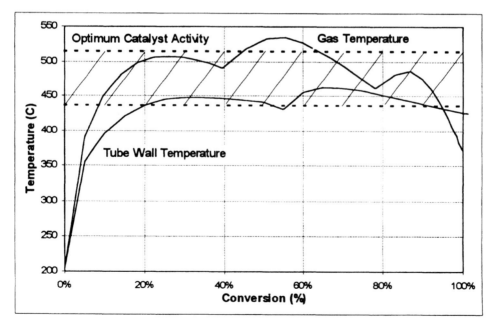

Figure 10.4 Reaction temperature (using LAR catalyst with graduated activity).

of the reaction tubes. A further 700 mm length containing catalyst only, is sufficient to replace the catalyst removed from the first 1000 mm of each tube.

Key Results

- Experimental data (Nikolov and Anastasov, 1989) was used to affirm the accuracy of the computer model of the reactor for the conventional PAN catalyst.
- Similar kinetic behaviour was assumed and the operating parameters were adjusted in order to simulate operation of the reactor with the LAR catalyst.
- The model identified heat dissipation as the greatest difficulty for operation with the new catalyst.
- Operating limits for the maximum gas and wall temperatures were established at 550°C and 500°C, respectively.
- It was determined that the overall heat transfer coefficient controlled the maximum reactor temperature, and that the flow of coolant predominantly controlled the temperature drop in the coolant.
- It was determined that catalyst dilution was required in order to spread the heat load along the reactor tubes, and to prevent the formation of hot-spots above the temperature limitations indicated above.
- A catalyst activity gradient created by including 80% of non-active particles in the first 500 mm and 60% in the following 500 mm, together with a flow of coolant of 2500 T/hr, would produce acceptable reactor performance.

10.6.4 Shell Configuration

Several operations with graduated catalyst activity were examined. Table 10.2 (similar to Table 10.1) indicates the effect of heat transfer coefficient on maximum reactor temperatures for cases where graduated catalyst activity is used. The required values of the overall heat transfer coefficient are much more easily realisable, and peak temperatures are displaced along the tubes by 400–500 mm. Furthermore, the rate of cooling below the optimum reaction temperatures is significantly slower.

The data in Table 10.2 establishes that the minimum allowable overall heat transfer coefficient for acceptable operation is about 180 W/m²/°C. Various shell

Table 10.2 Effect of heat transfer coefficient on maximum operating temperatures with graduated catalyst activity.

Heat Transfer Coefficient (W/m²/°C)	Maximum Gas Temperature (°C)	Maximum Wall Temperature (°C)
150	572	482
170	554	474
180	546	469
195	534	462
210	525	457
250	506	446

configurations can be investigated with this target in mind. The overall heat transfer coefficient for simple segmental baffles, in a one-pass shell configuration is indicated in Table 10.3 for varying numbers of baffles. Details of the calculation procedure are provided in Appendix B at the end of this chapter. It should be noted that the limiting heat transfer coefficient is the convection coefficient inside the tubes. The maximum heat transfer coefficient is approximately 208 W/m²/°C. A more complex shell-side configuration is not warranted.

Table 10.3 shows that three baffles (and possibly even one) are sufficient to provide adequate heat transfer. However, five baffles are preferred in order to maximise the heat transfer rate and allow an additional length of nearly 600 mm per pass which reduces the average velocity and minimises the pressure drop. An odd number of baffles is required to ensure that both the shell-side inlet and outlet are on the same side of the reactor. Although, four passes would appear adequate, the reduced flow in some areas will cause the minimum heat transfer coefficient to be somewhat less than the average value and hot-spots could form near the outside of the tube bundle. The temperature profile for a reactor which is operated with graduated catalyst activity and with five baffles is shown in Figure 10.3. Although, the heat transfer coefficient will decline with time (due to fouling), the aging of the catalyst should compensate for this effect and prevent the formation of hot-spots.

The tube diameter of 25 mm was found to be suitable from the heat transfer perspective. However, the exact tube diameter and layout have not yet been specified. A heavy tube gauge is required to minimise the effects of corrosion due to the difficulty of replacing tubes in the reactor, and 12 BWG was selected. Therefore, the tubes will have an outside diameter of 31.8 mm ($1\frac{1}{4}''$). A standard 39.7 mm ($1\frac{9}{16}''$) square pitch was selected to allow easy cleaning between the parallel rows of tubes. The large number of tubes in the bundle suggest that the bundle will be packed close to its ultimate capacity (i.e. the number of tubes per unit area at infinite vessel diameter). A packing efficiency of 99% was initially assumed for the reactor. The minimum inside diameter of the shell can, therefore, be calculated from:

$$D_{min} = \sqrt{\frac{(4/\pi)\, 13,500 \times 0.0397^2}{99\%}} = 5.23\, m \tag{10.10}$$

Table 10.3 Effect of number of baffles on overall heat transfer coefficient and average coolant velocity.

Number of Shell Baffles	Overall Heat Transfer Coefficient (W/m²/°C)	Average Coolant Velocity (m/s)
0	172	0.1
1	182	0.2
3	191	0.4
5	195	0.6
7	198	0.8
9	200	1.0

The actual reactor diameter will also depend on the mechanical vessel design because less than 100% of the cross-sectional area may be available for reaction gas flow. Various mechanical fittings and supports could reduce the available area by up to 15%. This is discussed in more detail in Section 10.12.

Key Results

- The required heat transfer coefficient for adequate reactor operation with the LAR catalyst (without activity gradients) is not realisable with a simple reactor and shell-and-tube configuration.
- The use of a graduated catalyst activity allows the required heat transfer coefficients to be reduced considerably, and maintains the reaction at the optimum temperature for significantly longer.
- The minimum heat transfer coefficient for acceptable operation is 180 W/m^2/°C.
- A shell configuration with five segmental baffles satisfies the heat transfer requirements without producing an excessive pressure drop.
- Heavy duty tubes were specified due to the risk of corrosion.
- A square pitch was selected for ease of cleaning and maintenance.
- The minimum reactor diameter is 5.23 m.
- The actual reactor diameter may be up to 15% larger than the minimum diameter.

10.6.5 Salt Cooler

The cooling duty for the heat transfer salt (Hitec) will be provided by a power-plant evaporator that produces high-pressure, superheated steam which is suitable for electricity generation. This exchanger should be integrated with the gas cooler to maximise the production of useful, high-grade energy. Hot water is nominally available at 165°C from the gas-cooler. A superheated steam production of 24 T/hr at 60 bar is achievable with this configuration.

The configuration of the salt-cooler was based on several key assumptions: 25 mm diameter tubes; 40 mm square pitch; 6.0 m straight tubes, without fins; and a tube-side velocity of 2.5 m/s. The overall heat transfer coefficient was conservatively estimated to be 270 W/m^2/°C under these conditions. The average heat transfer salt temperature is 378°C and the steam saturation temperature (at 60 bar) is 276°C, producing a very large average temperature driving-force of 102°C. The required heat transfer area was calculated to be 817 m^2. This requires 2270 tubes of the dimensions given above. The relevant calculations are detailed in Appendix B.

Key Results

- Salt cooling is performed by a power plant evaporator producing 24 T/hr of superheated steam at 60 bar which is to be used for electricity generation.
- The salt-cooler tube bundle will contain 2270 tubes (25 mm Ø × 6.0 m) on a square pitch. The overall diameter of the bundle will be 1400 mm.

10.6.6   Salt Circulation Pump

The heat transfer salt, Hitec, must be circulated through the reactor and salt-cooler in order to remove the heat of reaction from the reactor. The closed system must also include a storage vessel (accumulator) to hold the heat transfer salt during shut-downs, and to ensure that the pump suction pressure (or net positive suction head required, NPSHR) is maintained above the lower limit of operation for the pump (i.e. net positive suction head available, NPSHA).

The pump selection is primarily dependent on two quantities: head and flow rate. The pump inlet temperature, corrosivity of the fluid, and the required pressure rating (i.e. the maximum suction pressure plus the maximum developed head) are also important, and may influence the selection. The final pump specification should be completed in consultation with a pump manufacturer or supplier. An accurate match between the selected pump and the system charac-teristics is sought. If this is not available, a customised pump design is likely to be required.

The normal flow rate of the salt is 2500 T/hr (22,500 L/min). The pump should have at least 30% flow over-capacity to deal with emergency situations (e.g. a very high reactor outlet temperature which must be reduced quickly). Therefore, the maximum design flow rate should be 29,250 L/min. The required head must be calculated based on the pipe layout shown in Figure 10.5, and the correspond-ing system pressure drops are summarised in Table 10.4.

The piping layout includes sufficient space between the reactor and salt-cooler for the pump and driver, and a control valve. The salt storage drum will be located directly above the salt-cooler to conserve space, and to increase the NPSHA. There is sufficient clearance around each item of equipment to allow easy opera-tion and maintenance. The pipe diameter was estimated from an equation for the optimum pipe diameter (Harker, 1978), in SI units:

$$D = 8.41 \times W^{0.45}/\rho^{0.31} \tag{10.11}$$

where D = estimated economic pipe diameter (mm); W = mass flow (kg/hr); ρ = density (kg/m^3).

Table 10.4   Pressure losses in the heat transfer salt circuit.

Pipe Component	Number	Loss Coefficient	Number of Velocity Heads
Straight pipe	—	$f \times (L/D)$	0.9
Entrances	3	0.5	1.5
Exits	3	1.0	3.0
Elbows (90°)	10	0.9	9.0
Orifice plate	1	1.0	1.0
Block valves	6	0.25	1.5
Sub-total			16.9
Control valve	1	—	+30%
Total			24.1

Note: The friction factor (f) is 0.014 for this system.

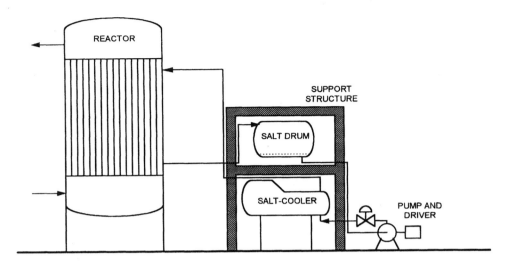

Figure 10.5 Heat transfer salt piping circuit.

The static head is the difference between the low liquid level in the salt storage drum and the reactor outlet. This value was estimated to be 5.0 m. The total length of straight pipe in the salt circuit was estimated to be 34 m. The dynamic pressure losses were calculated in terms of flow coefficients using standard values. This method is considered more accurate than the 'equivalent length' technique. The control valve should be sized to provide a moderate pressure drop for controllability, and 30% of the total pressure drop was assumed to be due to the control valve.

The static head and the dynamic pressure losses can be combined to create a system curve which can then be matched against pump curves. Figure 10.6 shows the calculated system curve and three possible pump curves. Curve A represents a pump that meets the system requirements but is unnecessarily large, which adds to both capital and operating costs. Curve B represents a suitable pump as the control valve would be close to fully open at the maximum flow condition, and approximately 70% open (based on a linear characteristic) at the normal operating point. Curve C shows a pump that is too small for this application.

An mixed-flow centrifugal pump was selected for this combination of high flow rate and moderate head, using recommendations from Perry (1984), Walas (1988) and Neerken (1978). The corrosivity of Hitec requires the pump casing and rotor to be constructed from a corrosion resistant material, such as stainless steel, type 316, due to the high fluid velocities (which are likely to increase the corrosion rate) and narrow clearances. Other components (e.g. fittings) should be constructed from low-alloy chromium steel. The operating temperature is high for a centrifugal pump and a customised design might be required.

The pump should be installed with a horizontal suction and a vertical discharge. The low liquid level in the salt storage drum was estimated to be 4.0 m above the pump suction. This results in a NPSHA of approximately 9.0 m (assuming the vapour pressure of Hitec is negligible at the operating temperature, and the suction

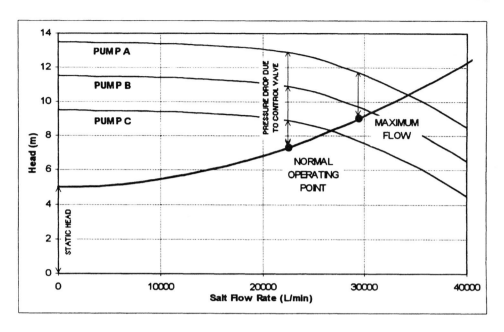

Figure 10.6 Pump and system curves for the heat transfer salt flow-loop.

line losses are 0.5 m). The NPSHR will be dependent on the vendor's final specification but it could be expected to be significantly less than the calculated NPSHA based on nomograms (Walas, 1988).

The optimum impeller diameter and the optimum rotational speed of the pump can be estimated from nomographs (Neerken, 1978). These have been constructed in terms of the pump specific speed, which is calculated from equation (10.12) below, and is shown in its usual form (with US common units):

$$N_s = N \times Q^{0.5}/h^{0.75} \tag{10.12}$$

where N_s = specific speed; N = rotational speed (rpm); Q = flow (US gpm); h = head (ft).

The maximum efficiency is obtained at a specific speed of 1500 to 7000. The optimum rotational speed at the normal operating point can be found by rearranging equation (10.12), and was calculated to be 285 rpm to 1335 rpm. This speed is much lower than the normal operating speed (3000 rpm) which is based on the local power supply. A variable speed drive will be required to avoid the large efficiency losses which would result from operating the salt pump at 3000 rpm. An operating speed of 600 rpm was selected, corresponding to a specific pump speed of 3150. At this speed, the optimum impeller diameter is 520 mm. The pump housing should be constructed to install impellers of up to 600 mm diameter to provide capacity for future expansions.

The specific speed and the impeller size can be used to provide an initial estimate of the efficiency of the pump via nomographs (Neerken, 1978). The salt pump efficiency was estimated to be 80% at 600 rpm with a 520 mm impeller. This efficiency estimate can be used to calculate the power requirement:

$$\text{Power} = (\rho \times g \times h) \times Q / \zeta \qquad (10.13)$$

where ρ = density (kg/m^3); g = acceleration due to gravity (m/s^2); h = head (m); Q = flow (m^3/s); ζ = efficiency.

The power required at normal operating conditions was calculated to be 94 kW. At the maximum flow conditions, the required power rating is 112 kW due to the increased flow rate and a slight loss in efficiency. A large capacity motor, of a standard size (150 kW), was specified for flexibility and for uncertainty in the efficiency estimate. A motor was preferred to a turbine driver in order to minimise the capital cost of the pump-plus-driver combination. The pump characteristic curve of NPSHR and operating efficiency cannot be finalised until consultations with the pump manufacturer or vendor have been completed, and a specific pump has been selected (or designed, if a customised model is required). At that stage, the motor rating should be reviewed to accommodate any changes.

The hazards associated with the heat transfer salt and its importance in controlling the reactor temperatures require a highly reliable seal. Double-faced mechanical seals are recommended due to the high operating temperature and the corrosivity of Hitec. It was considered that a packed gland would be ineffective under these conditions. Single-faced mechanical seals would be subject to deformation at the operating temperatures.

A removable insulated cover, constructed from mineral-fibre blocks with external sheet-metal lining, will be installed to reduce the heat losses from the pump. The cover should enclose the body of the pump and the suction and discharge, but not the driver. The insulation around the connecting pipes should extend to the edge of the removable pump cover.

Key Results

- A storage drum for the heat transfer salt is required to hold the circulating salt during shut-downs, and to ensure that an adequate suction head is always present for the salt pump.
- The storage drum will be located above the salt-cooler to conserve ground-level space, and to increase the NPSHA.
- A mixed-flow centrifugal pump was selected due to the combination of high flow and moderate head required.
- A customised design will probably be required to deal with the high operating temperature and the corrosivity of the heat transfer salt.
- A low rotational speed and a comparatively small impeller diameter were specified to maximise the pump efficiency.
- The pump driver will be an electric motor which is rated significantly higher than the estimated pump power consumption in order to provide flexibility and to compensate for uncertainty in the high efficiency estimate.
- The final pump specification must be confirmed with the pump manufacturer and vendor after a specific pump has been selected or designed.
- Double-surface mechanical seals were specified due to the extreme operating environment and hazards of the hot heat transfer salt.
- An insulated cover will be installed to reduce heat losses from the pump.

10.7 CHEMICAL ENGINEERING DESIGN SPECIFICATION

10.7.1 Reactor Specification

General

Type of Reactor	*Catalytic, tubular, fixed-bed*
Orientation	*Vertical*
Operation	*Continuous*

Feed

Reactants	*100% ortho-xylene, air*
Air to Hydrocarbon ratio	*9.5:1*
Hydrocarbon loading	*340 g/hr/tube*
Total Flow	*49.35 T/hr*
Temperature	*205°C*
Pressure	*175 kPa*

Catalyst

Type	*Vanadium pentoxide (V_2O_5)*
Conversion of o-Xylene	*100%*
Selectivity for PAN	*78.8%*
Yield of PAN	*1.10 kg/kg o-xylene*
Support	*6 mm ceramic spheres*
Holding Method	*Wire gauze and clamping ring*
Catalyst Dilution	*80% over first 500 mm and 60% over next 500 mm using inert balls*

Tubes

Number	13,500
Length	3.5 m
Catalyst Fill Height	3.4 m
Inside Diameter	25.4 mm
Outside Diameter	31.8 mm
Gauge	BWG 12 (heavy)
Heat Transfer Area	4710 m^2
Passes	1 (Upflow)
Tube Pattern	square
Tube Pitch	39.7 mm
Pressure Drop	6.2 kPa
Inlet Temperature	205°C
Outlet Temperature	370°C

Shell

Fluid	Hitec (heat transfer salt)
Composition of Fluid	40% $NaNO_2$, 7% $NaNO_3$, 53% KNO_3
Flow	2500 T/hr
Inlet Temperature	371°C
Outlet Temperature	385°C
Heat Removed per Tube	1193 W
Heat Transfer Coefficient	195 W/m^2/°C
Inside Diameter	5500 mm
Passes	1
Baffles	5 (Segmental)
Pressure Loss	18 kPa

10.7.2 Salt Cooler Specification

Type	Power-plant boiler
Duty	22.5 MW
Hot Fluid	Hitec (40% $NaNo_2$, 7% $NaNO_3$, 53% KNO_3)
Flow	2500 T/hr
Number of Tubes	2270
Tube Dimensions	25 mm Ø × 6.0 m
Heat Transfer Area	817 m^2
Tube Bundle Outside Diameter	1.40 m
Cold Fluid	Boiling water
Flow	24 T/hr
Temperature	Inlet: 165°C; outlet: 315°C
Steam Pressure	60 bar
Material	Low-alloy chromium steel

10.7.3 Salt Circulation Pump Specification

Type	Single-stage centrifugal
Flow Pattern	Mixed-flow centrifugal
Orientation	Vertical discharge
Speed	870 rpm
Flow Rate	Normal: 22,500 L/min; maximum: 29,250 L/min
Head	Normal: 11.0 m; shut-off: 11.5 m
Power	Normal draw: 94 kW; motor rating 150 kW
Efficiency	Normal: 80%; maximum flow: 75%
Impeller Diameter	520 mm
NPSHA	>9.0 m
Seals	Double-surface mechanical

PART II: MECHANICAL ENGINEERING DESIGN

10.8 MECHANICAL ENGINEERING DESIGN PARAMETERS

After the chemical engineering design has been completed, the mechanical details which are needed to construct the reactor must be finalised. The design parameters which require evaluation at this stage are:

a) material of construction
b) insulation
c) vessel dimensions
d) wall thickness
e) ends (type, thickness)
f) openings
g) joints (type, efficiency)
h) pressure relief valves and bursting disks
i) supports
j) foundations

Engineering drawings can only be produced after all the relevant details have been finalised, and these drawings can then be used to estimate the cost.

10.9 MECHANICAL ENGINEERING DESIGN METHODS

10.9.1 Shell Design

The reactor shell, ends, joints and openings were designed in accordance with Australian Standard AS1210: Unfired Pressure Vessels, in order to withstand the effects of an internal explosion creating a pressure rise of not greater than 1800 kPa. Openings required for process flow were sized from estimates of the economic pipe diameter. Two large manholes were incorporated into the design to allow regular renewal of the catalyst. One manhole was located below the bottom tube sheet to allow removal of the old catalyst, and one was located above the top tube sheet to allow new catalyst to be added.

The Australian Standard for Unfired Pressure Vessels (AS1210) lists and explains the equations needed to complete a satisfactory pressure vessel design. No further details are included here. The minimum thickness of the tube plate was calculated from a modified form of the flat-end equation which appears in BS5500. This equation is given here, and includes an estimate of 75% ligament efficiency, based on a standard square-pitch tube pattern:

$$t = C \times D \times (\Delta P/f\lambda) \qquad (10.14)$$

where t = minimum tube plate thickness; C = design constant (= 0.45 for welded plate); D = plate diameter; ΔP = differential pressure (shell to tube); f = tensile strength of the plate material; λ = ligament efficiency.

The dimensions of the openings required for process streams were based on estimates of the economic pipe diameter using equation (10.11), above. A bursting disk for the reactor and a pressure relief valve were specified from AS1210.

Key Results

- AS1210 (Australian Standard for Unfired Pressure Vessels) was the primary design code used to determine the vessel dimensions.
- The pressure relief valves and bursting disks were designed using the methods and equations specified in AS1210.
- Openings and nozzles were sized for entry and exit process streams, and manholes to allow the catalyst to be replaced when necessary.

10.9.2 Supports and Foundations

The supports and foundations were designed to meet the mechanical require-ments under three conditions: empty vessel; hydraulic testing of vessel; and vessel in operation. The maximum weight was used to determine the cross-sectional area required for the skirt and footing. The overturning moment, due to self-load and wind force, was calculated in order to determine the number and size of anchor bolts and the reinforcement requirements.

The vessel supports were also designed in accordance with AS1210, Section 3.24. Equation (10.15), below, was used to calculate the wind force moment, and equation (10.16) for the reinforcing requirements. Wind forces are assumed to predominate over earthquake forces for the selected site at Kemerton.

$$M_w = 0.66D \times P \times (h^2/2) \tag{10.15}$$

$$A_r = M_w/(a \times d) \tag{10.16}$$

where M_w = moment due to wind-force (N mm); D = diameter of vessel (mm); P = pressure of wind-force (Pa); h = height of vessel (mm); A_r = cross-sectional area of reinforcing (mm^2); a = characteristic material constant ($= 130.6$); d = depth of footing (m).

The dimensions of the base-plate ring were calculated from equations by Scheiman, in Sinott (1993):

$$L_b = F_b/f_c \tag{10.17}$$

$$t_b = L_r \times (3 \times f_c/f_r)^{0.5} \tag{10.18}$$

where L_b = width of the base ring (mm); F_b = compressive load on the base ring (MPa); f_c = allowable bearing pressure on the concrete foundation (MPa); t_b = minimum thickness of the base ring (mm); L_r = distance from the skirt to the outer edge of the base ring (mm); f_r = allowable bearing pressure in the base ring material (MPa).

Key Results

- The supports and foundations were designed using the approximate methods of Brown (1966, 1974) and Schwartz (1983) and checked against regulations given in AS1210.
- Three vessel conditions were considered so that the design is adequate for all possible operating scenarios. Both compressive stresses and overturning moments were considered.

10.10 MATERIALS OF CONSTRUCTION

The material selection for the reactor was based on six factors: resistance to crude phthalic and maleic anhydride vapours; resistance to Hitec (the heat transfer medium); operating temperature; mechanical strength; ease of fabrication; and cost. The shell is susceptible to attack from the cooling salt only, while the tubes are in direct contact with both the cooling medium and the reaction gases.

The preferred construction material for the reactor is ASTM A387 (5% Cr, 0.5% Mo) low alloy steel, which is described by the materials specification given in Table 10.5. Phthalic and maleic anhydride vapours are not particularly corrosive to most steels. The average corrosion rates are generally less than 0.05 mm/year. The cooling salt, Hitec, is strongly oxidising at high temperatures and a significant chromium content is required to provide adequate resistance. The addition of chromium also improves the mechanical properties at high temperature. Several stainless steels, notably type 316, satisfy all the material requirements. However, A387 is substantially cheaper and can be used with little penalty to the corrosion rate. At high pressures (and, hence, large wall thicknesses), cladding is normally recommended in order to reduce the vessel cost when alloy steels are used. However, the operating temperature for the reactor precludes the use of carbon steel in this case, and unclad A387 will be used.

Literature reports (e.g. Rabald, 1968; De Renzo, 1985) concerning the corrosivity of crude phthalic anhydride vapours (as found within reactors) were often found to be contradictory. Recommendations range from plain steel boiler tubes, to common stainless steels, to 60% nickel alloys! A material should not be specified without specific testing under all operating conditions. An alternative, more conservative specification for the material of construction is stainless steel type 316 (18Cr, 8Ni, 2.5Mo).

Table 10.5 Material specification for the reactor shell.

ASTM Code	A387: Pressure vessel plates, alloy steel, chromium–molybdenum
General description	Low-alloy chromium steel
Grade and class	A387-5-2
Composition	0.15% C, 0.25–0.66% Mn, 0.030% P, 0.030% S, 1.00% maximum Si, 4.0–6.0% Cr and 0.45–0.65% Mo
Tensile strength	515–690 MPa
Yield strength	310 MPa
Heat treatment	Tempering at 705°C

Key Results

- The choice of construction material was based on chemical resistance, mechanical properties at the operating temperatures, and overall cost of construction.
- A low alloy steel containing 5% Cr and 0.5% Mo (ASTM A387) is sufficient for both the mechanical and chemical requirements of the reactor.
- Corrosion testing is required for the low alloy steel, A387. After results are available it may be necessary to specify a stainless steel such as type 316.

10.11 PRESSURE VESSEL DESIGN (AS1210)

The normal operating pressure for the reactor is 200 kPa, up to a maximum of 220 kPa. However, the consequences of a gas leak due to over-pressure are such that the reactor must be capable of containing small explosions which increase the internal pressure by up to 1800 kPa. Therefore, the reactor will be designed for a pressure of 2200 kPa which is equivalent to the maximum operating pressure plus the explosion allowance plus a 10% safety margin.

ASTM A387 is a group D2 material under the AS1210 classification. This material grouping and vessel diameter of (at least) 5.23 m require Class 1 construction. Joints will be welded with a minimum efficiency of 90% and must undergo a 100% radiographic or ultrasonic examination. The longitudinal joints and all joints within or connecting the vessel ends are to be double-U, double-butt welds (joint efficiency of 100%). All of the other joints will be single-V, single-butt welds with a backing strip that remains in service (joint efficiency of 90%).

At the minimum reactor diameter, the required wall thickness is 54.3 mm (including a corrosion allowance of 0.1 mm/year for 20 years). The next nominal size above this thickness is 63.5 mm (nominal thickness: 2.5″) and this size was selected for construction. This wall thickness and reactor diameter result in a maximum allowable working pressure (MAWP) of 2490 kPa.

At high pressure and for vessels with large diameter to height ratios, ellipsoidal ends are preferred. To minimise the overall vessel height and reduce the amount of material required, 2.5 : 1 ends are specified. The minimum required thickness is 73.8 mm (including a corrosion allowance of 0.05 mm/year for 20 years). The specified material is 76.2 mm plate (nominal thickness: 3″). The MAWP for the ends, and also the overall MAWP, is 2220 kPa.

The minimum, unstayed tube plate thickness was calculated to be 320 mm (using AS1210). This thickness was considered to be unacceptable due to its weight and the difficulty associated with machining such a large item. Subject to a full mechanical analysis, a 100 mm tube plate with 150 × 150 mm RSJ supports on a hexagonal matrix, was specified. This reduces the cross-sectional area available to be used for reaction tubes by 11%. Consequently, the reactor diameter must be increased to 5.50 m from the calculated minimum diameter of 5.23 m.

The increased reactor diameter necessitates a revision to the calculated MAWP for both the vessel walls and the vessel ends. The new MAWP for the vessel walls is 2370 kPa. This pressure is still above the design pressure, and no changes are required to the specification. The new MAWP for the dished ends is 2110 kPa, which is below the design pressure. The end thickness must be increased to the next nominal size ($3\frac{1}{2}''$). A plate thickness of 88.9 mm was specified which results in a MAWP of 2470 kPa. The overall MAWP of the revised design is 2370 kPa. Calculations of the vessel wall and end thicknesses are included in Appendix B, and summarised in Table 10.6.

The reactor baffles will be subject to impact from the flow of the high density HTS. To counteract the potential damage resulting from extended reactor operation and to provide stability given the wide spans, each baffle will be 12.5 mm thick. The baffles will be held in place by welds to 18 supporting rods which will be connected to the tube plate, as shown in Figure 10.7.

The process stream connections were sized on the basis of selecting the nearest pipe size to the economic pipe diameter as recommended by Harker (1978). The inlet and outlet pipes for gas streams will be 1000 mm (nominal size 40") ducting. The openings for the coolant are to be 600 mm (nominal size 24") low-alloy chrome steel pipe. Two elliptical manholes, 450×400 mm, will be installed (one in each end)

Table 10.6 Plate thicknesses for the reactor.

Section	Thickness	Nominal Size
Vessel walls	63 mm	$2\frac{1}{2}''$
Ends	89 mm	$3\frac{1}{2}''$
Tube plate	100 mm	$4''$

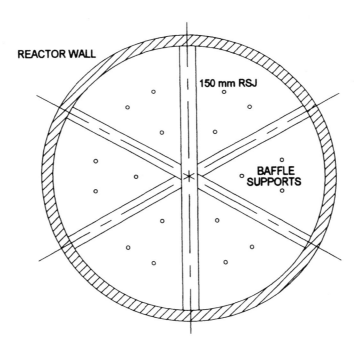

Figure 10.7 Reactor tube plate layout.

Table 10.7 Openings and reinforcement requirements.

Duty	Size	Reinforcement
Gas streams	1000 mm Ø	63,750 mm^2
Coolant streams	600 mm Ø	38,350 mm^2
Manholes	2 × 450 mm × 400 mm	28,600 mm^2 each

to allow access for catalyst charging and removal. Reinforcement, in the form of increased thickness, will be added around each opening. The sizes of the various openings and the reinforcement requirements are summarised in Table 10.7, while the relevant calculations are included in Appendix B.

The vessel will also be fitted with two protective devices: a pressure relief valve which discharges at 2000 kPa, and a bursting disk which discharges at 2150 kPa. The bursting disk requires a discharge capacity of 120 Nm^3/min which corresponds to a discharge area of 361 mm^2. A 50 mm diameter disk was specified. This provides a significant over-capacity but also provides protection against a possible malfunction due to blockage. Calculations are also included in Appendix B.

Key Results

- The design pressure for the reactor is 2200 kPa, including an 1800 kPa allowance for internal explosions.
- Class 1 construction is required according to AS1210. The joints will be either double-butt welds (longitudinal, ends) or single-butt welds with a backing strip remaining in service (all other joints).
- The wall thickness will be 63 mm for vertical sections and 89 mm for the 2.5 : 1 ellipsoidal ends. The maximum allowable working pressure with these thicknesses is 2260 kPa.
- The tube plate will be 100 mm thick and will be supported by 150 × 150 mm RSJs on a hexagonal matrix.
- The vessel openings will be 1000 mm ducting for the gas connections and 600 mm steel pipe for the coolant connections. The manholes will be elliptical, 450 × 400 mm, and should be installed in the top and bottom ends of the vessel.
- Two safety devices will be installed: a pressure relief valve which discharges at 2000 kPa, and a bursting disk which discharges at 2150 kPa.

10.12 INSULATION

Insulation was selected based on the type and size of vessel, the operating temperatures, safety considerations, and the insulation cost. The optimum thickness was calculated from the thermal characteristics of the insulation, typical cost data (including installation), and the value of energy saved by reducing the heat losses from the vessel.

The most suitable types of thermal insulation for large diameter vertical vessels are flexible blankets and rigid blocks (MICA, 1979). The reactor operating temperatures restrict the choice of materials to calcium silicate, expanded silica (or pearlite) and mineral fibre. Mineral wool blankets are widely available and economically attractive. An average price, including installation and a galvanised covering, was estimated at $500/m³ for a 100 mm thickness. Other materials would also be acceptable and are comparably priced.

The preferred thermal insulation consists of two blankets of mineral wool, each 75 mm thick, covering both the walls and the ends of the vessel. The insulation will extend at least 200 mm down the skirt. Pins, at 750 mm centres vertically and 400 mm centres horizontally, will be used to attach each blanket. A sheet metal jacket is to be used to protect the insulation from environmental and mechanical damage. Generally, rivets will be used for securing the jacket. Over areas where access may be required periodically, sheet metal screws will be used.

The thermal conductivity of the vessel wall and of the mineral wool were used to estimate the energy losses for different thicknesses of insulation. A free-convection heat transfer coefficient of 25 W/m²/°C was estimated for ambient conditions (Kreith and Bohn, 1986). A three year effective life for the insulation at 8000 operating hours/year was assumed for the economic basis. Energy was valued at $0.025 per kWhr (after electrical generation costs are considered).

The energy savings achieved by insulating the vessel are significant compared with the capital cost of insulation, and the insulation could be expected to last at least three years. Therefore, insulation should be viewed as an essential investment. The net economic benefit of insulation is considered for a range of thicknesses in Appendix B. The optimum thickness is around 150 mm. Although the energy savings continue to increase, the additional cost of insulation above this thickness becomes uneconomical. Standard blankets of mineral wool are available in thicknesses of 75 mm. A double blanket is considered to be the best alternative. If the value of energy increases, the optimum thickness will also increase. As an increase in energy value is considered more likely than a decrease, an over-estimate of the optimum insulation thickness is preferred.

Key Results

- Thermal insulation is considered essential due to the high operating temperature. A double blanket (2 × 75 mm) of mineral wool, with a protective sheet metal jacket, is the preferred alternative.
- Potential energy savings were estimated by calculating conduction and convection losses from the vessel wall with varying thicknesses of insulation.
- A suitable insulation material was selected to meet engineering requirements. The economic thickness was calculated from estimated costs of energy and the insulation, and an effective project life cycle (before the insulation must be replaced) of three years.
- The recommended thickness is greater than the current optimum, but the increasing value of energy warrants an over-estimate of the insulation requirement.

10.13 SUPPORTS AND FOUNDATIONS

The reactor supports were designed to withstand maximum wind loads of 1500 Pa (180 km/hr) at a height of 4.5 m. Earthquake loads were considered to be negligible by comparison due to the plant location at Kemerton. The maximum bending moment due to wind is 292 kNm (calculated over 66% of the total projected area). The maximum mass of the vessel, under the conditions of full operational loading, is 349 tonnes.

The reactor will be supported on a cylindrical skirt. The allowable axial stress in the skirt is 52.2 MPa (calculated from the larger of the two limits given in Clause 3.24.3.3 of AS1210). The combined axial stress from the wind-load and vessel mass is only 16.2 MPa (calculated from the same clause in AS1210). Equating the limiting stresses with the maximum operational stress, the minimum thickness of the skirt was determined to be 7.1 mm. A plate thickness of 12.7 mm (nominal thickness: $\frac{1}{2}''$) was selected to provide a corrosion allowance and safety factor. The skirt should be 2.25 m high to allow sufficient clearance below (1500 mm at the centre) for maintenance activities. A flush-weld will be used to join the vessel to the skirt. A 150 mm × 10 mm strap should be used to strengthen the joint.

The skirt is to be mounted on a base plate ring which is 35 mm thick. The base-ring should extend 50 mm outside the centre-line of the skirt support and 100 mm inside the skirt. This area is sufficient for anchor bolts to be placed around the inside circumference of the skirt, and to spread the load of the vessel to prevent over-stressing of the concrete footing. The base plate transmits a compressive stress of 2.65 MPa to the concrete footing. This is well below the maximum allowable stress of approximately 5.0 MPa for medium strength concrete. The layout of the skirt and base-plate is shown in Figure 10.8.

The reactor will be secured by 16 bolts of 25 mm (1'') diameter. Combined, they have a resistance well in excess of the minimum requirement of 150% of the overturning moment. The bolts will be placed inside the skirt to avoid the hazard

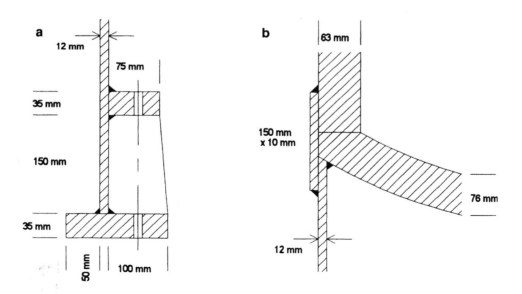

Figure 10.8 Reactor support: (a) joint between skirt and base-plate; (b) joint between skirt and reactor.

of exposed lugs and bolts. Double-gusset bolting chairs should be used at each point of securement.

The footing will be an octagonal concrete pad, 1.3 m deep and 6.8 m diameter, with a mass of 129 tonnes. The total soil bearing load will be 478 tonnes which will exert a compressive stress of 136 kPa. The limiting capacity for the soil which is classified as a medium-strength, sandy soil is about 300 kPa. The footing will be reinforced with a standard mesh of 6 mm rods on a 100 mm square pattern.

The maximum and allowable stresses in the various supports are summarised in Table 10.8. Details of the calculations for the design of the skirt, the base ring and the footing are included in Appendix B.

Key Results

• The reactor will be supported by a cylindrical skirt, 2.25 m high and 12.7 mm thick. At winds loads of up to 180 km/hr (at 4.5 m above grade), the maximum stress in the skirt is 31% of the allowable stress.
• The mass of the vessel under full operational load was estimated to be 349 tonnes. This weight will be transmitted to a concrete footing by a 38 mm base plate, with an area of 1.29 m². The maximum stress exerted on the concrete will be 53% of the allowable maximum.
• The vessel will be secured with 16 × 25 mm Ø bolts, placed inside the skirt.
• The foundation will be a reinforced, octagonal, concrete footing, 1.3 m deep and 6.8 m diameter, and weighing 129 tonnes. The maximum stress transmitted to the soil is 35% of the maximum allowable for sandy soil.

10.14   COSTING

A factored cost estimate for the reactor (accurate to about 20%) was completed using the method of Purohit (1983) for shell and tube heat exchangers (equation (10.19)) and the method of Mulet *et al.* (1981) for pressure vessels (equation (10.20)):

$$b = \frac{6.6}{1 - \exp{[(7-D)/27]}} \times p \times f \times r \times A \qquad (10.19)$$

$$\ln(c) = 8.6 - 0.21651 \times \ln(W) + 0.04576 \times \ln(W)^2 \qquad (10.20)$$

Table 10.8   Stresses in the support structures.

Section	Maximum Working Stress (MPa)	Allowable Stress (MPa)
Skirt	16.2	52.2
Bolts	26.5	125
Concrete pad	2.65	5.00
Soil	0.105	0.300

where b = base cost for heat exchanger (1983 US$); D = shell diameter (inches); A = heat transfer area (square feet); p = cost multiplier for tube layout and pitch; f = cost multiplier for front end; r = cost multiplier for rear end; c = base cost for pressure vessel; W = weight (kg).

Equation (10.19) is strictly only applicable for shell diameters less than 3.1 m. The estimate from equation (10.20) must be modified to account for the reactor internals. Thus, the estimated base cost is associated with a large uncertainty (at least 25%). The data is 8–10 years old and there is an additional uncertainty associated with the escalation to current prices (the Chemical Engineering Cost Index was used). The estimated total delivered cost, in 1997 US dollars, is $1.26 million with an accuracy of ±25%. The breakdown of costs is indicated in Table 10.9. At the prevailing exchange rate, this corresponds to A$1.59 million.

Several other costs must be added to the delivered cost of the reactor: the cost of construction and installation of the salt-cooler; the cost of the heat transfer salt pump; the cost of an initial charge of catalyst; the cost of the inventory of the heat transfer salt; the cost of associated pipes, valves and fittings; the cost of instrumentation and process control equipment (both purchase cost and installation); and, the cost of an inventory of essential spares. Manufacturer's (or vendor's) price quotations are required for each of these items but it is expected that the delivered costs of the two main ancillary capital items, the salt-cooler and the salt pump, would cost A$50,000–A$100,000, and A$300,000–A$400,000, respectively.

Key Results

- The cost of the reactor was estimated from correlations for shell and tube exchangers (Purohit) and pressure vessels (Mulet).
- The estimated total installed cost is US$1.26 million or A$1.59 million (±25%).
- The costs of salt-cooler, heat transfer salt pump, catalyst, coolant, process control equipment and an inventory of spares must be added to the final overall cost.

Table 10.9 Pre-construction cost estimate for the reactor.

	1997 US$
Base cost as a heat exchanger	$388,150
Base cost as a pressure vessel	$177,000
Estimated base cost	$320,000
Cost multipliers	
Tube length	0.24
Design pressure	0.52
Tube material	1.15
Shell material	0.50
Tube gauge	0.07
Sum of cost multipliers	2.48
Ratio of cost indices	1.14
Total cost	US$1,264,000

10.15 ENGINEERING SPECIFICATION

Mechanical Design of Reactor Vessel

Overall Height	9.5 m
Outside Diameter	5.65 m
Total Weight (empty)	241 tonnes
Construction	AS1210, Class 1
Design Pressure	2200 kPa
MAWP	2100 kPa
Wall Thickness	63 mm
Type of Ends	Ellipsoidal, 2.5:1
End Thickness	89 mm
Gas Connections	1000 mm ducting
HTS Connections	600 mm steel pipe
Manholes	2 × 450mm × 400 mm
Bursting Disk	2150 kPa, 50 mm

Supports and Foundations

Type of Support	Cylindrical skirt
Height of Support	2.25 m
Thickness of Support	12.7 mm
Base Ring Thickness	38 mm
Bolts	16 × 25 mm Ø
Shape of Footing	Octagonal
Depth of Footing	1.3 m
Diameter of Footing	6.8 m
Weight of Footing	129 tonnes
Reinforcing	Mesh, 6.4 mm × 100 mm squares

Material Specification

ASTM Code	*A387: Pressure Vessel Plates, Alloy Steel, Chromium–Molybdenum*
General Description	*Low-alloy chromium steel*
Grade and Class	*A387-5-2*
Nominal Composition	*5% Cr, 0.5% Mo*
Tensile Strength	*515–690 MPa*
Yield Strength	*310 MPa*
Heat Treatment	*Tempering at 705°C*

Insulation

Type	*Mineral fibre double blanket*
Thermal Conductivity	*0.036 W/m/°C*
Thickness	*150 mm (2 × 75 mm)*

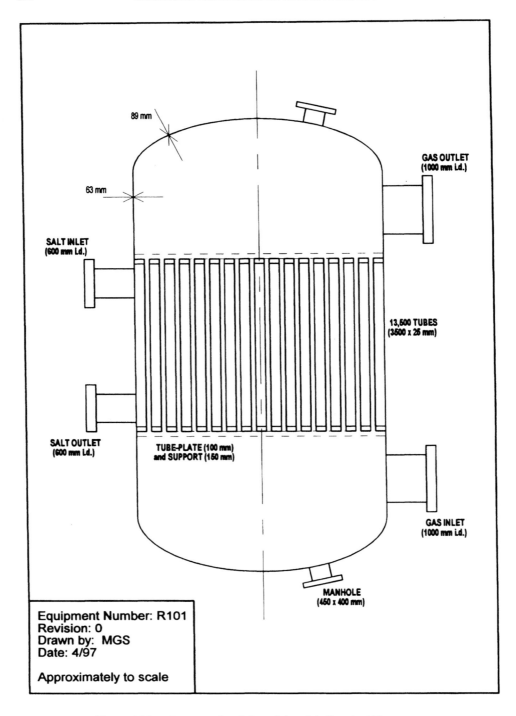

Figure 10.9a Cross-sectional view of the phthalic anhydride reactor.

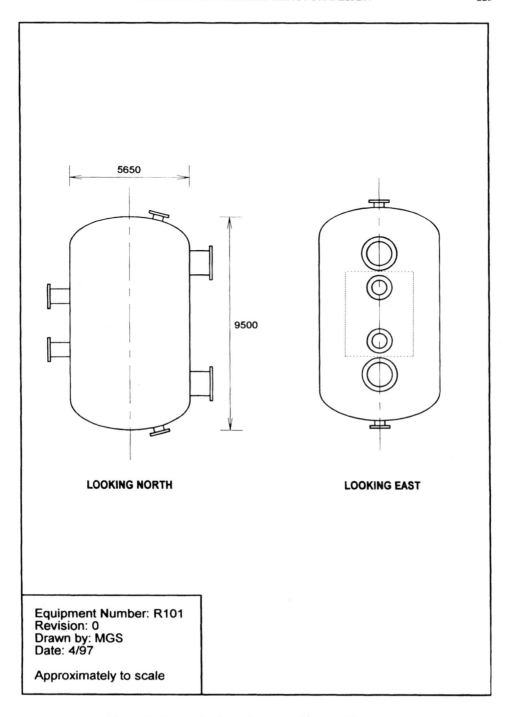

5650

9500

LOOKING NORTH

LOOKING EAST

Equipment Number: R101
Revision: 0
Drawn by: MGS
Date: 4/97

Approximately to scale

Figure 10.9b Profile views of the phthalic anhydride reactor.

PART III: OPERATIONAL CONSIDERATIONS

10.16   HAZOP ANALYSIS

A preliminary HAZOP analysis of the reactor was completed. Full details are included in Appendix D at the end of this chapter. There exists a wide range of possible failures and operating conditions that could lead to hazardous situations. Hopefully, most of these events have been identified and safety measures have been included to counteract them, either in terms of additional equipment (e.g. gravity-feed cooling water tank mounted above the salt-cooler), or the control system which is described in Section 10.21. Any omissions will be identified in the next stage of the design process which should include a formal review by a multi-disciplinary team of professionals.

The process design review team should include representatives from the departments responsible for process operations, maintenance, process engineering and process control. The complete and final (to that point!) P&ID should then be reviewed to incorporate modifications arising from the review process. Once this step has been completed, a formal HAZOP study should be conducted prior to commissioning.

10.17   PROCESS HAZARDS

Both o-xylene and phthalic anhydride form explosive mixtures with air at the concentrations present in the reactor. Therefore, any leak from a line or the vessel would be extremely hazardous. Consequently, all ignition sources will be banned from the site. Regular operator patrols of gas lines will be established to detect leaks before they initiate disasters. A leak of phthalic anhydride would be visible as a white cloud. A leak of o-xylene would be detectable by smell.

The cooling salt, Hitec, is a powerful oxidant and could cause an explosion if it contacts a hydrocarbon or any other organic material. A leak in the reaction tubes would initiate contact between Hitec and hydrocarbon and possibly cause an explosion which might start another reaction between the salt and the metal. Under such conditions, the reactor could be severely damaged. In some cases, an explosion may not result but the salt-metal reaction could destroy many tubes before any damage becomes visible externally. The reactor tubes will be inspected at every shut-down and catalyst changeover.

A cooling system failure could also be the source of a serious hazard. If the reaction is allowed to proceed after cooling has ceased, the temperature will rise quickly with disastrous effects. The partial oxidation reaction (producing phthalic anhydride) will become unfavourable and the complete oxidation, which releases nearly four times as much heat, will predominate. The temperature will then increase uncontrollably which will further increase the risk of tube failure. The salt-metal reaction would then also become likely which would damage the vessel internally.

Clearly, emergency safeguards are required for the cooling system. A gravity-feed water tank could be placed on standby near the salt cooler to provide an emergency supply of coolant in the event of an equipment failure. Two pumps (in parallel) will be installed on the boiler feed-water line and on the salt circuit. Any of the above failures will necessitate an immediate emergency shut-down.

Key Results

- A leak from a connecting line or directly from the reactor would release an explosive mixture to the atmosphere. Therefore, all sources of ignition will be banned from the site and regular operator patrols implemented to provide early leak detection.
- Damaged reactor tubes could cause contact between the cooling salt (a strong oxidant) and hydrocarbon gases, starting an explosion and/or a reaction between the salt and metal. Periodic inspection of the tubes is required.
- A failure in the cooling system could cause the reactor temperature to increase uncontrollably. An emergency cooling system will be installed.

10.18 SAFETY

An operational policy favouring quality rather than quantity will be emphasised by the plant manager. A high-purity product reduces the load on the recovery and purification equipment and reduces the likelihood of serious accidents. The correct operation of all process equipment is central to such a policy. No equipment should be overloaded, except under conditions of strict supervision and, then, only for short periods. Operating at above the rated capacity increases the risk of equipment failure and reduces the life of the equipment. Overloading the catalyst could cause premature deactivation and an increased level of impurities in the product. Cooling equipment will be designed with a 25% over-capacity for abnormal operating conditions. A regular maintenance program is essential and will be implemented for all critical items of equipment.

To make the reactor 'safe' following an abnormal situation, both the hydrocarbon feed and air feed should be reduced or stopped while coolant (HTS) flow is continued. The boiler feed-water flow should be manipulated to control the reactor temperature. The response should be automatic but manual overrides must be present as a contingency for controller failure. Fires may be contained with water, carbon dioxide, dry chemical or foam extinguishers. Operations should be suspended until all vapours have been dispersed and the source of the accident has been detected and fixed.

Key Results

- The operating policy will stress quality and safety.
- All equipment will be operated to minimise the risk of failure. Cooling equipment will have a 25% over-capacity for emergency situations.
- The reactor can be made safe by reducing the hydrocarbon feed and adjusting the boiler feed-water flow in order to control the reactor temperature. The response should be automatic with a manual override backup.

10.19 OPERABILITY

There are three primary requirements for satisfactory operation of the reactor. First, the feed quality must be high. The hydrocarbon should be fully vaporised and contain no impurities that could be catalytically oxidised to form unwanted by-products. Stocks of o-xylene will be checked and tested on arrival to avoid contamination of the storage tank with impure material. Under normal conditions, the air feed is not critical. However, the catalyst life can be extended by occasionally adding small quantities of sulphur dioxide, and this will be done periodically under controlled conditions.

Second, the catalyst should be maintained in good condition. Damage incurred when loading the catalyst could restrict the surface area available for reaction and, thereby, reduce the catalyst life. Poor catalyst loading techniques could also increase the pressure drop through the reactor. If the catalyst is stored for prolonged periods in an unsuitable environment it can become deactivated before it is used. Therefore, correct handling prior to installation is also required.

Third, the process must be controlled correctly. The flow and composition of the feed and the flow of coolant are all critical. An appropriate testing program will be established to ensure that the feed composition is monitored continuously so that suitable operating adjustments can be made to respond to changes in the feed.

Key Results

- The feed quality should be carefully controlled and monitored to restrict the formation of by-products in the reactor.
- Poor storage conditions and/or a poor loading technique of the catalyst could reduce the length of the catalyst cycle (i.e. result in premature deactivation) and increase the pressure drop through the reactor.
- The reactor control system should be carefully monitored to ensure satisfactory operation.

10.20 ENVIRONMENTAL CONSIDERATIONS

The results of the HAZOP study can also be used to assess the potential environmental impact of the reactor operation. During normal operations, the process is fully enclosed and there should be no releases from the reactor to the environment, in either the solid, liquid or vapour phase. The process should also be relatively quiet, except for the noise of the main feed pump, the circulating coolant pump and the feed-water pump to the salt-cooler.

However, a reaction runaway that led to an explosion is clearly a major environmental concern due to the release of toxic organic components (such as PAN and MAN), flammable and explosive components (hydrocarbons such as o-xylene), and highly corrosive components (such as the cooling salt). The reactor is fitted with two pressure relieving devices that should limit the release of organic vapour, but these may be inadequate to handle some explosions. The control system will be required to provide further protection against dangerous operating conditions. Routine thickness testing of the various components of the reactor system will be required during shut-downs to monitor corrosion and ensure that adequate vessel and pipe wall thicknesses remain throughout.

The danger of a phthalic anhydride dust cloud release from the reactor is lower than at other points in the process as the concentration of PAN is always less than 15%. However, explosive mixtures of PAN dust and air could still form after a process breach. Respirators will be provided to allow process operators to safely investigate any suspected process breach before it becomes a major hazard.

Key Results

- There are no emissions from the reactor or associated equipment during normal operation.
- A reaction runaway followed by an explosion is the most serious environmental risk. Pressure relief equipment and an advanced control system provide protection against reaction runaway.
- Regular equipment thickness testing is required to monitor corrosion.
- An explosive PAN dust cloud could form after a process breach.
- Respirators will be provided due to the toxic nature of some components in the reaction gas mixture (especially PAN and MAN).

10.21 CONTROL AND INSTRUMENTATION

Tight process control of the reactor temperature is essential in order to maintain consistent product quality. An advanced control system will be established to regulate the flow of coolant through the reactor shell. The system consists of four control loops which control: (a) the reactor feed rate; (b) the reactor outlet temperature; (c) the coolant supply temperature; and (d) the salt cooler level.

A simple PI controller will be used to control the reactor feed rate. Set-point ramping will be used to smooth changes to the reactor and downstream units. An air-to-open control valve will be used so that the valve will fail shut if the plant instrument air system fails. This will prevent a runaway reaction and return the plant to a safe condition in the event of an emergency.

A feedback–feedforward controller will be used to control the reactor outlet temperature. Feedforward control action will be utilised to detect changes in the reactor feed rate before they impact on the temperature control loop. Feedback control is used so that a temperature set-point can be sustained without offset as the feedforward relationship will be inexact. A cascade loop will also be used to improve the dynamic responsiveness of the loop which is critical to the operation of the reactor and the entire plant.

The peak reactor temperature will be monitored using a selector relay and inputs from at least 15 temperature sensors located within the reactor. If the peak reactor temperature exceeds the process limit, the coolant control valve and the water-supply to the salt-cooler control valve will be fully opened. Both of these valves will be air-to-close (or fail open) control valves.

The coolant supply temperature to the reactor will be controlled via the water flow to the salt-cooler. The salt-cooler pressure controller could be incorporated into a cascade control system to yield tighter control of the salt inlet temperature. However, that type of system has the potential to transmit disturbances to the downstream power generation equipment which is likely to be detrimental to other process areas.

The fourth control loop will also use feedforward–feedback control. This controller will stabilise the water level in the salt-cooler and will manipulate the steam production rate according to changes in the water supply rate. An air-to-close valve will again be used to maximise the intrinsic safety of the process. The overall control scheme for the reactor is shown in Figure 10.10.

Pressure indicators will be installed at the inlet and outlet of the gas streams to provide a continuous record of the pressure drop over the catalyst, and to detect any abnormal pressure buildups. A high-level alarm (i.e. requiring immediate action) will be added to warn of dangerous pressure buildups and initiate emergency shut-down procedures if required. A pressure indicator will also be installed on the salt-cooler to warn of potential PRV discharges (steam venting to atmosphere).

Temperature indicators will be positioned at several positions along the reactor tubes to monitor the temperature profile and the peak temperature. Several sets of temperature indicators will be required as tubes close to the centre will behave slightly differently to those near the outside wall of the reactor. Both the average

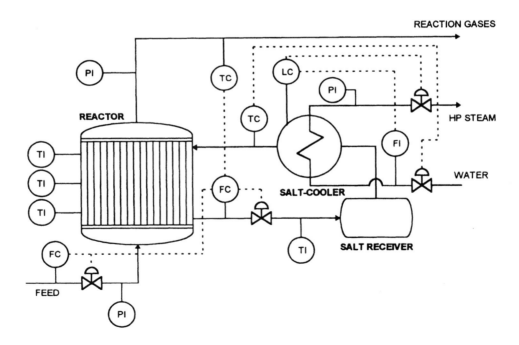

Figure 10.10 Control scheme for the reactor.

(weighted) reactor outlet temperature and the peak reactor temperature will be calculated online.

Temperature indicators will also be installed to monitor the coolant temperatures. High or low salt temperatures could affect the control of the reactor temperature and cause a greater proportion of the reaction to occur at undesirable temperatures, even with appropriate feedforward control action. Low-level alarms (i.e. warning of a change in current operating conditions that might *lead to* a situation which would require immediate action) will be connected to the indicators to provide warning of excessive cooling that may lead to local solidification which could create a blockage. High-level alarms are required to warn of potential salt–metal reactions which can occur at elevated temperatures.

The majority of the alarms which will be installed on the process will require operator intervention to initiate changes to the process. This reduces the risk of a process shut-down being triggered by instrument failure and allows the process operator to contribute their judgement and experience to an emergency response situation. However, the following alarm systems will immediately initiate changes in the process as violation of these conditions indicates that an emergency situation has already occurred:

— very high reactor outlet temperature,
— very high peak reactor outlet temperature,
— very high coolant temperature.

Key Results

• An advanced control system, comprising four main control loops, will be used to control the reactor and salt-cooler.
• Control valves have been specified as either air-to-open or air-to-close in order to maximise the intrinsic safety of the process.
• The average reactor outlet temperature and the peak reactor temperature will both be monitored online. High-level alarms should be installed to warn of the formation of reactor hot-spots.
• Pressure indicators at the reactor inlet and outlet and high-level alarms are required to monitor pressure buildup, and to provide adequate warning of sudden pressure increases.
• The reactor temperature profile will be monitored at several points using a series of temperature sensors and transmitters.
• Cooling salt temperatures will also be monitored to provide warnings of salt solidification at low temperatures and a possible salt–metal reaction at high temperatures.

10.22 OPERATING CONSIDERATIONS

10.22.1 Operation Under Normal Conditions

Air and *o*-xylene (9.5% by weight) enter the reaction tubes at about 205°C and are heated to the reaction temperature by the circulating heat transfer salt

(HTS) which is around 370°C at the inlet. The highly exothermic oxidation reaction starts almost immediately and the reactant temperature rises above the salt temperature within the first 100 mm. The temperature continues to rise (dependent on the temperature difference between the gas and the salt) until the rate of heat dissipation to the salt equals the rate of heat generation from the reaction. When the catalyst is fresh, the reaction stops within about 2 m as the supply of o-xylene is exhausted, and the temperature drops off rapidly after that point. With an older catalyst, the reaction may continue until close to the end of the tubes. Under these conditions, the hot-spot will be reduced in intensity. The residence time in the reactor is only about 1.3 seconds. The product gases contain no o-xylene and about 10.6% (by weight) phthalic anhydride vapour.

A single operator could control the reactor, preheating equipment, salt cooler and gas cooler. Indeed, there are advantages to one person controlling all of those items. The units are all interconnected and a single operator would be best able to respond to changes as they occur, and these will affect all the units together. The aspects of operation that require the most attention, and are critical to the overall operation of the reactor, are the temperature profiles in the reactor and salt. The flow of boiler feed-water through the gas-cooler and salt-cooler must be controlled to ensure that the salt temperature remains essentially constant. This is critical to controlling the reactor temperature. Although an increase in the flow of salt will reduce the reactor temperature, the change will only be small. It will often be necessary to make a small (temporary) reduction in the hydrocarbon feed to the reactor to reduce the reactor temperature significantly.

The control system, if correctly tuned and operated, should be capable of automatically controlling most of the plant. However, the consequences of an accident are such that the operator has a responsibility to oversee all adjustments to the unit. If control of the reactor temperatures is lost, the first action should always be to reduce the hydrocarbon feed rate.

Key Results

- The circulating HTS is initially used to heat the reactants and then to remove the heat of reaction. After reaching a peak temperature (at the hot-spot), the products, containing 10.5% phthalic anhydride cool to the salt temperature.
- One operator should be responsible for control of the reactor, preheaters, salt-cooler and gas-cooler. Particular attention should be given to the temperatures in the reactor.
- The first response to abnormal situations should be to make the reactor safe by reducing the hydrocarbon feed rate.

10.22.2 Commissioning

The reactor must be hydrostatically tested, to prove that it is mechanically sound, before it can enter service. After preliminary checks have been conducted, the catalyst can be loaded into the vessel to complete pre-commissioning. Hot air (250°C or more) enters through the gas inlet and will be used to heat the reactor tubes to around 200°C. At this temperature, HTS can be pumped into the circuit from a heated tanker. The flow of hot air should continue to flow until the reactor temperature approaches 250°C. The next phase of the start-up procedure is to introduce a small quantity of o-xylene into the air stream (about 1%, by weight) to boost the temperatures further via heat liberated from the reaction. The flow of boiler feed-water to the salt cooler should be started simultaneously.

The reactor should be operational within 8–12 hours after the start-up has begun. At this stage, testing can commence. The o-xylene feed rate should be increased slowly (in increments of 0.5% from 1.0% to the normal operating level of 9.5%). The cooling water flow to the salt cooler must be increased simultaneously with each change. Steady-state and satisfactory operating temperatures should be attained at each feed concentration before a further increase is made. The full commissioning process could be expected to take up to a week before satisfactory operating conditions are determined exactly for the first time. Subsequent start-ups which occur after planned shut-downs should only take 24 hours. Table 10.10 summarises the key steps in the commissioning procedure.

Key Results

- The mechanical testing of the reactor and catalyst charging are the main components of pre-commissioning.
- HTS should be introduced after hot air has heated the reactor tubes to around 250°C.
- Hydrocarbon should be introduced slowly to the feed. The boiler feed-water flow should be started as soon as the reactor heats up.
- Testing should be performed at a range of o-xylene concentrations to ensure safe operation at all conditions.

Table 10.10 Reactor commissioning procedure.

Step	Time	Action
Pre-Commissioning		
1		Preliminary mechanical tests.
2		Introduce catalyst.
Commissioning		
3	0:00 hrs	Start hot air flow.
4	1:00 hrs	Introduce HTS once the reactor temperature reaches 200°C. Establish circulation of HTS at 75% of design rate. Start water flow to salt-cooler at minimum.
5	1:30 hrs	Start o-xylene flow (1% w/w only).
6	2:00 hrs	Set all controllers to automatic when reactor outlet temperature reaches 300°C.
7	3:00 hrs	Increase the o-xylene concentration in 1% steps until the design concentration is reached. Initiate sampling program for the reactor outlet gases. Ensure each sample is tested immediately by the on-site laboratory.
8	15:00 hrs	Stable operation at 100% design rate.
9	36:00 hrs	Commence testing program to establish the optimum operating conditions and temperature set-points.

10.22.3 Shut-Down

A minor shut-down can be initiated by reducing the concentration of *o*-xylene in the feed in order to slow the rate of reaction. As temperatures fall, the boiler feed-water flow to the salt cooler can be reduced. The hot air flow should be sustained after the hydrocarbon feed has stopped. The air flow should then be reduced slowly as the rate of circulation of HTS is reduced. When the HTS flow has been reduced to 100–500 T/hr (0.02–0.10 m/s), the HTS pump should be stopped. Natural circulation will continue for a short time. The flow of air should then be slowed until the HTS starts to solidify in the reactor. A small air flow should be maintained until the solidification process is complete and the reactor has cooled satisfactorily.

A major shut-down requires the system to be purged of HTS. This is achieved at a temperature of about 300°C, with hot air (but no *o*-xylene) flowing through the reactor. When the salt is fully removed, the air flow can be reduced as the reactor cools. Care should be taken to ensure that condensation does not occur in the cold reactor. Although water should not affect the catalyst, contaminants (for example, chlorides) in the water may have a detrimental affect. The basic procedure for a major shut-down is given in Table 10.11.

Key Results

- Minor shut-downs can be achieved by reducing the flow of *o*-xylene through the reactor, followed by slow cooling. The HTS can be left in the reactor unless a full inspection is required.
- Major shut-downs require the HTS to be purged from the reactor at about 300°C (i.e. above the solidification point of the HTS).

10.22.4 Start-up

Start-up after a major shut-down should follow the same procedure as that used for commissioning. After a minor shut-down, hot air should be used to melt the HTS stored in the reactor. The HTS pump should be started when the majority of the salt has melted. Then, slowly introduce the *o*-xylene and start the boiler feed-water flow. Reactor product should be recycled until satisfactory operation has

Table 10.11 Major shut-down procedure for the reactor.

Step	Action
1	Reduce *o*-xylene feed rate.
2	Reduce cooling water flow to salt cooler.
3	Stop *o*-xylene flow.
4	Slow HTS flow and stop cooling water flow.
5	Purge system of HTS (at about 300°C).
6	Reduce air flow as reactor cools.
7	Stop air flow.

been attained. Waste gases can be recycled through the recovery equipment or vented if the hydrocarbon concentration is low.

Key Results

- The start-up procedure is similar to the commissioning procedure.
- When HTS remains in the reactor, hot air should be used to melt the salt before other operations commence.

10.22.5 Regular Maintenance

The LAR catalyst will slowly deactivate with continuous use due to the presence of very small amounts of poisons in the feed and the formation of hot-spots caused by non-optimal operation. Once deactivation has started, the reactor temperatures will be lower than for normal operation and, consequently, a greater proportion of by-products will be produced. After 18–24 months of operation, the catalyst will be incapable of fully oxidising the *o*-xylene feed and will need to be replaced.

Figure 10.11 shows the temperature and conversion profiles that might be expected from catalyst that has reached the end of its cycle and needs to be replaced. The data was generated using the FORTRAN program that was used to design the reactor. Relatively widespread deactivation (95% in the first 500 mm, 90% in the next 500 mm and 80% over the remainder of the reactor) of the catalyst was assumed. In an industrial environment, the deactivation profile is likely to be

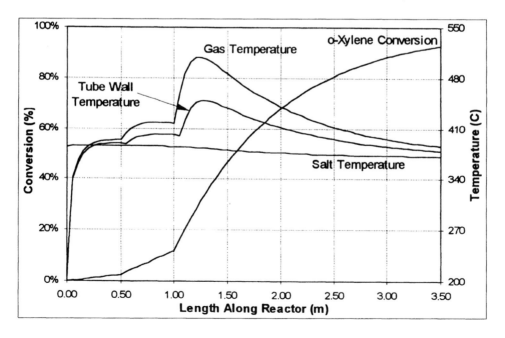

Figure 10.11 Temperature and conversion profiles with deactivated catalyst.

smoother. However, it would be expected to show a similar trend as the catalyst near the reactor entrance will always deactivate preferentially, as it has the initial contact with the catalyst poisons that enter with the reactor feed.

Clearly, the gas and the tube-wall temperatures are much lower than the case with fresh catalyst (see Figure 10.4). The presence of around 3% o-xylene in the reactor product will significantly reduce the final product purity and reduce the effectiveness of the recovery and purification stages of the process. Furthermore, the loss of o-xylene in the waste streams from the plant represents a significant economic penalty which could substantially reduce profitability. This loss of profitability should be calculated and optimised with respect to both the cost of replacing the reactor catalyst and the cost of lost production while the plant is shut-down, in order to determine when a catalyst re-charge is required.

Used (deactivated) catalyst can be removed from the reactor through the bottom of the tubes by removing the gauze support with a specially designed tool. Chutes can be provided to transport the catalyst particles to drums for re-use as inert (or semi-inert) support or for disposal. If the catalyst is to be re-used, its mechanical properties must be checked. Visual inspection would normally be sufficient.

Charging the new catalyst is a difficult process that must be performed quickly (to minimise plant down time), cheaply and in a manner that does not damage the catalyst and, therefore, reduce the yields for the whole of the next cycle. A common rule of thumb is that the particles should not have a free fall of more than 500–1000 mm. The catalyst vendor normally provides specific information concerning the best method of charging. However, two methods are suggested here. The particles can be pre-loaded into flexible socks with drawstring ties, and each sock would contain a measured amount of catalyst for one tube. An operator would pass the sock down a tube, release the tied end by pulling the drawstring, and then withdraw the sock as catalyst particles fall to form the bed. An alternative loading scheme would utilise an air stream (flowing from the base of the reactor through the reactor tubes) to cushion the fall of the catalyst particles as they are loaded from above.

When the reactor is free of catalyst, the mechanical integrity of the tubes should be tested. Initially, the testing can be performed hydrostatically, but further testing may also be required. Tubes that are extensively corroded or damaged should be blocked off and taken out of service to avoid contact between salt and hydrocarbon. The HTS pumps should be serviced regularly to minimise failure during normal plant operation. The reactor insulation should be replaced or repaired after about three years, or whenever reactor energy losses become significant.

Key Results

- The catalyst should be replaced after 18–24 months, or when the yield shows a sustained and significant reduction.
- Used catalyst can be discharged through the bottom of the tubes to storage or disposal.
- New catalyst should be charged through the top of the tubes in a manner recommended by the catalyst proprietor.
- The mechanical integrity of the reactor tubes should be tested during a shut-down to prevent serious accidents from occurring during normal operating periods.

PART IV: CONCLUSIONS, RECOMMENDATIONS AND REFERENCES

10.23   CONCLUSIONS

Chemical Engineering Design

- A tubular fixed-bed catalytic reactor has been designed to produce 5000 kg/hr of phthalic anhydride from o-xylene.
- LAR catalyst (primarily vanadium pentoxide) was chosen as it can process significantly higher hydrocarbon loadings without deactivation.
- A high temperature (salt-based) coolant will be used to dissipate the substantial heat of reaction.
- A computer model of the reactor was written in FORTRAN to determine the reactant conversion and temperature profiles. The model was tested on experimental data and then modified to simulate the LAR catalyst.
- A range of configurations for the reactor were tested to find a set of operating conditions that were able to dissipate the heat of reaction effectively and produce the desired conversion.
- Up to 24 tonne/hr of HP (6000 kPa) steam will be produced from cooling the circulating salt stream in a power-plant evaporator.
- The reaction is highly temperature dependent. At lower temperatures, the reaction is much slower and dependent on the partial pressure of oxygen and the PAN yield is reduced. At higher temperatures, the catalyst can deactivate irreversibly.
- The cooling salt is strongly oxidising and will react violently with any reactive hydrocarbons.
- Shut-down and start-up are made more difficult by the high melting point of the cooling salt (143°C).
- The operating limits for the maximum gas and wall temperatures were established at 550°C and 500°C, respectively.
- Catalyst dilution is required to spread the heat load along the reactor tubes and prevent the formation of hot-spots near the reactor inlet.
- A shell configuration with five segmental baffles satisfies the heat transfer requirements without producing an excessive pressure drop.
- Heavy-duty tubes were specified due to the risk of corrosion.
- A square pitch was selected for ease of cleaning during maintenance.
- The minimum reactor diameter is 5.20 m. The actual reactor diameter may be up to 15% larger than the minimum diameter.
- The salt-cooler tube bundle will contain 239 tubes (25 mm Ø × 1600 mm) on a square pitch. The overall diameter of the bundle will be 600 mm.
- A storage drum for the heat transfer salt is required to hold the circulating salt during shut-downs, and to ensure that an adequate suction head is always present for the salt pump.
- The storage drum will be located above the salt-cooler to conserve ground-level space, and to increase the NPSHA.
- An mixed-flow centrifugal pump was selected for the combination of high flow and moderate head required.

- A customised design will probably be required to deal with the high operating temperature and corrosivity of the heat transfer salt.
- A low rotational speed and a comparatively small impeller diameter were specified to maximise the pump efficiency.
- The pump driver will be an electric motor which is rated significantly higher than the estimated pump power draw in order to provide flexibility and to compensate for uncertainty in the high efficiency estimate.
- The final pump specification must be confirmed with the pump manufacturer and vendor after a specific pump has been selected or designed.
- Double-surface mechanical seals were specified due to the extreme operating environment and hazards of the hot heat transfer salt.
- An insulated cover will be installed to reduce heat losses from the pump.

Mechanical Engineering Design

- The reactor vessel shell was designed to withstand an internal pressure of 1800 kPa that could occur due to an internal explosion.
- AS1210 (Australian Standard for Unfired Pressure Vessels) was the primary design code used to determine the vessel dimensions.
- The pressure relief valves and bursting disks were designed using the methods and equations specified in AS1210.
- Two safety devices will be installed: a pressure relief valve which discharges at 2000 kPa, and a bursting disk which discharges at 2150 kPa.
- Openings and nozzles were sized for the flow rates of the entry and exit process streams, and for manholes to allow the catalyst to be replaced when necessary.
- The supports and foundations were designed using the approximate methods of Brown (1966, 1974) and Schwartz (1983) and checked against regulations given in AS1210.
- The choice of construction material was based on chemical resistance, mechanical properties at the operating temperatures, and overall cost of construction. A low-alloy steel containing 5% Cr and 0.5% Mo (ASTM A387) is sufficient for both the mechanical and chemical requirements of the reactor.
- Class 1 construction is required according to AS1210. All joints will be either double-butt welds (longitudinal joints and end joints) or single-butt welds with a backing strip remaining in service.
- The wall thickness will be 63 mm for vertical sections and 76 mm for the 2.5 : 1 ellipsoidal ends. The maximum allowable working pressure with these thicknesses is 2260 kPa.
- The tube plate will be 100 mm thick and will be supported by 150 × 150 mm RSJs on a hexagonal matrix.
- The vessel openings will be 1000 mm ducting for the gas connections and 600 mm steel pipe for the coolant connections. The manholes will be elliptical, 450 × 400 mm, and should be installed in the top and bottom ends of the vessel.
- Thermal insulation is considered essential due to the high operating temperature. A double blanket (2 × 75 mm) of mineral wool, with a protective sheet metal jacket, is the preferred option.

- The reactor will be supported by a cylindrical skirt, 2.25 m high and 12.7 mm thick. At winds loads of up to 80 km/hr (at 4.5 m above grade), the maximum stress in the skirt is 34% of the allowable stress.
- The weight of the vessel under full operational load was estimated to be 349 tonnes. This weight will be transmitted to the foundation by a 35 mm base plate, with an area of 2.30 m². The maximum stress exerted on the foundation will be 29% of the allowable maximum.
- The vessel will be secured by 16 × 12 mm Ø bolts placed inside the skirt.
- The foundation will be a reinforced, octagonal, concrete footing (1.3 m deep and 8.5 m diameter) and weighing 201 tonnes. The maximum stress transmitted to the soil is 28% of the maximum allowable for sandy soil.
- The estimated total installed cost of the reactor is US$1.26 million ± 25%.
- The costs of catalyst, coolant, process control equipment, and an inventory of spares must be added to the final overall cost.

Operational Considerations

- A leak from a connecting line or directly from the reactor would release an explosive mixture to the atmosphere. Therefore, all sources of ignition will be banned from the site and regular operator patrols implemented to provide early leak detection.
- Damaged reactor tubes could cause contact between the cooling salt (a strong oxidant) and hydrocarbon gases, starting an explosion and/or a reaction between the salt and metal. Periodic inspection of the tubes is required.
- A failure in the cooling system could cause an uncontrollable increase in the reactor temperature. An emergency cooling system will be installed.
- The operating policy will stress quality and safety.
- All equipment will be operated in order to minimise the risk of failure. Cooling equipment will have a 25% over-capacity for emergency situations.
- The reactor can be returned to a safe condition by reducing the hydrocarbon feed, and adjusting the boiler feed-water flow in order to control the reactor temperature. The response will be automatic with a manual override backup.
- The feed quality should be carefully controlled and monitored to restrict the formation of by-products.
- Poor storage conditions and/or a poor loading technique of the catalyst could reduce the catalyst life, and increase the pressure drop through the reactor.
- There are no emissions from the reactor or associated equipment during normal operation.
- A reaction runaway followed by an explosion is the most serious environmental risk. Pressure relief equipment and an advanced control system provide protection against reaction runaway.
- Regular thickness testing is required to monitor corrosion.
- An explosive PAN dust cloud could form after a process breach.
- Respirators will be provided due to the toxic nature of some components in the reaction gas mixture (especially PAN and MAN).

- The reactor control system will be designed to provide tight control of key process variables, and it will utilise a range of advanced control techniques.
- Control valves have been specified as either air-to-open or air-to-close to maximise the intrinsic safety of the process.
- The average reactor outlet temperature and the peak reactor temperature will both be monitored online. High-level alarms should be connected in order to warn of the formation of reactor hot-spots.
- Pressure indicators at the reactor inlet and outlet, and high-level alarms, are required to monitor pressure buildup and to provide adequate warning of sudden pressure increases.
- The reactor temperature profile will be monitored at several points using a series of temperature sensors and transmitters.
- Cooling salt temperatures will be monitored to provide a warning of salt solidification at low temperatures and a possible salt–metal reaction at high temperatures.
- During normal operation, the circulating cooling salt is initially used to heat the reactants and then to remove the heat of reaction. After reaching a peak temperature (at the hot-spot), the products, containing 10.5% phthalic anhydride cool to the salt temperature.
- One operator should be responsible for the control of the reactor, preheaters, salt-cooler and gas-cooler. Particular attention should be given to the temperatures in the reactor.
- The first response to abnormal situations should be to make the reactor safe by reducing the hydrocarbon feed rate.
- The mechanical testing of the reactor, and the catalyst charging, are the main components of pre-commissioning.
- During commissioning, the cooling salt should only be introduced after hot air has heated the reactor tubes to around 250°C.
- Hydrocarbon should be introduced slowly into the feed. The boiler feed-water flow should be started as soon as the reactor heats up.
- Testing should be performed over a range of o-xylene concentrations to ensure safe operation at all conditions.
- Minor shut-downs can be achieved by reducing the flow of o-xylene through the reactor, followed by slow cooling. The cooling salt can be left in the reactor unless a full inspection is required.
- Major shut-downs require the cooling salt to be purged from the reactor at a temperature above its melting point (143°C).
- The catalyst should be replaced after 18–24 months, or when the yield shows a sustained and significant reduction.
- Used catalyst can be discharged through the bottom of the tubes to storage or for disposal.
- New catalyst should be charged through the top of the tubes in a manner recommended by the catalyst proprietor.
- The mechanical integrity of the reactor tubes should be tested during a shut-down to prevent serious accidents from occurring during normal operating periods.

10.24 RECOMMENDATIONS

The reactor design presented here is full and complete. It does not require any additional design effort but the following list of recommendations indicate the areas of greatest uncertainty and suggests where additional resources, if available, might best be allocated:

- Laboratory tests should be conducted to determine interaction parameters for VLE models, especially the affinity between PAN and water.
- Laboratory tests should be conducted to determine the reaction product composition distribution at varying temperatures with both new and old catalyst.
- The effect of feed composition on the reactor product composition should be investigated.

This design is NOT sufficient for construction of a phthalic anhydride reactor. The following items and tasks must be completed before construction begins:

- Isometric drawings of the reactor, salt-cooler and all connecting equipment and pipes.
- Installed costs must be confirmed with suitable vendors.
- Contracts with a construction company(s) must be negotiated and signed.
- The requisite funds for equipment construction, installation and commissioning must be secured.
- Relevant government approvals for construction must be obtained.
- The necessary engineering, operating and maintenance staff must be hired.

10.25 REFERENCES

ASTM Standards, *A387: Pressure Vessel Plates, Alloy Steel, Chromium–Molybdenum*, (1989).

ASTM Standards, *A354: Quenched and Tempered Alloy Steel Bolts, Studs and Other Externally Threaded Fasteners*, (1989).

Blachschmitt, K., Reuter, P., Wirth, F., Buerger, M. and Seubert, R., *United States Patent No. 4,036,783, Supported Catalyst Containing Vanadium and Titanium* (1977).

Brown, A.A., Tank Foundation Design, *Hydrocarbon Process.*, October, 153–156 (1974).

Calderbank, P.H., Kinetics and Yields in the Catalytic Oxidation of o-Xylene to Phthalic Anhydride with V_2O_5, Catalysts, in *Chemical Reaction Engineering — II*, American Chemistry Society, Washington (1974).

Calderbank, P.H., The Prediction of the Performance of Packed Bed Catalytic Reactors in the Air Oxidation of o-Xylene, *Chem. Eng. Sci.*, 32, 1435–1443 (1977).

Coulson, J.M. and Richardson, J.F., *Chemical Engineering Volume 2* (4th ed.), Pergamon, New York (1991).

DeRenzo, D.J., *Corrosion Resistant Materials Handbook* (4th ed.), Noyes Data Corp, New Jersey (1985).

Daugherty, R.L., Franzini, J.B. and Finnemore, E.J., *Fluid Mechanics with Engineering Applications* (SI ed.), McGraw-Hill, New York (1989).

Evans, L.S., *Selecting Engineering Materials for Chemical and Process Plant*, Wiley, New York (1974).

Everett, A., *Materials*, Batsford, London (1981).

Graham, J.J., The Fluidized Bed Phthalic Anhydride Process, *Chem. Eng. Prog.*, 66(9), 54–58 (1970).

Harker, J.H., Finding an Economic Pipe Diameter, *Hydrocarbon Process.*, March, 74–76 (1978).

Kern, D.Q., *Process Heat Transfer*, McGraw-Hill, New York (1950).

Kreith, F. and Bohn, M.S., *Principles of Heat Transfer* (4th ed.), Harper & Row, New York (1986).

McCabe, W.L., Smith, J.C. and Harriott, P., *Unit Operations of Chemical Engineering* (4th ed.), McGraw-Hill, New York (1985).

McKetta, J.J. (ed.), Phthalic Anhydride, in (McKetta) *Encyclopedia of Chemical Processes and Design*, Marcel Dekker, New York (1990).

MICA, *Commercial and Industrial Insulation Standards* (1979).

Mulet, A., Corripio, A.B. and Evans, L.B., Estimate Costs of Pressure Vessels via Correlations, *Chem. Eng. (N.Y.)*, 5 October, 145–150 (1981).

National Association of Corrosion Engineers, *Corrosion Data Survey* (6th ed.), NACE, Houston (1974).

Neerken, R.F., Selecting the Right Pump, *Chem. Eng. (N.Y.)*, 3 April, 122–128 (1978).

Nikolov, V.A. and Anastasov, A.I., A Study of Coolant Temperature in an Industrial Reactor for *o*-Xylene Oxidation, *AIChE J.*, 35(3), 511–513 (1989).

Nikolov, V.A., Anastasov, A.I. and Elenkov, D., Oxidation of *o*-Xylene into Phthalic Anhydride with Two Fixed Beds of Vanadium–Titania Catalyst, *Chem. Eng. Prog.*, 25(3), 127–132 (1989).

Perry, R.H. and Green, D. (eds), *Perry's Chemical Engineers' Handbook* (6th ed.), Chapters 4,10,11, McGraw-Hill, New York (1984).

Peters, M.S. and Timmerhaus, K.D., *Plant Design and Economics for Chemical Engineers* (4th ed.), McGraw-Hill, New York (1991).

Purohit, G.P., Estimating Costs of Shell-and-Tube Heat Exchangers, *Chem. Eng. (N.Y.)*, 22 August, 56–67 (1983).

Rabald, E., *Corrosion Guide* (2nd ed.), Elsevier, Amsterdam (1968).

Schwab, R.F. and Doyle, W.H., Hazards in Phthalic Anhydride Plants, *Chem. Eng. Prog.*, 66(9), 49–53 (1970).

Schwartz, M., Supports for Process Vessels and Storage Equipment, *Chem. Eng. (N.Y.)*, 3 September, 119–123 (1983).

Singh, J., Selecting Heat Transfer Fluids for High Temperature Service, *Chem. Eng. Prog.*, 88(11), 53–58 (1981).

Sinnott, R.K., *Chemical Engineering Volume 6* (2nd ed.), Pergamon, New York (1993).

Sittig, M., *Handbook of Catalyst Manufacture*, Noyes Data Corp, New York (1978).

Standards Association of Australia, *AS1210:1989 Unfired Pressure Vessel Code*, Sydney (1989).

Standards Association of Australia, *AS1358:1989 Bursting Disks and Bursting Disk Devices*, Sydney (1989).

Twigg, M.V., *Catalyst Handbook* (2nd ed.), Wolfe Publishing, London (1989).

Verde, L. and Neri, 'A., Make Phthalic Anhydride with Low Air Ratio Process, *Hydrocarbon Process.*, 63(11), 83–85 (1984).

Wainwright, M.S. and Hoffman, T.W., The Oxidation of Ortho-Xylene on Vanadium Pentoxide Catalysts II — The Influence of Catalyst Support Material on Product Distribution, *Can. J. Chem. Eng.*, 55(5), 557–564 (1977).

Walas, S.M., *Chemical Process Equipment: Selection and Design*, Butterworths, Boston (1988).

APPENDIX B: CALCULATIONS FOR PHTHALIC ANHYDRIDE REACTOR DESIGN

B.1 REACTIONS

The phthalic anhydride synthesis reaction is a network of three parallel reactions: the partial oxidation of o-xylene to phthalic anhydride; the partial oxidation of o-xylene to maleic anhydride; and the complete oxidation of o-xylene to carbon dioxide and water. Each reaction is highly exothermic but the complete oxidation reaction releases approximately four times as much heat as the partial oxidation reactions. The stoichiometry of the reactions is:

Reaction (1): $C_8H_{10} + 3O_2 \Rightarrow C_8H_4O_3 + 3\ H_2O$
Reaction (2): $C_8H_{10} + 7O_2 \Rightarrow C_4H_4O_3 + 4\ CO_2 + 3\ H_2O$
Reaction (3): $C_8H_{10} + 10\frac{1}{2}O_2 \Rightarrow 8CO_2 + 5\ H_2O$

The LAR catalyst normally favours the first reaction, but the exothermic reaction is increased at elevated temperatures. The normal distribution of reactions is:

Reaction (1): 78.8%
Reaction (2): 6.5%
Reaction (3): 14.7%

Therefore, the combined reaction can be written as:

$$C_8H_{10} + 4.36O_2 \Rightarrow 0.79C_8H_4O_3 + 0.065C_4H_4O_3 + 1.44CO_2 + 3.29H_2O$$

B.2 DERIVATION OF SIMULATION MODEL EQUATIONS

B.2.1 Conversion Profile (Calderbank, 1974)

Assume first-order reaction kinetics with a rate constant independent of temperature:

$$\frac{dx}{dL} = -k \cdot (1-x) \tag{B.1}$$

$$\Rightarrow x = 1 - \exp(-\alpha L) \tag{B.2}$$

$$\alpha = \frac{\rho_B \cdot S \cdot \bar{M}}{G} \tag{B.3}$$

where ρ_B = bulk density of the catalyst bed = 1500 kg/m^3; S = pre-exponential factor, determined experimentally by Calderbank (1974) = 0.8; $\bar{M}$ = average molecular weight of the feed; G = superficial gas flow rate (kg/m^2/hr).

$$\bar{M} = 0.905 \, (MW_{air}) + 0.095 \, (MW_{ox})$$
$$= 0.905 \, (29) + 0.095 \, (106)$$
$$= 30.9$$

$$G = 49{,}350 \, \text{kg/hr}/13{,}500 \text{ tubes}$$
$$= 3.66 \, \text{kg/hr/tube}$$
$$= 7440 \, \text{kg/m}^2/\text{hr}$$

$$\alpha = 1500 \times 0.8 \times 29.8/7440$$
$$= 5.06$$

Using the conventional catalyst, $\bar{M} = 29.8$, $G = 10{,}170 \, \text{kg/m}^2/\text{hr}$ and $\alpha = 3.52$ (Nikolov and Anastasov, 1989).

B.2.2 Temperature Profiles

Calderbank (1974) derived the energy balance over an interval of a reaction tube:

$$T - T_w = \frac{\beta}{\gamma - \alpha} (e^{-\alpha L} - e^{-\gamma L}) \qquad (B.4)$$

where:

$$\alpha = \frac{\rho_B \cdot S \cdot \bar{M}}{G} \qquad (B.5)$$

$$\beta = \frac{\Delta H_r \cdot y_o \cdot \alpha}{c_p} \qquad (B.6)$$

$$\gamma = \frac{2 \cdot U}{R \cdot G \cdot c_p} \qquad (B.7)$$

T = reaction gas temperature (°C); T_w = tube-wall temperature (°C); ΔH_r = heat released per mole of o-xylene = 13,395 kJ/mol; y_o = mole fraction of o-xylene; c_p = specific heat of reactants = 1.19 kJ/kg/°C; R = tube radius = 0.0127; U = overall heat transfer coefficient ($\approx 70 \, \text{W/m}^2$/°C).

The energy balance over an interval of a reaction tube, including the circulating heat transfer salt, is given by:

$$\int_{T_1}^{T_r} \dot{m} \cdot c_p dT + \dot{m} \cdot y_o \cdot \Delta H_r \cdot (x_2 - x_1) = \int_{T_r}^{T_2} \dot{m} \cdot c_p dT + U \cdot A \cdot \Delta T \qquad (B.8)$$

$$\Rightarrow \dot{m} \cdot y_o \cdot \Delta H_r \cdot (x_2 - x_1) = \dot{m} \cdot c_p \cdot (T_2 - T_1) + U \cdot A \cdot \Delta T \qquad (B.9)$$

where $\dot{m}$ = mass flow = 49,350 kg/hr = 0.0010 kg/s/tube; T = mean tube temperature (°C) $\approx [0.8 T_{gas} + 0.2 T_{wall}]$; A = heat transfer area (m^2); U = overall heat transfer coefficient (W/m^2/°C); $\Delta T = T - T_{salt}$ (determined iteratively).

The energy balance over the same interval can be used to calculate the change in the salt temperature:

$$U \cdot A \cdot \Delta T = \dot{m}_s \cdot c_{ps} \cdot (T_{s1} - T_{s2}) \tag{B.10}$$

where $\dot{m}_s$ = mass flow of salt (kg/s); c_{ps} = specific heat of the heat transfer salt = 1.61 kJ/kg/°C;

B.3   TUBE-SIDE HEAT TRANSFER COEFFICIENT

Kreith and Bohn (1986) provide a correlation for tube-side flow, based on the Nusselt number:

$$Nu = 0.203 \, (Re \cdot Pr)^{0.33} + 0.220 \, Re^{0.8} Pr^{0.4}$$

where:

$$Re = \frac{G \cdot D}{\mu \cdot (1 - \varepsilon)} \tag{B.11}$$

$$= \frac{(2.04)(0.0254)}{(0.334 \times 10^{-6})(1 - 0.405)}$$

$$= 2561$$

$$Pr = \frac{c_p \cdot \mu}{k} \tag{B.12}$$

$$= \frac{(1.204)(0.334 \times 10^{-6})}{0.0499}$$

$$= 0.806$$

$$\Rightarrow Nu = 110$$

$$\Rightarrow h_i = \frac{Nu \cdot k}{D} \tag{B.13}$$

$$= \frac{(110)(0.0499)}{0.0254}$$

$$= 220 \, W/m^2/°C$$

B.4   SHELL-SIDE HEAT TRANSFER COEFFICIENT

The method of Kern (1950) was used to determine the shell-side heat transfer coefficient via the Reynolds number and the Prandtl number:

$$A = \frac{D_i \cdot C' \cdot B}{P_T} \tag{B.14}$$

$$= \frac{(5.50)(0.0400 - 0.0318)(3.5/6)}{0.040}$$

$$= 0.640 \, m^2$$

where A = shell-side cross-flow area (m²); D_i = shell inside diameter (m); C' = clearance between tubes (m); B = baffle spacing (m); P_T = tube-pitch (m).

$$G = \frac{\dot{m}}{A} \tag{B.15}$$

$$= \frac{694.4}{0.640}$$

$$= 1085 \, \text{kg/m}^2\text{/s}$$

$$u = \frac{G}{\rho} \tag{B.16}$$

$$= \frac{1085}{1851}$$

$$= 0.59 \, \text{m/s}$$

$$\text{Pr} = 6.29 \, @ \, 377°C$$

$$\text{Re} = \frac{G \cdot u}{\mu} \tag{B.17}$$

$$= \frac{(1085)(0.59)}{1.997 \times 10^{-3}}$$

$$= 3.21 \times 10^{-5}$$

$$\Rightarrow \quad j_H = 76 \quad \text{(Kern, 1950)}$$

$$\Rightarrow \quad \text{Nu} = j_H \cdot \text{Pr}^{0.33} \tag{B.18}$$

$$= 140$$

$$D_e = 0.99 D_o$$

$$= 31.4 \, \text{mm}$$

where
D_e = effective tube diameter, based on the actual tube diameter and the tube-pitch; D_o = actual tube diameter.

$$\Rightarrow \quad h_o = \frac{\text{Nu} \cdot k}{D_e} \tag{B.19}$$

$$= \frac{(140)(0.515)}{0.0314}$$

$$= 2320 \, \text{W/m}^2\text{/°C}$$

B.5 OVERALL HEAT TRANSFER COEFFICIENT (CLEAN)

The overall, 'clean' heat transfer coefficient is calculated from the heat transfer resistances of: the tube-side heat transfer (h_i); the shell-side heat transfer (h_o);

and conduction through the tube-wall (h_k).

$$h_k = \frac{2k}{d \cdot \ln(d_o/d_i)} \tag{B.20}$$

$$= \frac{2(25)}{0.0286 \cdot \ln(31.8/25.4)}$$

$$= 7370 \, W/m^2/^\circ C$$

$$\frac{1}{U} = \frac{1}{h_i} + \frac{1}{h_o} + \frac{1}{h_k} \tag{B.21}$$

$$= \frac{1}{220} + \frac{1}{2320} + \frac{1}{7370}$$

$$\Rightarrow U = 195 \, W/m^2/^\circ C$$

B.6 TUBE COUNT

Kern (1950) gives a graphical correlation between the tube-bundle diameter and the tube packing efficiency. At a tube-bundle diameter of 600 mm, the packing efficiency is 87% due to wall effects. At a diameter of 1000 mm, the packing efficiency is 95%. Extrapolating to a tube-bundle diameter of approximately 5.0 m, the packing efficiency is at least 99%.

The reactor requires 13,500 tubes which will be arranged on a 39.7 mm square pitch. Therefore, the ultimate capacity is given by:

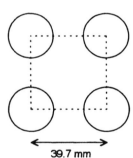

$$1 \, tube/(0.0397 \, m)^2 = 634 \, tubes/m^2$$

$$\text{Minimum bundle area} = (13,500/634)/99\%$$

$$= 21.5 \, m^2$$

$$\Rightarrow \text{Minimum bundle diameter} = 5.23 \, m$$

39.7 mm

B.7 TUBE-SIDE PRESSURE DROP

B.7.1 Catalyst Pressure Drop: Method 1 — Darcy Equation for Packed Beds

$$\frac{\Delta P}{L} = \left(\frac{150 \, (1 - \varepsilon) \cdot \mu}{D_p} + 1.75G \right) \frac{1 - \varepsilon}{\varepsilon} \cdot \frac{G}{D_p \cdot \rho \cdot g} \tag{B.22}$$

where:

$$\varepsilon = \text{porosity of packed bed} = 0.405;$$

$$\mu = \text{gas-phase viscosity} = 3.34 \times 10^{-5} \, Pa \, s;$$

$$D_p = \text{particle diameter} = 6.0\,\text{mm};$$

$$G = \text{mass flux} = 2.04\,\text{kg/m}^2\text{/s};$$

$$\rho = \text{average gas-phase density} = 0.77\,\text{kg/m}^3;$$

$$L = \text{height of packed bed} = 3.5\,\text{m}.$$

$$\Rightarrow \Delta P = 5.8\,\text{kPa}$$

B.7.2 Catalyst Pressure Drop Method 2 — Nomograph (McCabe *et al.*, 1985)

$$G = 2.04\,\text{kg/m}^2\text{/s}$$

$$= 1600\,\text{lb/ft}^2\text{/hr}$$

$$\Rightarrow \Delta P/m = 1.2\,\text{inches H}_2\text{O/ft}$$

$$= 1\,\text{kPa/m}$$

$$\Rightarrow \Delta P = 3.5\,\text{kPa}$$

B.7.3 Total Pressure Drop

The tube-side pressure drop has three components: the catalyst; the gauze and holding grid; and the entrance and exit losses.

Catalyst pressure drop: use 5.8 kPa (more conservative estimate)
Gauze and holding grid: assume 3 kPa/m
  $\Rightarrow \Delta P = 0.3\,\text{kPa}$
Entrance and exit losses: assume 4 velocity heads
  $\Rightarrow \Delta P = 0.1\,\text{kPa}$
Total tube-side pressure drop $= 5.8 + 0.3 + 0.1 = 6.2\,\text{kPa}$

B.8 SHELL-SIDE PRESSURE DROP

Kern (1950) provides a correlation for the shell-side pressure drop in shell-and-tube heat exchangers:

$$\Delta P = \frac{f \cdot G_s^2 \cdot D_s \cdot (N+1)}{2 \cdot \rho \cdot g \cdot D_e \cdot \phi} \tag{B.23}$$

$$= 17.7\,\text{kPa}$$

where f = friction factor (determined graphically from Re) $= 0.0018 \times 144$; G_s = mass flux of salt $= 1530\,\text{kg/m}^2\text{/s}$; D_s = shell inside diameter $= 5.50\,\text{m}$; N = number of baffles $= 5$; ρ = salt density $= 1851\,\text{kg/m}^3$; D_e = effective tube diameter $= 0.0312\,\text{m}$; ϕ = number of shells $= 1$.

B.9 SALT-COOLER DESIGN

B.9.1 Basic Exchanger Configuration

The salt-cooler should be a power-plant evaporator which produces high-pressure steam on the shell-side from superheated water. The required duty is 22.5 MW. A standard tube arrangement should be used: 19.1 mm diameter tubes on a 25.4 mm square pitch. Similarly, a standard tube length of 6.0 m should also be used to minimise the initial capital cost of the exchanger. A tube-side velocity of 2.5 m/s is typical of many installations, and is a compromise between pressure drop, heat transfer and the fouling tendency.

B.9.2 Temperature Driving Force

Salt outlet temperature $= 385°C$

Average salt temperature $= 378°C$

Steam saturation temperature ($@6000$ kPa) $= 276°C$

Temperature driving force, $\Delta T = 102°C$

B.9.3 Heat Transfer Coefficient

The tube-side heat transfer coefficient can be found from:

$$h_i = \frac{k}{D}(0.023 \cdot Re^{0.8} \cdot Pr^{0.4}) \qquad (B.24)$$

$$= 6710 \, W/m^2/°C$$

where $k =$ thermal conductivity of the heat transfer salt $= 0.515$ W/m/°C; $D =$ tube diameter $= 0.0191$ m; $Pr =$ Prandtl number of the heat transfer salt $= 6.29$; $Re =$ tube-side Reynolds number $= 44,030$.

The heat transfer regime on the shell-side is convective boiling and the heat transfer coefficient can be expected to be very high ($>20,000$ W/m²/°C). The uncertainty in this value will have only a limited effect on the accuracy of the overall heat transfer coefficient as the tube-side heat transfer and the fouling factor are limiting.

The overall heat transfer coefficient is calculated from:

$$\frac{1}{U} = \frac{1}{h_i} + \frac{1}{h_o} + R_d \qquad (B.25)$$

$$= \frac{1}{6710} + \frac{1}{20,000} + 0.0035$$

$$\Rightarrow U = 270 \, W/m^2/°C$$

where U = overall heat transfer coefficient (W/m²/°C); R_d = conservative fouling factor = 0.0035 m² °C/W.

B.9.4 Bundle Dimensions

The minimum required heat transfer area is calculated from:

$$A = \frac{Q}{U \cdot \Delta T} \tag{B.26}$$

$$= \frac{22.5 \times 10^6}{(270)(102)}$$

$$= 817 \, \text{m}^2$$

$$\text{Number of tubes} = \frac{A}{\pi \cdot d \cdot L} \tag{B.27}$$

$$= \frac{817}{\pi (0.0191)(6.0)}$$

$$= 2270$$

The tube-bundle diameter can be estimated from the ultimate packing capacity of the tubes by assuming a packing efficiency of 95%:

$$D = \sqrt{\frac{4}{\pi} \left(\frac{N \cdot \text{pitch}^2}{95\%} \right)} \tag{B.28}$$

$$= 1.40 \, \text{m}$$

This diameter is near the maximum allowable tube-bundle diameter (1.52 m by the ASTM Standards) but is acceptable for the salt-cooler. If required, two shells could be used, pending a more rigorous design.

B.10 SHELL DESIGN (AS1210)

B.10.1 Operating and Design Conditions

Temperatures: normal   385°C (maximum salt temperature)
               peak   450°C
          ⇒   tensile strength of ASTM A387, f = 107 MPa

Pressures: normal   220 kPa
               peak   350 kPa
               design   2200 kPa (internal explosion)

Joint Efficiency:
    100% double-welded, double-V butt-joints (longitudinal joints and end joints)
    90% single-welded, single-V butt-joints (circumferential joints)

Test Efficiency:
  100% full radiographic or ultrasonic examination (all joints)

B.10.2 Wall Thickness

The thickness of the cylindrical section must be the larger of the thicknesses calculated to resist circumferential stresses and longitudinal stresses. The minimum thickness required to resist the maximum circumferential stress is given by:

$$t = \frac{P \cdot D}{2 \cdot f \cdot \eta - P} \tag{B.29}$$

$$= 57.1 \, \text{mm}$$

where t = minimum wall thickness (mm); D = shell inside diameter (mm); P = design pressure = 2.2 MPa; f = tensile strength of wall material = 107 MPa; η = joint efficiency = 100%.

The minimum thickness required to resist the maximum longitudinal stress is given by:

$$t = \frac{P \cdot D}{4 \cdot f \cdot \eta - P} \tag{B.30}$$

$$= 31.6 \, \text{mm}$$

where η = 90% for longitudinal joints; and other variables are defined above.

The corrosion allowance for Hitec was 0.1 mm/year for 20 years. Therefore, the minimum wall thickness for the reactor is:

$$t = 57.1 + 20(0.1)$$

$$= 59.1 \, \text{mm}$$

The next larger nominal plate-steel thickness (63.5 mm) will be used to fabricate the reactor shell. The maximum allowable working pressure (MAWP) for this wall thickness is given by:

$$P = \frac{2 \cdot f \cdot \eta \cdot t}{D + t} \tag{B.31}$$

$$= 2.37 \, \text{MPa}$$

where t = (actual thickness − corrosion allowance) = 61.5 mm.

B.10.3 Ends

Torispherical ends are not suitable at the required working pressures. Either ellipsoidal or spherical ends could be used but ellipsoidal ends are preferred to minimise the overall height of the vessel.

The minimum thickness for 2.5 : 1 ellipsoidal ends is given by:

$$t = \frac{P \cdot D \cdot K}{2 \cdot f \cdot \eta - 0.2 \cdot P} \tag{B.32}$$

$$= 77.5\,\text{mm}$$

where $K = 1.37$ for 2.5 : 1 ends, by nomograph from AS1210; and other variables are defined above.

The next larger nominal plate-steel thickness is 88.9 mm, which provides a substantial corrosion allowance. The maximum allowable working pressure (MAWP) for this wall thickness is given by:

$$P = \frac{2 \cdot f \cdot \eta \cdot t}{D \cdot K - 0.2 \cdot t} \tag{B.33}$$

$$= 2.47\,\text{MPa}$$

This pressure is greater than the MAWP for the cylindrical section. Therefore, the overall MAWP is 2.37 MPa.

B.11 TUBE-PLATE DESIGN

B.11.1 General Requirements (AS1210)

Tube pitch: must be more than 25% larger than the tube diameter
 ⇒ pitch/diameter $= 25.4/19.1 = 1.33$ ⇒ OK
Tube holes: machined finish required
Attachment: expansion fitting, followed by beading

B.11.2 Tube-plate Thickness

The minimum required tube-plate thickness is not covered by AS1210. The method recommended in the British Standard BS5500 uses:

$$t = C \cdot D_e \sqrt{\frac{P}{f \cdot \lambda}} \tag{B.34}$$

where $t =$ tube-plate thickness (mm); $C = 0.45$ for welded plate; $D_e =$ tube-plate diameter (mm); $P =$ maximum tube-plate differential pressure $= 1.8\,\text{MPa}$; $f =$ tensile strength $= 125\,\text{MPa}$; $\lambda =$ ligament efficiency $= 0.75$.

At the minimum tube-plate diameter (equal to the minimum shell diameter) of 5230 mm, the required tube-plate thickness is 320 mm. The weight of a tube-sheet of these dimensions virtually precludes its fabrication. A shell-side supporting structure (e.g. several RSJ beams mounted radially) would occlude some tube holes but should allow the overall tube-plate thickness to be reduced. However, a slightly larger tube-plate diameter is required to compensate for the loss of tube area. The recommended tube-plate design consists of a 100 mm tube-plate and

$6 \times 150mm \times 150mm$ RSJ supporting beams. This design is subject to a detailed mechanical analysis.

B.12 VESSEL OPENINGS

B.12.1 Nozzle Diameters

The Harker equation (1978), modified for SI units, for the optimum pipe (and nozzle) diameter is:

$$D_{opt} = 8.41 \frac{W^{0.45}}{\rho^{0.31}} \qquad (B.35)$$

where D_{opt} = pipe diameter (mm); W = mass flow rate (kg/hr); ρ = fluid density (kg/m^3).

The reactor feed flow rate is 49,350 kg/hr and its density is 1.04 kg/m^3. Therefore, the optimum pipe diameter is given by:

$$D_{opt} = 8.41 \frac{(49,350)^{0.45}}{(1.04)^{0.31}}$$

$$= 1074 \, mm$$

The nearest standard pipe size *lower* than the optimum was selected since the construction materials are more expensive than normal carbon steel. Therefore, a 1000 mm diameter nozzle was specified for the reactor feed gas. Similarly, the optimum pipe diameters and selected sizes for the other inlet and outlet streams are given in Table B.1.

B.12.2 Reinforcement (AS1210)

The overall strength of the vessel is slightly reduced by the opening and must be compensated with reinforcing. The required area of reinforcing is given by:

$$A_r = D \times t_s \qquad (B.36)$$

where A = area of reinforcing in any plane (mm^2); D = diameter of the hole plus twice the corrosion allowance (mm); t_s = thickness of the shell (mm). Therefore, the reinforcing area around the reactor feed gas inlet and reactor product gas

Table B.1 Reactor nozzle diameters.

Stream	W	ρ (kg/m^3)	D_{opt} (mm)	D (mm)
Reactor feed gas	49,350 kg/hr	1.04	1074	1000
Reactor product gas	49,350 kg/hr	0.85	1145	1000
Heat transfer salt inlet	2500 T/hr	1856	617	600
Heat transfer salt outlet	2500 T/hr	1846	618	600

outlet is given by:

$$A_r = (1000 + 4) \times 63.5$$

$$= 63,750 \, \text{mm}^2$$

Similarly, the heat transfer salt nozzles require 38,350 mm² of reinforcement.

B.12.3 Manholes

Two manholes are required for inspection of the reactor, and to load and unload the catalyst. The manholes should be located in the dished ends, above and below the tube bundle. Elliptical manholes are preferred for strength in large vessels (AS1210, 1989). The reactor contents have the potential to be toxic and relatively small manholes are preferred (AS1210, 1989). Each manhole requires 28,600 mm² of reinforcing, based on the long axis, using equation (B.36). The manhole covers should be dished, gasketed and fixed at two points (AS1210, 1989).

B.13 PROTECTIVE DEVICES (AS1210)

Pressure relief is required to protect the reactor from damage due to over-pressure. A two-level protective system is preferred. A pressure relief valve (PRV) should operate at 2000 kPa and contain a minor over-pressure. A rupture disk will also be fitted to provide relief for emergency situations where the pressure continues to rise after the PRV has lifted, or if the PRV fails. The rupture disk should be set to fail at a pressure of 2100 kPa. The reactor will not contain liquefied gases and is not subject to any other special consideration. Therefore, the required discharge capacity is determined from the reactor volume only:

$$Q = 0.0018 \times W_c \qquad \text{(B.37)}$$

$$= 119 \, \text{Nm}^3/\text{min}$$

where Q = discharge capacity (Nm³/min); W_c = water capacity (volume) of the reactor = 66,000 L.

The required minimum area for the pressure relieving devices is determined from the required discharge capacity:

$$A = \frac{60 \cdot Q}{\rho} \left(\frac{T \cdot Z}{M}\right)^{0.5} \left(\frac{1}{10 \cdot C \cdot P}\right) \qquad \text{(B.38)}$$

$$= \frac{60(119)}{1.35} \left(\frac{658(1.0)}{42}\right)^{0.5} \left(\frac{1}{10(2.9)(2.000)}\right)$$

$$= 361 \, \text{mm}^2$$

where A = minimum flow area of the pressure relieving device (mm²); Q = discharge capacity (m³/min); ρ = gas density = 1.35 kg/m³; T = maximum operating temperature = 658 K; Z = compressibility factor = 1.0; M = molecular weight of the discharge gas ≈ 42; $C =$ a constant from Tables in AS1210 ≈ 2.9; $P =$ relieving pressure (MPa). A 50 mm diameter was selected for both the PRV and the rupture disk. This provides a large over-capacity (the flow area for a 50 mm diameter valve or rupture disk is 1963 mm^2), and reduces the potential for a blockage which might occur with a smaller diameter.

B.14 INSULATION

The optimum thickness of insulation is determined by balancing the economic value of the overall heat loss from the reactor against the annualised cost of installing an incremental layer of insulation. The heat loss is estimated from equations (B.39) to (B.41):

$$Q = U \times A \times \Delta T \tag{B.39}$$

$$\frac{1}{U} = \frac{1}{h_c} + \frac{1}{h_i} + \frac{1}{h_k} \tag{B.40}$$

$$h_i = \frac{2 \cdot k}{\bar{r} \cdot \ln (d_o/d_i)} \tag{B.41}$$

where $Q =$ overall heat loss (W); $U =$ overall heat transfer coefficient (W/m^2/°C); $A =$ heat transfer area (m^2); $\Delta T =$ (average reactor temperature−ambient temperature) ≈ 360°C; $h_c =$ heat transfer coefficient for conduction through the vessel wall ≈ 400 W/m^2/°C; $h_i =$ heat transfer coefficient for conduction through the insulation; $h_c =$ heat transfer coefficient for free convection from the insulation ≈ 25 W/m^2/°C; $k =$ thermal conductivity of the insulation $= 0.036$ W/m/°C; $\bar{r} =$ average radius of the reactor, including the insulation (m); $d_o =$ outside diameter of the reactor, including the insulation (m); $d_i =$ outside diameter of the reactor, excluding the insulation (m).

Table B.2 indicates the value of the incremental energy saved and the incremental cost of the additional insulation for varying thicknesses of insulation from 25 mm to 250 mm. Heat loss was valued at A\$0.04 per kWh. The insulation cost, including installation, was estimated at A\$500 per m^3, and it was apportioned

Table B.2 Optimum insulation thickness.

Insulation Thickness (mm)	Total Heat Loss (kW)	Incremental Energy Saving (A\$)	Cost of Incremental Insulation (A\$)	Net Benefit of Incremental Insulation (A\$)
0	1423	—	0	—
25	82	1.3 million	2120	1.3 million
50	43	38,210	2160	36,050
75	29	13,250	2190	11,060
100	22	6720	2230	4490
125	18	4070	2270	1800
150	15	2720	2300	420
175	13	1950	2340	−390
200	11	1470	2380	−910

over three years (the estimated life of the insulation before it needs to be replaced). Each incremental addition of insulation, up to a total thickness of 150 mm, provides a net benefit. Any additional insulation incurs a loss as the cost of the insulation is higher than the value of the energy which is saved. Therefore, the optimum thickness of insulation is 150 mm.

B.15 SUPPORTS

B.15.1 Weight of the Reactor

The overall weight of the reactor, and associated fittings and internals, must be determined for three conditions: the empty vessel, the vessel under hydrostatic test conditions; and the vessel under full operational load. The various components of the vessel and their respective weights, assuming a steel density of 7850 kg/m^3, are tabulated in Table B.3.

B.15.2 Skirt (AS1210)

A cylindrical skirt was selected to support the reactor. It will extend 2.25 m from the point it joins the reactor to provide a clearance of 1.5 m below the lowest point of the reactor base. The axial (compressive) stress exerted on the skirt comes from the weight of the vessel and the wind moment, and it can be calculated from:

$$S_a = \frac{1}{t}\left(\frac{4 \cdot M}{\pi \cdot D_o^2} + \frac{W}{\pi \cdot D_o}\right) \tag{B.42}$$

where S_a = axial stress (MPa); M = bending moment (N mm); W = maximum weight of vessel (N); D_o = outside diameter of skirt (mm); t = thickness of skirt (mm).

Table B.3 Reactor weight (kg).

Component	Empty Vessel	Hydrostatic Test	Operating Load
Vessel walls	96,000	96,000	96,000
Tube-plate	21,000	21,000	21,000
Tubes	115,000	115,000	115,000
Baffles	5000	5000	5000
Catalyst			36,000
External fittings	1000	1000	1000
Attached pipes	2500	2500	2500
Insulation	500	500	500
Process (salt + gases)			62,000
Water		108,000	
Live loads			1000
Impact loads			2000
Total	241,000 kg	349,000 kg	342,000 kg

The bending moment is a product of the wind speed and the effective area of the vessel. The wind is assumed to act at the mid-height of the vessel and to have a maximum speed of 180 km/hr (equivalent to a pressure of 1500 Pa).

The effective diameter of the vessel is given by:

$$D_e = D_o + 2t_i + t_p \tag{B.43}$$

$$= 5630 + 2(150) + 600$$

$$= 6530 \, \text{mm}$$

where D_o = outside diameter of the vessel = 5630 mm; t_i = thickness of the insulation = 150 mm; t_p = length of the projected piping $\approx$ 600 mm.

The total height of the vessel is 9.5 m. Therefore, the total wind force is given by:

$$F = 0.66 \times P \times D_e \times h \tag{B.44}$$

$$= 0.66 \times 1500 \times 6.53 \times 9.5$$

$$= 61.4 \, \text{kN}$$

The wind-moment is the product of the force and the mid-height of the reactor:

$$M = F \times (h/2) \tag{B.45}$$

$$= 61,400 \times (9500/2)$$

$$= 292 \times 10^6 \, \text{N mm}$$

There are two limits to the allowable compressive stress, which are functions of the yield strength and the Young's modulus (E) of the construction material, respectively. The pertinent relationships are (B.46) and (B.47):

$$S_a < 0.5 \times f \times \eta \tag{B.46}$$

$$S_a < \frac{0.125 \times E \times t}{D_o} \tag{B.47}$$

where S_a = compressive stress in the skirt support (MPa); f = yield strength of the construction material (plain carbon steel) = 125 MPa; η = joint efficiency = 85%; E = Young's modulus of the construction material = 0.185 $\times$ 10^6 MPa; t = thickness of the skirt support (mm); D_o = reactor outside diameter = 5630 mm.

Substituting equation (B.42) into equations (B.46) and (B.47) produces inequalities (B.48) and (B.49):

$$\frac{1}{t} \left(\frac{4 \cdot M}{\pi \cdot D_o^2} + \frac{W}{\pi \cdot D_o} \right) < 0.5 \times f \times \eta \tag{B.48}$$

$$t > 3.9 \, \text{mm}$$

$$\frac{1}{t} \left(\frac{4 \cdot M}{\pi \cdot D_o^2} + \frac{W}{\pi \cdot D_o} \right) < \frac{0.125 \times E \times t}{D_o} \tag{B.49}$$

$$t > 7.1 \, \text{mm}$$

With the addition of a small corrosion allowance (1.0 mm over 20 years), the minimum thickness of the skirt support is, therefore, 8.1 mm. A plate thickness of 12.7 mm (nominal $\frac{1}{2}''$) was selected. The maximum compressive stress in the reactor skirt was determined from equation (B.42) to be 16.2 MPa. The allowable compressive stress was determined to be 52.2 MPa; the lower of the stresses calculated from equations (B.46) and (B.47).

B.15.3 Base-Ring

The compressive load on the base-ring is given by:

$$F_b = \frac{4 \cdot M}{\pi \cdot D_o^2} + \frac{W}{\pi \cdot D_o} \tag{B.50}$$

$$= 207 \text{ N/m}$$

The minimum width of the base-ring can be determined from the compressive load by:

$$L_b = \frac{F_b}{f_c} \tag{B.51}$$

$$= \frac{207}{5.0}$$

$$= 41 \text{ mm}$$

where L_b = base-ring width (mm); f_c = maximum allowable bearing pressure of the concrete pad = 5 MPa.

The calculated base-ring width is not practical as it would not allow sufficient room for fixing, either inside or outside the skirt. A width of 150 mm was specified, comprising 50 mm outside the centre-line of the skirt and 100 mm inside the centre-line of the skirt. The stress which is transmitted to the concrete pad can be determined from:

$$\sigma = \frac{\text{weight}}{\text{area}} \tag{B.52}$$

$$= \frac{3.49 \times 10^5 \times 9.81}{(\pi/4)(D_o^2 - D_i^2)}$$

$$= 2.65 \text{ MPa}$$

where σ = compressive stress (MPa); D_i = inner diameter of the base-ring = 5400 mm; D_o = outer diameter of the base-ring = 5550 mm.

The base-ring must be thick enough to resist deformation due to the weight of the vessel. The minimum thickness can be determined from:

$$t_b = L_r \times \left(\frac{3 \cdot f_c}{f_r}\right)^{0.5} \tag{B.53}$$

$$= 94 \times \left(\frac{3 \times 5}{135} \right)^{0.5}$$

$$= 31 \, \text{mm}$$

where t_b = thickness of the base-ring (mm); L_r = distance from the edge of the skirt to the outer edge of the ring = $100 - \frac{1}{2}(12.7) = 94 \, \text{mm}$; f_r = tensile strength of the base-plate = 135 MPa. Again, a standard plate thickness was selected: 38.1 mm ($1\frac{1}{2}''$ nominal).

B.15.4 Bolts

The reactor and skirt will be secured with bolts which must support 150% of the maximum overturning moment due to wind-force. A total of 16 bolts will be used. They will be mounted inside the skirt support to eliminate the hazard of protruding bolts. The bolt area required is calculated from the overturning moment by assuming an equal load distribution:

$$A_b = \frac{1}{N_b \times f_b} \left(\frac{4 \times M}{D_b} \right) \tag{B.54}$$

$$= \frac{1}{16 \times 125} \left(\frac{4 \times 2.92 \times 10^8}{5450} \right)$$

$$= 107 \, \text{mm}^2$$

where A_b = area of one bolt at the root of the thread (mm²); N_b = number of bolts; f_b = tensile strength of the bolt material = 125 MPa; M = wind-moment = $2.82 \times 10^8 \, \text{N mm}$ (see equation (B.45)); D_b = diameter of the bolt circle = 5450 mm.

The calculated bolt area corresponds to a bolt diameter of only 11.7 mm. This is below the practical limit, and 25 mm bolts were specified in accordance with Sinnott (1993) which recommends that bolts smaller than 25 mm should not be used for securement.

B.16 FOUNDATION

B.16.1 Concrete Pad

An octagonal concrete pad was considered to be the most appropriate foundation for the reactor. Using rules-of-thumb, the pad will be located 300 mm above grade and 1000 mm below grade (total depth of pad will be 1300 mm). The compressive load on the soil comprises two parts: the load from the reactor; and the load from the foundation. This load is resisted by the soil bearing capacity which was estimated to be 300 kPa (a typical value for the sandy soil which is found in the local area). A 50% safety factor was applied to the estimate of the soil bearing capacity due to the absence of specific geological data measurements. The total compressive load on the soil can be determined from equation (B.55), which can be rearranged to calculate the minimum footing area required (equation (B.56)):

$$\sigma = \frac{W_v}{A_f} + \gamma \tag{B.55}$$

$$A_f = \frac{W_v}{\sigma_{max} - \gamma} \tag{B.56}$$

$$= \frac{3.42}{0.150 - 0.031}$$

$$= 28.7 \, m^2$$

where A_f = minimum pad area (m²); W_v = maximum weight of the reactor = 3.42 MN; γ = specific weight of the concrete footing = 0.031 MPa; σ = compressive load on the soil (MPa); σ_{max} = maximum soil bearing capacity = 0.150 MPa.

The minimum (short) diameter of the octagonal pad can be calculated from its area by:

$$D_f = \left(\frac{9}{8} A_f\right)^{0.5} \tag{B.57}$$

$$= 5.7 \, m$$

The calculated footing diameter is only slightly greater than the outside diameter of the base-ring. A clearance of 600 mm around the outside of the base-ring was considered appropriate, and a footing diameter of 6.8 m was specified. The estimated weight of the footing is 129 tonnes.

Access routes to the reactor could be laid directly onto the surrounding ground. However, some sub-surface strengthening might be required for heavy machinery or heavy vehicles (e.g. a crane to lift the reactor into place).

B.16.2 Reinforcing

The reinforcing requirement can be determined from the maximum overturning moment:

$$A_r = \frac{M}{a \times d} \tag{B.58}$$

$$= \frac{2.92 \times 10^5}{130.6 \times 1.3}$$

$$= 1720 \, mm^2$$

where A_r = cross-sectional area of reinforcing required (mm²); M = overturning moment = 2.92 N m; a = a constant = 130.6 for standard reinforcing; d = depth of footing (m).

Standard reinforcing mesh with 6 mm rods located at 100 mm centres has a cross-sectional area of approximately 40 mm² per 100 mm length, and comfortably satisfies the minimum requirements. The reinforcing should be laid at the *bottom* of the concrete pad.

APPENDIX C: FORTRAN PROGRAM FOR
PHTHALIC ANHYDRIDE REACTOR SIMULATION

```
C      Program to model temperature and composition profiles
C      in catalytic tubular fixed bed reactor for the partial
C      oxidation of ortho-xylene to phthalic anhydride.
C
C-----------------------------------------------------------
C
C      Variable names refer to equations derived in Appendix B.
C
       DOUBLE PRECISION X(0:90), T(0:90), TS(0:90), TW(0:90),
      .LENGTH(0:90), FLOW, S_FLOW,AREA,CP,S_CP,DEL_T(2),TR1,
      .TR2, DENS,MW,G,Y0,RADIUS,TOTLEN,EFFLEN,ALPHA,BETA,
      .GAMMA,FACTOR,S,HC,U,HEAT_R
       INTEGER I,K
C
C      Define output file for results.
       OPEN (FILE="REACTOR.DAT",UNIT=4,STATUS="OLD")
C
C-----------------------------------------------------------
C
C      Physical properties common to both catalysts.
       DENS=1500
       RADIUS=0.0125
       AREA=0.003927
       HEAT_R=13395
       S=0.8
       HC=210/1000
       CP=1.19
       S_CP=1.61
C
C      Physical properties specific to the LAR catalyst.
       FLOW=0.00100
       G=7330
       Y0=0.0952
       MW=30.9
C
C      Physical properties specific to traditional catalysts.
C      FLOW=0.00139
C      G=10170
C      Y0=0.0356
C      MW=29.8
C
C      Define constants.
       ALPHA=DENS*S*MW/G
       BETA=HEAT_R*Y0*ALPHA/CP
       GAMMA=2*3600*HC/(RADIUS*G*CP)
```

```
C
      PRINT*,'  LAR CATALYST'
      PRINT*,'  COUNTER-CURRENT SALT FLOW'
      PRINT*,'  GRADUATED CATALYST ACTIVITY'
      PRINT*,
C
C     Parameters to be inputted by the user.
      PRINT*,'  LENGTH OF TUBULAR REACTOR (m):          '
      READ*,TOTLEN
      PRINT*,'  TOTAL SALT FLOW (T/HR):                 '
      READ*, S_FLOW
      PRINT*,'  HEAT TRANSFER COEFFICIENT (W/m2/C):    '
      READ*,U
      PRINT*,'  CORRECTION FACTOR FOR FLOW ARRANGEMENT: '
      READ*, FACTOR
C
C     Corrections for units.
      S_FLOW=S_FLOW / (13500*3600/1000)
      U=U/1000
C
C     Set conditions at entrance to reactor.
      LENGTH(0)=0
      EFFLEN=0
      X(0)=0
      T(0)=300
      TS(0)=385
      TW(0)=T(0)
C
C     Initialize variables.
      DO 30 J=1,60
      X(J)=0.5
      T(J)=400
      TS(J)=400
      TW(J)=400
   30 CONTINUE
      K=INT(TOTLEN/0.05)+1
C
C-----------------------------------------------------------
C
C     Titles for results output.
      PRINT*,'    ----------------------------------------'
      PRINT*
      PRINT*,'   LENGTH  CONV  TEMP  TEMP-W  TEMP-S'
      PRINT*
      PRINT1,LENGTH(0),X(0),T(0),TW(0),TS(0)
      WRITE(4,1) LENGTH(0),X(0),T(0),TW(0),TS(0)
C
```

```
C       Start calculations at entrance to reactor tubes and
C       integrate in steps of 0.05m.
        DO 10 I = 1, K
        LENGTH(I) = I * 0.05
C
C       Define activity profile for catalyst.
        IF (I .LE. 10) THEN
        EFFLEN = EFFLEN + 0.01
        GAMMA = (2*3600*U/(RADIUS*G*CP)) * 5
        ELSEIF (I .GT. 10 .AND. I .LE. 20) THEN
        EFFLEN = EFFLEN + 0.02
        GAMMA = (2*3600*U/(RADIUS*G*CP)) * 2.5
        ELSE
        EFFLEN = EFFLEN + 0.05
        GAMMA = 2*3600*U/(RADIUS*G*CP)
        ENDIF
C
C       Define activity profile for deactivated catalyst.
C       IF (I .LE. 10) THEN
C       EFFLEN = EFFLEN + 0.01
C       GAMMA = (2*3600*U/(RADIUS*G*CP)) * 100
C       ELSEIF (I .GT. 10 .AND. I .LE. 20) THEN
C       EFFLEN = EFFLEN + 0.02
C       GAMMA = (2*3600*U/(RADIUS*G*CP)) * 25
C       ELSE
C       EFFLEN = EFFLEN + 0.05
C       GAMMA = (2*3600*U/(RADIUS*G*CP))*10
C       ENDIF
C
C       Equation (1)
        X(I) = 1 - EXP( -ALPHA*EFFLEN)
C
C       First estimate for temperature difference.
        DEL_T(1) = 40
C
C       Change thermodynamic properties with temperature.
   20   CP = 1.180 + (X(I)/2)*0.025
        S_CP = 1.610 + 0.00118*(387 TS(I))
C
C       Equation (3)
        TR1 = ((X(I) - X(I-1))*HEAT_R*Y0*FLOW - U*AREA*DEL_
        T(1))/(FLOW*CP)
        T(I) = T(I-1) + TR1
C
C       Equation (2)
        TR2 = (BETA/(GAMMA - ALOG(1 - X(I))))*
       .(EXP( -ALPHA*EFFLEN) - EXP( -GAMMA*EFFLEN))
```

```
      TW(I)=T(I)-TR2
C
C     Equation (4)
      TS(I)=TS(I-1)-(U*AREA*DEL_T(1))/(S_FLOW*S_CP)
C
C     Calculate   temperature   difference   and   iterate   if
      necessary.
      DEL_T(2)=FACTOR * (TW(I)+TW(I-1)-TS(I)-TS(I-1))/2
      IF (DEL_T(2)-DEL_T(1).GT.0.5 .OR. DEL_T(1)-DEL_T(2)
      .GT.0.5) THEN
      DEL_T(1)=DEL_T(2)
      GOTO 20
      ENDIF
C
C     Output results at each interval along the reactor tubes.
      PRINT1,LENGTH(I),X(I),T(I),TW(I),TS(I)
      WRITE(4,1) LENGTH(I),X(I),T(I),TW(I),TS(I)
    1 FORMAT (8X,F4.2,6X,F5.3,6X,F5.1,6X,F5.1,6X,F5.1)
C
   10 CONTINUE
      STOP
      END
```

APPENDIX D: HAZARD AND OPERABILITY STUDY FOR PHTHALIC ANHYDRIDE REACTOR

Table D.1 Reactor feed gases from preheater at 49.65 T/hr and 205°C.

Key Word	Deviation	Possible Causes	Consequences	Action Required
NONE	No flow	1. Flow stopped upstream.	Process stops. HTS gradually cools (no heat input) and may solidify.	a) Ensure continuity of flow upstream and good communications from feed storage.
		2. Catalyst bed blocked.	Pressure build-up in reactor (no hazard until 2100 kPa).	b) Install pressure gauge in reactor. Emergency shutdown if blockage does not clear itself.
		3. Line breakage.	As for 1. Release of explosive mixture to atmosphere.	c) Regular operator patrol of all lines. Emergency shutdown procedures.
MORE OF	More flow	4. Blower operating above rated capacity.	Reaction not complete if catalyst is old. May cause premature deactivation. Increased load on coolant.	Covered by a).
				d) Ensure over-capacity of coolant.
	More temperature	5. Preheater operating above rated capacity.	Small decrease in coolant duty. Inefficient use of energy upstream.	e) Ensure good operating practices upstream. (No hazard.)
	More pressure	6. Partial blockage of line (possibly due to partially closed valve).	Increased pressure downstream. Excess load on pumps leading to reduction in flow through reactor.	f) Install pressure gauge at inlet of reactor. Clear blockage or open valve.
LESS OF	Less flow	7. Reduced feed rate.	Reduced PAN and steam production.	No action required.
		8. Leaking flange.	As for 8. Release of explosive mixture to atmosphere.	Covered by c).
	Less temperature	9. Preheater operating below rated capacity.	Small decrease in cooling duty. Risk of by-product formation at low reaction temperatures.	Covered by e).
				g) Install temperature gauge at inlet of reactor.
PART OF	Low xylene concentration	10. Xylene feed pump operating below rated capacity.	Reduced production of PAN. Reduced duty on coolant.	Covered by e).
MORE THAN	High xylene concentration	11. Ratio controller not operating correctly.	Possible deactivation of catalyst.	h) Sample feed regularly and monitor flow of product.
AS WELL AS	Impurities in xylene	12. Poor quality feed.	Formation of by-products in reactor. Low purity crude product.	i) Test quality of feed regularly.
OTHER	Maintenance	13. General equipment failure or catalyst changeover in reactor.	Process stops.	j) Good practices in construction and operation. Ensure shutdown and start-up procedures are well detailed.

Table D.2 Reactor product gases at 49.65 T/hr and 370°C.

Key Word	Deviation	Possible Causes	Consequences	Action Required
NONE	No flow	1. Blockage in reactor.	Pressure build-up in reactor (no hazard until 2100 kPa).	a) Shutdown if blockage does not clear itself.
		2. Rupture of reactor.	As for 1. Release of explosive mixture to atmosphere.	b) Emergency shutdown.
MORE OF	More flow	3. High feed rate.	Increased duty downstream.	c) Limit operations to design capacity.
	More temperature	4. Inadequate cooling.	Coolant temperature rises. Increased duty downstream.	d) Install controller to regulate flow of coolant based on flow through reactor.
	More pressure	5. Partial blockage of reaction tubes.	Increased pressure downstream. Hot spots may form in unblocked tubes.	e) Shutdown and clean reactor. f) Install temperatures indicators in reactor.
LESS OF	Less flow	6. Low feed rate.	Process inconvenience but no hazard.	No action required.
		7. Leak within reactor.	Salt contacts hydrocarbon — EXPLOSION.	g) Emergency shutdown. Reactor repairs required before start-up. Covered by f).
	Less temperature	8. Low coolant temperature.	More by-products from reaction. Lower pressure steam produced. Reduced duty downstream.	h) Install controller on salt cooler to regulate cooling.
AS WELL AS	Low purity of crude PAN	9. Formation of by-products.	Increased duty in recovery and purification units.	i) Good operating practices.
OTHER	Maintenance	10. General equipment failure or catalyst changeover in reactor.	Process stops.	j) Good practices in construction and operation. Ensure shutdown and start-up procedures are well detailed.

Table D.3 Cold Hitec salt at 2500 T/hr and 370°C.

Key Word	Deviation	Possible Causes	Consequences	Action Required
NONE	No flow	1. Line breakage or vessel rupture in salt cooler.	Release of dangerous material to atmosphere.	a) Emergency shutdown.
		2. Hitec solidified in salt cooler.	No cooling in reactor. EXPLOSION likely.	Covered by a).
MORE OF	More flow	3. Pump working at above rated capacity.	Risk of equipment or line failure increased. No effect on process performance.	b) Good operating practice.
	More temperature	4. Inadequate cooling.	Temperature difficult to control. Risk of hot spots in reactor.	c) Ensure surplus capacity for coolant circuit.
LESS OF	Less flow	5. Pump working at below rated capacity.	Cooling may be inadequate.	Covered by c). d) Regular maintenance of pump.
		6. Leaking flange or ruptured line.	As for 5. Release of dangerous material to atmosphere.	Covered by a). e) Replace coolant lost from circuit.
	Less temperature	7. Low temperature at inlet of salt cooler.	Gradual cooling of HTS if flow to salt cooler is not reduced.	f) Install controller to regulate flow of cooling water to salt cooler.
		8. High water flow to salt cooler.	As for 7.	Covered by f).
AS WELL AS	Air	9. Leak in salt cooler.	As for 5.	g) Install air vent at high point in line.
OTHER	Maintenance	10. General equipment failure or catalyst changeover in reactor.	Process stops.	h) Good practices in construction and operation. Ensure shutdown and start-up procedures are well detailed.

Table D.4 Hot Hitec salt at 2500 T/hr and 385°C.

Key Word	Deviation	Possible Causes	Consequences	Action Required
NONE	No flow	1. Hitec solidified in reactor.	Process stops. Risk of accident during start-up. No steam generation.	a) Install temperature indicator.
		2. Line breakage.	Release of dangerous material to atmosphere. No steam generation.	b) Emergency shutdown.
MORE OF	More flow	3. Pump working at above rated capacity.	Risk of equipment or line failure increased. No effect on process performance.	c) Good operating practice. Do not overload equipment.
	More temperature	4. Excess heat generated in reactor (possibly from complete oxidation reaction).	Gradual warming of coolant circuit. Reaction temperature difficult to control.	d) Ensure surplus capacity for coolant circuit.
LESS OF	Less flow	5. Pump working at below rated capacity.	Less steam produced.	Covered by d). e) Regular maintenance of pump.
		6. Leaking flange or ruptured line.	As for 5. Release of dangerous material to atmosphere.	Covered by b). f) Replace coolant lost from circuit.
	Less temperature	7. Less heat produced within reactor (possibly from reduced feed rate).	Gradual cooling of HTS if flow to salt cooler is not reduced.	g) Install controller to regulate flow of water to salt cooler.
AS WELL AS	Air	8. Leak in salt cooler.	As for 5.	h) Install air vent at high point in line.
OTHER	Maintenance	9. General equipment failure or catalyst changeover in reactor.	Process stops.	i) Good practices in construction and operation. Ensure shutdown and start-up procedures are well detailed.

Table D.5 Water at 24 T/hr and 165°C.

Key Word	Deviation	Possible Causes	Consequences	Action Required
NONE	No flow	1. Line rupture.	No steam production. No cooling of HTS — increase in reactor temperatures.	a) Emergency shutdown.
		2. Valve shut incorrectly.	As for 1.	b) Open valve. Covered by a).
		3. Pump failure.	As for 1.	
MORE OF	More temperature	4. Increased duty upstream (gas cooler).	Cooling may be inadequate.	c) Install controller to regulate flow.
LESS OF	Less flow	5. Pump working at below rated capacity.	Cooling may be inadequate.	Covered by c). d) Regular maintenance of pump.
		6. Leaking flange or ruptured line.	As for 5.	e) Increase cooling water flow until line or flange can be repaired.
	Less temperature	7. Decreased duty upstream (gas cooler).	Improved cooling capacity.	No action required.
		8. Weather conditions.	As for 7.	No action required.
AS WELL AS	Suspended solids	9. High suspended solids concentration in water supply.	Increased fouling in salt cooler.	f) Cooler must be cleaned regularly.
OTHER	Maintenance	10. General equipment failure or catalyst changeover in reactor.	Process stops.	g) Good practices in construction and operation. Ensure shutdown and start-up procedures are well detailed.

Table D.6 High pressure steam at 24 T/hr and 215°C.

Key Word	Deviation	Possible Causes	Consequences	Action Required
NONE	No flow	1. Line or vessel rupture.	No power generation. Salt cooler failure.	a) Emergency shutdown.
		2. Valve shut incorrectly.	As for 1.	b) Open valve. Covered by a).
		3. Pump failure.	As for 1.	
LESS OF	Less flow	4. Leaking flange or ruptured line.	Reduced power generation. Process hazard (HP steam).	c) Repair and/or shutdown.
	Less temperature	5. Fouling reducing heat transfer.	Inadequate cooling.	d) Ensure surplus capacity of cooling water.
OTHER	Maintenance	6. General equipment failure or catalyst changeover in reactor.	Process stops.	e) Good practices in construction and operation. Ensure shutdown and start-up procedures are well detailed.

11. CASE STUDY — PHTHALIC ANHYDRIDE AFTER-COOLER DESIGN

OVERALL SUMMARY

A desuperheater–cooler–condenser, which functions as the after-cooler in the LAR process for phthalic anhydride production, was designed to recover approximately 50% of the PAN vapour which is present in the reaction gases. During normal operation, the unit processes 49.35 T/hr of reaction gases and recovers 2.53 T/hr of crude liquid product at a purity of 98.0%.

The after-cooler is a shell-and-tube heat exchanger with condensation on the outside of 280×25 mm o.d. horizontal tubes. The exchanger shell has an inside diameter of 1000 mm in order to accommodate large vapour nozzles, and to provide sufficient clearance above and below the tube-bundle. Five internal shell-side baffles are used to increase the heat transfer rate, and a sloped insert is used in the base of the exchanger to collect the liquid condensate product. The cooling duty is satisfied by diathermic oil (71 m^3/hr) which enters the after-cooler at 100°C. Various exchanger configurations were evaluated using a specially written FORTRAN model of a desuperheater–cooler–condenser. The final design was obtained after several iterations and fulfils all of the design requirements.

The after-cooler will be constructed from stainless steel type 317 to minimise the corrosion rate from hot, liquid phthalic anhydride. The mechanical design was developed using pressure vessel guidelines (AS1210 and BS5500) in order to withstand all feasible operating conditions. The vessel supports and concrete foundation are typical of those used for shell-and-tube exchangers. Thermal insulation was specified to reduce energy losses from the unit. The total installed cost of the after-cooler was estimated at A$89,000 with an uncertainty of $\pm 25\%$.

A HAZOP study was completed around the after-cooler to provide recommendations for the safe operation of the after-cooler, and for the control and instrumentation requirements. Procedures for the commissioning, shut-down and start-up of the after-cooler have been outlined. A tentative maintenance schedule has also been included.

PART I: CHEMICAL ENGINEERING DESIGN

11.1 GENERAL DESIGN CONSIDERATIONS

In conventional processes, phthalic anhydride is sublimed directly from the reaction gases in large switch-condensers. The LAR process differs from the

Updated Material and Energy Balance for the After-Cooler (E105)

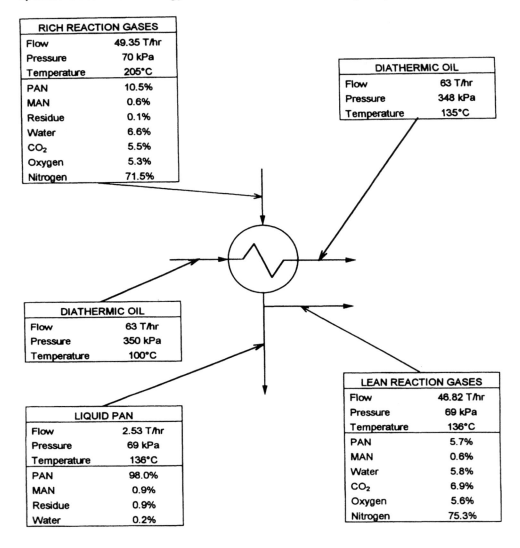

RICH REACTION GASES	
Flow	49.35 T/hr
Pressure	70 kPa
Temperature	205°C
PAN	10.5%
MAN	0.6%
Residue	0.1%
Water	6.6%
CO_2	5.5%
Oxygen	5.3%
Nitrogen	71.5%

DIATHERMIC OIL	
Flow	63 T/hr
Pressure	348 kPa
Temperature	135°C

DIATHERMIC OIL	
Flow	63 T/hr
Pressure	350 kPa
Temperature	100°C

LIQUID PAN	
Flow	2.53 T/hr
Pressure	69 kPa
Temperature	136°C
PAN	98.0%
MAN	0.9%
Residue	0.9%
Water	0.2%

LEAN REACTION GASES	
Flow	46.82 T/hr
Pressure	69 kPa
Temperature	136°C
PAN	5.7%
MAN	0.6%
Water	5.8%
CO_2	6.9%
Oxygen	5.6%
Nitrogen	75.3%

conventional process mainly in relation to the reactor catalyst which allows much higher concentrations of *o*-xylene in the feed. Consequently, the reaction gases contain a higher concentration of phthalic anhydride and approximately half of the product can be recovered by condensation (rather than sublimation) upstream of the switch-condensers. The condensation process is performed in the after-cooler. The LAR process is the only process which uses an after-cooler so the available design information is limited.

The principal design requirement for the after-cooler is to recover liquid phthalic anhydride from the reaction gases by desuperheating, condensing and cooling the

Engineering Specification Sheet for the After-Cooler (E105)

Chemical Engineering Design

Functional Description	*Desuperheater — cooler — condenser*		
Type of Exchanger	*Shell-and-tube*		
Orientation of Tubes	*Horizontal*		
Operation	*Continuous*		
General Properties			
Duty	*1290 kW*		
Heat Transfer Area	*124 m^2*		
Overall HT Coefficient	*210 W/m^2/°C*		
Shell-Side			
Fluid	*Reaction gases / crude phthalic anhydride*		
	Flow	Temperature	Composition
Inlet Gases	*49.35 T/hr*	*205°C*	*10.5 % PAN*
Outlet Gases	*46.82 T/hr*	*136°C*	*5.7 % PAN*
Liquid Product	*2.53 T/hr*	*136°C*	*98.0 % PAN*
Tube-Side			
Fluid	*Diathermic oil*		
	Flow	Temperature	Composition
Inlet	*63 T/hr*	*100°C*	*100%*
Outlet	*63 T/hr*	*135°C*	*100%*
Notes			
1. Yield of phthalic anhydride: 50%			

inlet gases. The flow of gases to the unit is 49.35 T/hr, of which phthalic anhydride is 10.5%. It is not necessary to condense all the product as the remainder will be recovered as a solid in the switch-condensers. A recovery efficiency of *50% was assumed in developing the process flow diagram.*

To prevent the formation of solids in the after-cooler, the minimum film temperature should be maintained above the melting point of pure phthalic anhydride (130.8°C). This produces limitations on the design and may even necessitate the use of a cooling fluid other than water.

Mechanical Engineering Design

Design Code		AS1210 / BS5500	
Shell-Side		**Tube-Side**	
Type	TEMA type E	Type	Plain, seamless
Material	Stainless steel, type 317	Material	Stainless steel, type 317
Diameter	1000 mm i.d.	Number	280
Passes	1	Passes	2
Baffles	5	Length	6.0 m
Baffle Type	Horizontal-cut, seg.	Diameter	25.4 mm o.d.
Front Head	Fixed (type A)	Gauge	14 BWG (medium)
Rear Head	Floating (type S)	Pattern	Square
Insulation	Mineral fibre (100 mm)	Pitch	32 mm centres
Nozzles	1 inlet x 760 mm,	Nozzles	1 inlet x 150 mm NB,
	1 outlet x 760 mm,		1 outlet x 150 mm NB
	1 outlet x 40 mm NB		
Pressure Drop	1.0 kPa	Pressure Drop	1.7 kPa
Vessel		**Supports and Foundation**	
Wall Thickness	12 mm	Supports	Steel saddles (2),
Ends Thickness	51 mm		concrete piers (2)
Tube-Plate Thickness	76 mm	Foundation	Concrete slab
Corrosion Allowance	1.0 mm		
Operating Weight	10,180 kg	Combined Weight	12,530 kg

The principal design parameters which must be determined or evaluated at this stage include all the operating conditions, the heat transfer conditions (and, therefore, the basic exchanger configuration) and other parameters which directly impact on the process flowsheet:

(a) tubes: diameter, length, number, layout;
(b) shell: diameter, baffles;
(c) composition of the liquid product;
(d) coolant flow rate;

Schematic Drawing of the After-Cooler

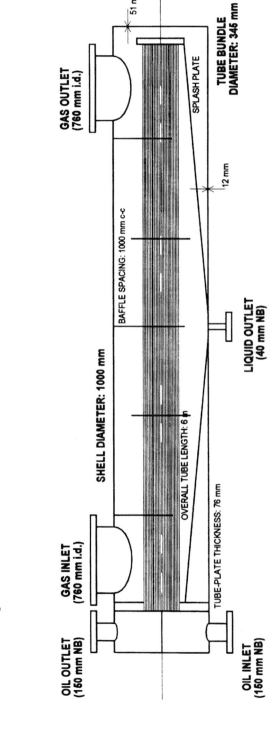

(e) temperature profiles of the gas, condensate film and coolant;
(f) pressure drop (shell-side and tube-side);
(g) heat transfer area;
(h) heat transfer rate.

11.2 DESIGN STRATEGY AND CRITERIA

Several preliminary design decisions must be made before specific condenser design methods can be applied to the design of the after-cooler. First, the type of condenser must be selected for the specific conditions and design requirements of the after-cooler. Second, a coolant must be selected that allows all of the process requirements to be met, and can also be integrated into the full process with maximum energy efficiency. The cost of the coolant and any safety issues should also be considered. Water is an obvious choice as it is cheap and can be easily integrated into the process, but it might not necessarily allow the process requirements to be met.

After the preliminary design decisions have been made, the procedure used to determine and evaluate the necessary design parameters can begin. The various parameters are mostly inter-related and cannot, therefore, be determined sequentially. An iterative approach is required to achieve a balance between, for example, the heat transfer rate and the temperature profiles, or between the heat transfer area and the pressure drop. A simulation model was considered to be the best tool to perform this iterative process. No suitable simulation programs were available so a FORTRAN program was written.

The first requirement of the FORTRAN program was physical property data for all materials and components involved. HYSIM was used to generate most of the necessary data. The density, specific heat, thermal conductivity, viscosity and average molecular weight of the reaction gas mixture and the phthalic anhydride liquid film were calculated at temperatures between 135°C and 200°C. The properties of the first-choice coolant (i.e. water) were obtained from steam tables while the properties of other potential coolants were taken from a suitable data source (Kreith and Bohn, 1986). To simplify the calculations required during each iteration, the enthalpy and the composition of the liquid product were also determined at temperatures of 135–200°C using HYSIM.

The second requirement of the computer model was to incorporate equations that accurately describe the operation of a desuperheater–cooler–condenser. The equations should be valid for the full range of conditions and configurations to be considered. The basic method used was to divide the exchanger into duty intervals (e.g. 50 kW each), based on counter-current flow. Heat transfer resistances were calculated for each interval and then the film temperatures were iterated until the heat balance equations were satisfied. Two algorithms were used: one to model the desuperheater section (no condensation) and one to model the cooler–condenser section. A third algorithm was used to calculate the heat transfer area from short cut calculations (ignoring the effects of diffusion of the condensing component through the non-condensable gases) to provide a comparison with the rigorous solution. The tube-side coolant velocity was chosen to be within a range that offers a compromise between good heat transfer, pumping duties and fouling rates, based on literature recommendations (Kern, 1950).

The program was written to accept inputs which described the condenser configuration, and the coolant inlet temperature, and the target temperature for the outlet gases. The required heat transfer area, length of tubes, coolant flow rate, temperature profiles and heat transfer profile were produced as output from the program. This allowed an iterative approach to be used so that a suitable configuration could be found which incorporated standard tube sizes, a customary pitch and layout, and which minimised the heat transfer area and the coolant flow rate. Most importantly, a design could be selected which resulted in condensate and water temperatures which were within the operable limits. Some surplus heat transfer area was provided to allow for inaccuracies in the heat transfer correlations (rarely accurate to better than 10%), a possible expansion in the overall PAN production rate, and a possible transfer of duty between the after-cooler and switch-condenser that might occur as a result of transient or dynamic effects in the process.

The remaining design parameters were calculated after the basic exchanger configuration was determined using the computer model. These parameters include the pressure drops across both the shell-side and the tube-side of the exchanger. These were estimated conservatively from literature correlations.

Key Results

- The first phase of the design process was the selection of an appropriate exchanger type.
- The second phase of the design process was coolant selection.
- The third phase of the design process was an iterative determination of the exchanger configuration and heat transfer area, using a specially developed computer model which was written in FORTRAN.
- Physical property data, cooling duties and compositions were obtained from HYSIM.
- The fourth phase of the design process was the calculation and evaluation of the remaining design parameters using various literature correlations.

11.3 PRELIMINARY DESIGN DECISIONS

11.3.1 Condenser Type

Several types of condensers were considered. Shell-and-tube exchangers are the first choice for most applications as they are low-cost, easily maintainable, readily available and widely used for all types of heat transfer services. Plate exchangers are becoming increasingly common, but the use of small plate clearances to increase the heat transfer rate make them highly susceptible to fouling, especially when there is a risk of solid formation. Consequently, a plate exchanger would be unsuitable for the after-cooler.

Direct-contact condensers are used in some services where a suitable coolant is available. However, the properties of phthalic anhydride vapours preclude the use of most common coolants. A cold stream of air could not be used as temperature profiles would be difficult to control if a small volume of air was used, and a larger volume would reduce the partial pressure of the phthalic anhydride and prevent any condensation. Water could not be used as it reacts with the anhydrides (both phthalic and maleic) in the liquid product to form acids which degrade the product. Any other liquid that might be used must be easy to separate from phthalic anhydride in the purification columns, must have a high specific heat to allow its temperature to be controlled effectively, must have a high boiling point so that a very high pressure is not required, and it must be relatively inexpensive. No material meeting all of these requirements was identified.

A shell-and-tube condenser was selected on the basis of the above analysis. Two tube arrangements were considered suitable for this process: *horizontal tubes with condensation on the outside*, or vertical tubes with condensation on the inside. The former was preferred as it provides a significantly higher heat transfer coefficient, easier separation of non-condensable gases, and easier cleaning of coolant scale. The disadvantages of horizontal tubes are an increased heat loss and wider exposure to corrosion from the phthalic anhydride vapours and condensate.

Key Results

- A shell-and-tube condenser was selected from standard exchanger designs.
- A configuration with condensation outside horizontal tubes was selected on the basis of expected heat transfer coefficients, vapour-liquid disengagement and ease of cleaning.

11.3.2 Coolant

The PFD presented in Figure 3.1 provided the basis for the mass and energy balances (Section 7.4) and the subsequent P&ID (Figure 8.1). This basis *assumes* that water, with an ambient inlet temperature (i.e. 25°C) is the coolant in the after-cooler. Water is attractive as a coolant from an economic perspective due to its low cost, and it is widely used throughout the process and so it could easily be integrated with other units to maximise the energy efficiency of the overall plant. However, preliminary heat transfer calculations suggested that the coolant inlet temperature would have to be regulated at around 120°C to allow the condensate film temperature to be maintained above the PAN melting point. This would be difficult to achieve with water and would create further problems in effectively integrating the heat exchanged in the after-cooler with other units. With these considerations, an alternative coolant may be preferred.

Three utilities other than water are already used within the plant: steam, dia-thermic oil and foundry salt. Steam could be used but the heat transfer coefficient from the steam to the tube wall would be very low and large heat transfer areas would be required. Accurate control of the inlet temperature would be achievable via pressure regulation but the operating pressure would be close to atmospheric and, therefore, a large volumetric flow rate of steam would be required. Diathermic oil has already been specified in the switch-condensers and would provide good temperature control and high heat transfer rates. Furthermore, the diathermic oil could be cooled by cooling water so that the previously developed heat integration scheme could be maintained without change to the process flowsheet. Foundry salt is suitable only for use at high temperatures and could not be used for this service.

Heat transfer coefficients in the boiling regime are very high and the pool temperature can be easily regulated by pressure so that a coolant with a boiling point around 130°C might be advantageous. However, it is difficult to distribute the cooling load evenly in an exchanger with both condensing and boiling occurring simultaneously. Regardless of this difficulty, no suitable boiling-coolant could be found to meet the process requirements in an economical manner. *Diathermic oil was selected as the preferred coolant*. This requires the *addition of another heat exchanger to the process layout: an oil-cooler*. The revised P&ID for the PAN recovery system is shown in Figure 11.1, which is directly analogous to the original scheme presented in Figure 8.1(b).

Key Results

- Diathermic oil is the preferred coolant as it allows the inlet temperature to be tightly controlled so that condensing phthalic anhydride does not solidify.
- The use of diathermic oil makes the control of the coolant inlet temperature independent of other process units so that their operation is not compromised via heat integration.
- No changes to the existing process flowsheet and heat integration scheme are required.
- An auxiliary exchanger is required to exchange heat between the hot oil and coolant.

11.4 CHEMICAL ENGINEERING DESIGN METHODS

11.4.1 Heat Transfer Coefficient

A desuperheater–cooler–condenser accepts a gas mixture feed above its dew point and condenses all or part of the condensable component while cooling the non-condensables to maintain the vapour–liquid equilibrium. The analysis of this type of exchanger divides readily into two sections: those with and without condensation.

Analysis of the desuperheating section is relatively simple. Four heat transfer resistances must be considered: coolant film resistance; scale deposits; tube wall resistance; and gas-phase resistance. These resistances are calculated from equations (11.1) to (11.4), below. Where banks of tubes are used (as opposed to a single tube), equation (11.4) provides a conservative estimate of the heat transfer coefficient. The actual heat transfer coefficient is likely to be at least 20% higher.

Convection through water film:

$$h_1 = (k_c/d_i) \times 0.027 \times Re^{0.8} \times Pr^{0.33} \qquad (11.1)$$

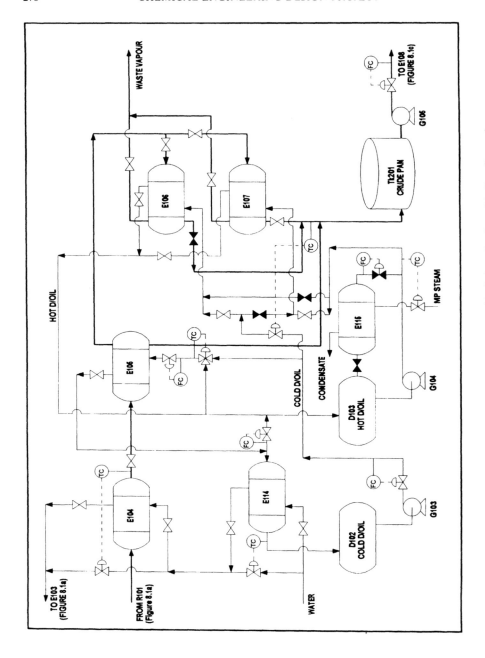

Figure 11.1 P&ID for the PAN recovery system with diathermic oil used in the after-cooler.

Resistance of scale deposits:

$$h_2 = (2 \times k_{st}/d_o)/\ln(d_o/d_i) \tag{11.2}$$

Conduction through tube wall:

$$h_3 = 1/R_d \tag{11.3}$$

Convection through gas phase:

$$h_4 = (k_g/d_o) \times 0.40 \times \mathrm{Re}^{0.6} \times \mathrm{Pr}^{0.36} \tag{11.4}$$

where: h = heat transfer coefficient (W/m^2/°C); k = thermal conductivity (W/m/°C); d_i = inside diameter of tubes (m); d_o = outside diameter of tubes (m); Re = Reynolds number; Pr = Prandtl number; R_d = scale resistance (typically 3.5×10^{-4} m^2 °C/W). *Subscripts:* c = coolant (i.e. diathermic oil); st = steel (tube walls); g = gas mixture.

The overall heat transfer coefficient, based on the outside diameter of the tubes, can be calculated from:

$$1/U = (d_i/d_o)/h_1 + 1/h_2 + 1/h_3 + 1/h_4 \tag{11.5}$$

where: U = overall heat transfer coefficient (W/m^2/°C).

The required heat transfer area can be calculated by using the log mean temperature difference:

$$A = Q/(U \times \Delta T) \tag{11.6}$$

where: A = heat transfer area (m^2); Q = total heat transferred (W); ΔT = log mean temperature difference (°C).

The analysis of the condensing section is more complicated. In addition to the resistances considered above, the effects of mass transfer diffusion by the condensable component(s) through the non-condensable gases are significant. Only phthalic anhydride needs to be considered as the contribution of maleic anhydride (and the residue which also condenses) to the heat transfer resistance is negligible, because the condensate is 98% pure in phthalic anhydride. Three heat flows are considered: sensible heat transferred from the bulk gas to the condensate film (equation (11.7), below); latent heat given up by diffusing vapour molecules (equation (11.8)); and the heat supplied from the coolant through the pipe wall and scale deposits (equation (11.9)):

$$q_v = h_4 \times (T_g - T_c) \tag{11.7}$$

$$q_l = k_G \times \lambda \times (P_g - P_s) \tag{11.8}$$

$$q_c = h_0 \times (T_c - T_o) \tag{11.9}$$

where: q_v = sensible heat transfer (W); q_l = heat transfer by diffusion (W); q_c = heat supplied by coolant (W); T_v = gas temperature (°C); T_c = condensate temperature (°C); T_o = coolant temperature (°C); k_G = mass transfer coefficient; λ = latent heat of condensate (J/kg); P_g = partial pressure of diffusing vapour (kPa); P_s = partial pressure at surface condensate temperature (kPa); $h_0 = 1/(1/h_1 + 1/h_2 + 1/h_3)$ (W/m^2/°C).

The mass transfer coefficient in equation (11.8) can be calculated from the following equations:

$$D = 10^{-7} \times T_g^{1.75}/(P \times [V_A^{0.33} + V_B^{0.33}]^2) \times [1/M_A + 1/M_B]^{0.5} \qquad (11.10)$$

$$Sc = \mu_g/(\rho_g \times D) \qquad (11.11)$$

$$k_G = (h_4/(c_{pg} \times P_{gln})) \times (Pr/Sc)^{0.667} \qquad (11.12)$$

where: D = diffusivity of PAN vapour through the reaction gases (m²/s); P = total gas pressure (kPa); V = molecular volume; M = mean molecular weight; Sc = Schmidt number; μ_g = mean viscosity of gas (Pa s); ρ_g = mean density of gas (kg/m³); c_{pg} = mean specific heat of gas (J/kg/°C); P_{gln} = log mean partial pressure of non-condensable gas (kPa).

An overall heat balance must be solved iteratively at various points along the condenser surface. The iterative procedure should be based on an estimate of the film temperature which is adjusted until the balance is satisfied. In general, h_0 remains nearly constant over the length of the condenser but k_G and the temperature differences change significantly. The overall heat balance is given by:

$$q_v + q_l = q_c = Q. \qquad (11.13)$$

The heat transfer area should also be calculated in successive intervals using the mean temperature difference and the duty increment at each interval. The overall heat transfer coefficient for the whole condenser can be calculated from the sum of the heat transfer and area increments, and the log mean temperature difference based on inlet and outlet conditions.

Key Results

- The operation of a desuperheater–cooler–condenser should be considered in two sections: with and without condensation.
- There are four heat transfer resistances occurring in the desuperheating section: coolant film, scale deposits, tube wall and gas phase.
- There are five heat transfer resistances occurring in the cooling–condensing section: coolant film, scale deposits, tube wall, gas phase and diffusion.
- The analysis of the cooler–condenser is completed by iteratively solving heat balances over intervals of the exchanger.

11.4.2 After-Cooler Simulation

The heat transfer equations outlined in Section 11.4.1 form the basis of the FORTRAN program written to model the operation of the after-cooler. The basic exchanger configuration and geometry must be inputted to the program so that the required heat transfer area, the coolant flow rate and the condensate film temperature can be calculated. Estimates of the shell diameter and tube-side nozzle diameter are also determined by the program so that the exchanger geometry can be optimised.

Eight, approximately equal, intervals of heat transfer were considered. The physical property data was specified separately and supplied for both the gas and condensate at each heat transfer interval using data generated by HYSIM. The first stage of the calculation routine is to determine all the heat transfer coefficients at each heat transfer interval. The second stage is then to iteratively solve the energy balance (equation (11.13)) by estimating the film temperature, and adjusting the estimate up or down depending on whether the net heat transfer is to or from the condensate film. The calculated results at each point are retained so that the gas, condensate and coolant temperature profiles can be outputted. The full FORTRAN model is included in Appendix F at the end of this chapter.

Key Results

- The computer model calculates the heat transfer area required, the coolant flow rate, and the condensate film temperature for specified condenser geometries using the method outlined in Section 11.4.1.
- Temperatures profiles for the gas, condensate and coolant are saved and outputted from the program.
- The shell diameter and tube-side nozzle diameter are estimated by the program so that the exchanger geometry can be optimised.

11.4.3 Pressure Drop

The pressure drop across the shell-side and tube-side of the after-cooler were calculated from equations which have been modified from Kern (1950), and corrected to SI units:

Tube-side:

$$\Delta P = f \times (\rho \times u^2/2) \times (L/D_i) \times N_P \qquad (11.14)$$

Shell-side:

$$\Delta P = f \times (\rho \times u^2/2) \times (D_s/D_e) \times (N_B + 1) \qquad (11.15)$$

where: ΔP = pressure drop (kPa); f = friction factor; ρ = average density (kg/m^3); u = average velocity (m/s); L = length of tubes (m); D_i = inside diameter of tubes (m); N_P = number of tube passes; D_s = outside diameter of tube bundle (m); D_e = effective diameter of tubes (m); N_B = number of baffles.

The pressure drop is normally significantly lower when condensation occurs, but no correction is included here as only 5% of the inlet gases are condensed. Neither is a viscosity correction included as inlet and outlet viscosities are essentially unchanged for both the gas stream and the coolant. The necessary physical property data were calculated at the average conditions in the vessel.

The estimates from equations (11.14) and (11.15) are considered to be potentially low as no allowance is included for entrance and exit (nozzle) pressure losses. Therefore, the final estimate of the shell-side pressure drop should be increased by two velocity heads (a velocity head is equal to the product of half the density and the square of velocity) to account for the nozzle losses. Kern (1950) recommends that the tube-side pressure drop should be increased by four velocity heads per pass.

Key Results

- Standard correlations for baffled shell-and-tube exchangers were used to provide initial estimates of the shell-side and tube-side pressure drops.
- No correction was made for condensation as it accounts for only 5% of the total gas flow rate.
- The calculated pressure drops were corrected for entrance and exit losses on both the shell-side and the tube-side.

11.5   DETAILED DESIGN

11.5.1   General Considerations

The after-cooler will be a shell-and-tube desuperheater–cooler–condenser. Horizontal tubes, with condensation on the outside of the tubes, will be used in a horizontal shell. The tube-side (cold) stream will be diathermic oil which will enter the after-cooler at 100°C at the rate of 71 m³/hr. The reaction gases from the gas-cooler will enter the shell-side of the condenser at 205°C at the rate of 49.35 T/hr. After cooling, phthalic anhydride will condense out of the gas stream and collect in the base of the exchanger. The reaction gases will exit the condenser at 136°C after 2.53 T/hr of crude phthalic anhydride has been recovered with a purity of 98.0%. The condensate film temperature is maintained at least 3°C above the PAN melting point so that solid formation is minimised. The thermodynamic duty of the condenser during normal operation is 1290 kW. This is equivalent to 1870 kJ per kg of pure phthalic anhydride recovered. The unit will be operated continuously.

11.5.2   Simulation Results

The computer model of the after-cooler was used to optimise the exchanger configuration and to minimise the coolant flow rate. A provisional design was developed to estimate an appropriate starting point for the optimisation of the design using the computer simulation model. This design is detailed in Appendix E at the end of this chapter. This design was based on several key assumptions: standard tube dimensions (25.4 mm o.d., 14 BWG, 6.0 m long); standard tube pitch (32 mm); log mean temperature difference (LMTD) of 10°C; heat transfer coefficient of 800 W/m²/°C; basic exchanger configuration with two tube-side passes and one internal shell-side baffle; coolant inlet temperature of 120°C; film temperature that was relatively close to the exchanger outlet temperature (already fixed at 136°C through optimisation of the process layout and PFD); and a tube-side velocity of 1.5 m/s. The film temperature is important to prevent solidification on the tubes which would significantly reduce the heat transfer rate.

The tube-side velocity is an important variable to iterate as it directly affects the heat transfer coefficient, the pressure drop and the coolant flow rate. These assumptions effectively define the starting point, shown in Table 11.1, in the iterative search for the optimum after-cooler design.

The results meet the main criteria (sufficient heat transfer area and condensate film temperature above the PAN melting point) but are unsatisfactory in two main aspects. First, the exchanger is much larger than it needs to be, primarily due to the original low estimate of the mean temperature difference. The over-design adds unnecessary cost to the exchanger construction. Second, the coolant flow rate is very high and the estimated nozzle diameter is probably too large by comparison with the estimated shell diameter. The film temperature is approximately 3°C higher than the PAN melting point and this is considered satisfactory, although a greater differential would be preferred.

The tube-side velocity effectively determines the coolant flow rate (if the tube count and diameter are constant) and is one of the controlling factors determining the heat transfer area (via the heat transfer coefficient which is strongly dependent on the Reynolds number). The velocity is also important in fixing the condensate film temperature as it influences the relative tube-side and shell-side heat transfer rates. Therefore, the second iteration of the after-cooler design kept that same exchanger configuration but with a significantly lower tube-side velocity (0.5 m/s). The exchanger configuration and simulation results are summarised in Table 11.2.

This modification significantly improved the condenser design but decreased the heat transfer coefficient to such an extent that the original design, which was significantly larger than required, is now too small. To compensate for the reduction in tube-side heat transfer coefficient which resulted from the lower tube-side velocity, the shell-side heat transfer was increased for the next iteration of the design by adding baffles to the shell. An odd number of baffles are required if both shell-side vapour nozzles are on the same side of the shell (clearly preferred). Furthermore, if horizontal-cut baffles are to be used (preferred for this type of exchanger) an even number of baffles on each side of the centre-point are preferred so that the central baffle opening is at the low point of the exchanger.

Table 11.1 Initial conditions and results used to determine the optimum after-cooler design.

Tube diameter	25.4 mm o.d.; 21.2 mm i.d.
Tube length	6.0 m
Tube pitch	32 mm
Number of tubes	335
Exchanger configuration	2 tube-side passes, 1 shell-side baffle
Inlet coolant temperature	120°C
Tube-side velocity	1.5 m/s
Heat transfer area required	88 m^2
Actual area/Required area	182%
Coolant flow rate	318 m^3/hr
Estimated minimum shell diameter	611 mm Ø
Estimated tube-side nozzle diameter	212 mm Ø
Minimum condensate film temperature	133.9°C

Thus, five baffles were specified for the next iteration. A secondary effect of increasing the shell-side heat transfer coefficient is that the film temperature should increase. Consequently, a lower coolant inlet temperature can be used (say, 110°C). This is beneficial as it reduces the overall heat transfer area required. This new set of conditions are summarised in Table 11.3 which describes the third iteration of the design.

Once again, there is a surplus of heat transfer area. This can be used to further reduce the coolant flow rate and also to reduce the number of tubes. The condensate film temperature remains acceptable so that the coolant inlet temperature can again be reduced. The after-cooler configuration arising from these changes, and the corresponding simulation results, are shown in Table 11.4.

This design is highly acceptable. Adequate heat transfer area is available and the coolant flow rate is relatively low. The exchanger diameter is moderate and the tube-side nozzles can be constructed of standard 100 mm NB (4″) pipe. The condensate film temperature is still approximately 3°C above the PAN melting point so that solidification is unlikely. There is an 8% over-design which provides flexibility against changes in the operating conditions and uncertainty in the heat

Table 11.2   Second iteration of the after-cooler design.

Tube diameter	25.4 mm o.d.; 21.2 mm i.d.
Tube length	6.0 m
Tube pitch	32 mm
Number of tubes	335
Exchanger configuration	2 tube-side passes, 1 shell-side baffle
Inlet coolant temperature	120°C
Tube-side velocity	0.4 m/s
Heat transfer area required	181 m²
Actual area/Required area	88%
Coolant flow rate	106 m³/hr
Estimated minimum shell diameter	611 mm Ø
Estimated tube-side nozzle diameter	122 mm Ø
Minimum condensate film temperature	134.8°C

Table 11.3   Third iteration of the after-cooler design.

Tube diameter	25.4 mm o.d.; 21.2 mm i.d.
Tube length	6.0 m
Tube pitch	32 mm
Number of tubes	335
Exchanger configuration	2 tube-side passes, 5 shell-side baffles
Inlet coolant temperature	110°C
Tube-side velocity	0.5 m/s
Heat transfer area required	125 m²
Actual area/Required area	128%
Coolant flow rate	106 m³/hr
Estimated minimum shell diameter	611 mm Ø
Estimated tube-side nozzle diameter	122 mm Ø
Minimum condensate film temperature	134.9°C

Table 11.4 Fourth iteration (and accepted result) of the after-cooler design.

Tube diameter	25.4 mm o.d.; 21.2 mm i.d.
Tube length	6.0 m
Tube pitch	32 mm
Number of tubes	280
Exchanger configuration	2 tube-side passes, 5 shell-side baffles
Inlet coolant temperature	100°C
Tube-side velocity	0.4 m/s
Heat transfer area required	124 m²
Actual area/Required area	108%
Coolant water flow rate	71 m³/hr
Estimated minimum shell diameter	560 mm Ø
Estimated tube-side nozzle diameter	100 mm Ø
Minimum condensate film temperature	134.8°C

transfer correlations. While it is feasible that this design could be improved still further, it is considered acceptable at this point.

The computer model also calculates the required heat transfer area by short-cut methods. Sinnott (1993) claims that when the fraction of non-condensables exceeds 70% (in this case, there are 90% non-condensables) the heat transfer area can be calculated by adding the condensation heat load and the cooling heat load together, and computing heat transfer coefficients by assuming forced convection only. However, the estimate obtained for the heat transfer area is 22% lower (more optimistic) than that calculated using the rigorous equations described in Section 11.4.1. Thus, the additional calculation rigour and computation time were justified.

The simulation results also show how the heat transfer rate varies through the exchanger. At the junction of the desuperheating and the condensing zones, there is also a distinct change in heat transfer rates. Figure 11.2 shows this effect and the relationship between the total heat transferred and the heat transfer area. Figure 11.3 indicates the temperature profiles through the gas, condensate film and coolant.

Key Results

- An iterative approach was used to find the optimum condenser design.
- A series of provisional assumptions were made in order to locate an appropriate starting point in the search for a satisfactory design.
- Preliminary assumptions were rejected or replaced until a satisfactory design was found.
- The tube diameter, gauge and length were selected from commonly available sizes.
- The final solution satisfied all the design requirements.
- It was found that the short-cut methods produced an overly optimistic design, i.e. the predicted heat transfer area required was much less than was found using a more rigorous approach.
- The heat transfer rate in the desuperheating zone was significantly lower than in the condensing zone.

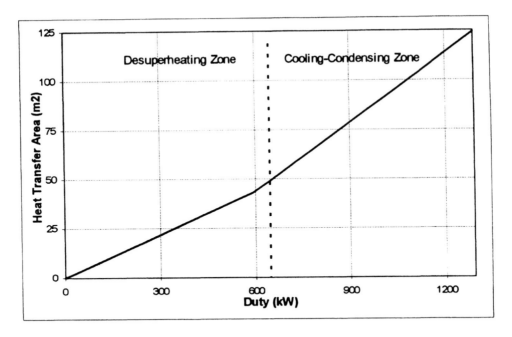

Figure 11.2   Heat transfer in the after-cooler.

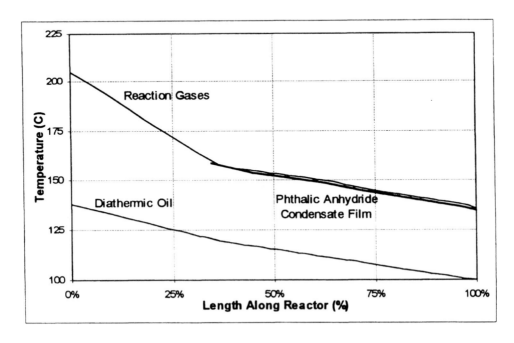

Figure 11.3   Temperature profiles.

11.5.3 Vessel Configuration

The basic configuration of the tube-bundle was determined by the iterative process described in Section 11.5.2. The tube length was fixed at 6.0 m, the longest standard size (approximately 20 feet) and the longest length that can be transported in Australia on the back of a truck without special arrangements. This length was specified directly, rather than iterating the length, to minimise the overall installed cost of the exchanger.

The tube diameter was also selected and not iterated. Again, 25 mm o.d. tubes are a standard size and are convenient to fabricate and install. Smaller tubes could have been used to increase the heat transfer area and the heat transfer coefficient inside the tubes, but the increased efficiency is marginal and comes at the expense of increased cleaning difficulty. Cleaning and maintenance were determined to be important considerations for this exchanger due to the possibility of PAN solidification on cooling and the corrosiveness of liquid PAN.

The shell-side inlet and outlet streams present some unusual problems for the design of the vessel. First, the vapour streams (both inlet and outlet) are at relatively low pressure and, therefore, occupy large volumes. Consequently, the nozzle diameters must be large compared with the diameter of the shell. The large nozzles occlude a significant amount of space that could otherwise have been used for the tube-bundle. Second, the outlet liquid flow is very small. As the crude phthalic anhydride liquid product is corrosive, hold-up in the exchanger should be minimised. Sloped inserts in the base of the exchanger alleviate this problem but, again, occlude space that would otherwise have been used for the tube-bundle. Thus, the shell diameter must be larger than the minimum calculated diameter which was based on uniform tube spacing inside the shell.

The optimum nozzle diameters were estimated from the optimum pipe diameter, which were calculated using equation (11.16), below, from Harker (1978) (converted to SI units). These estimates are shown in Table 11.5 together with the actual nozzle diameters specified. Both vapour nozzles were specified to be significantly smaller than the estimates obtained because of the size of the exchanger shell.

$$D = 8.41 \times W^{0.45}/\rho^{0.31} \tag{11.16}$$

where: D = estimated economic pipe diameter (mm); W = mass flow (kg/hr); ρ = density (kg/m^3).

The liquid transfer line to the crude PAN storage tank was carefully sized due to the possibility of solid formation which could potentially block the line. A small

Table 11.5 After-cooler nozzle diameters.

Opening	Calculated Optimum Size	Specified Inside Pipe Diameter
Gas inlet	953 mm	760 mm
Gas outlet	1026 mm	760 mm
Condensate	32 mm	40 mm (NB)
Diathermic oil	148 mm	150 mm (NB)

diameter increases the flow velocity and reduces residence time but is difficult to clean if a blockage occurs. A larger line diameter increases the risk of solidification but could be cleaned more easily. An intermediate size (40 NB) was specified. Standard pipe sizes, in schedule 40 pipe, were specified throughout. The liquid product outlet nozzle was sized to match the transfer line size.

Several tube-sheet configurations, each with a standard diameter in order to minimise fabrication costs, were considered and are shown in Figure 11.4. The smallest tube-sheet diameter, 890 mm, produces a clearance of only 44 mm between the inlet nozzle and the top of the tube-bundle, and a slope of 4.9° towards the liquid outlet nozzle. The inlet vapour velocity will be close to 20 m/s so that the tubes near the nozzle inlet will be subjected to large forces, and a heavy duty impingement plate is necessary to prevent damage to the tube-bundle. This requires a much larger clearance so that the largest shell diameter, 1000 mm, is preferred. The additional space allows the inlet vapours to expand uniformly across the entire bundle. On the outlet side, the clearance provides disengage-ment area for the condensate droplets, before they fall onto the sloped base-plate insert and hence flow toward the centrally located liquid outlet nozzle. The horizontal-cut segmental baffles should extend to the last row of tubes and be at least 10 mm thick for stability.

The tube dimensions were essentially fixed by the computer model. The wall thickness of 2.1 mm (BWG 14, medium gauge) was selected to provide a reasonable resistance to corrosion from the phthalic anhydride condensate. The working pressures did not impose any limitation on the tube thickness. A square tube-pattern (rather than triangular which is a more efficient method of packing tubes) was selected in order to simplify cleaning of the tube-bundle during maintenance. Figure 11.5 shows the interior layout of the after-cooler, and indicates important dimensions to ensure that the chemical engineering design requirements are satisfied.

The shell-side pressure drop across the after-cooler was calculated to be 1.05 kPa, which is typical of many condensers and is considered to be conservative.

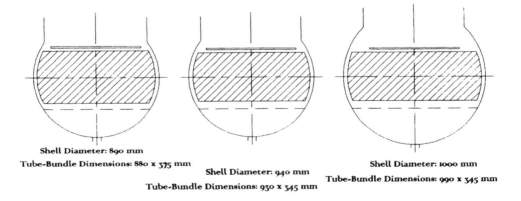

Shell Diameter: 890 mm
Tube-Bundle Dimensions: 880 x 375 mm

Shell Diameter: 940 mm
Tube-Bundle Dimensions: 930 x 345 mm

Shell Diameter: 1000 mm
Tube-Bundle Dimensions: 990 x 345 mm

Figure 11.4 Alternative tube-sheet configurations.

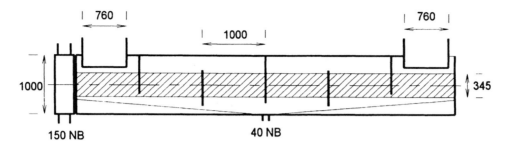

Figure 11.5 Interior layout of the after-cooler (see Table 11.5; approximately to scale).

The tube-side pressure loss was calculated to be 1.70 kPa and although this result is low, it is also conservative due to the low tube-side velocity which was used.

Key Results

• The shell diameter was selected to provide adequate clearance above and below the tube-bundle.
• A sloped base was installed to encourage the condensate to flow directly to the outlet nozzle in order to minimise the risk of corrosion.
• The vapour nozzles were sized below the economic pipe diameter due to the relatively small exchanger shell diameter.
• The liquid-product outlet nozzle was sized to minimise the potential for pipe blockages caused by solidification.
• An impingement plate was specified due to the high vapour velocity at the inlet.
• Medium-gauge tubes were used to provide adequate corrosion resistance.
• A square tube-pattern layout with a standard pitch (32 mm centres) was used to simplify tube-bundle cleaning during maintenance.
• Pressure drops were calculated to be 1.05 kPa on the shell-side and 1.70 kPa on the tube-side.

11.6   CHEMICAL ENGINEERING DESIGN SPECIFICATION

General

Functional Description	*Desuperheater–cooler–condenser*
Type of Exchanger	*Shell-and-tube*
Orientation of Tubes	*Horizontal*
Operation	*Continuous*
Yield of Phthalic Anhydride	50%
Duty	1290 kW
Heat Transfer Area	124 m²
Overall HT Coefficient	210 W/m²/°C

Shell

Fluid	*Reaction gases/crude phthalic anhydride*		
	Flow	Temperature	Composition
Inlet Gases	*49.35 T/hr*	*205°C*	*10.5% PAN*
Outlet Gases	*46.82 T/hr*	*136°C*	*5.7% PAN*
Liquid Product	*2.53 T/hr*	*136°C*	*98.0% PAN*
Diameter	*1000 mm i.d.*		
Baffles	*Segmental, horizontal-cut (5)*		
Nozzles	*Vapour nozzles: 760 mm (2)*		
	Liquid nozzle: 40 mm NB (1)		
Pressure drop	*1.05 kPa*		

Tubes

Fluid	*Diathermic oil*		
	Flow	Temperature	Composition
Inlet	*71 m³/hr*	*100°C*	*100% oil*
Outlet	*71 m³/hr*	*135°C*	*100% oil*
Diameter	*25.4 mm o.d.*		
Length	*6.0 m*		
Number	*280*		
Passes	*2*		
Pattern	*Square*		
Pitch	*32 mm centres*		
Nozzles	*150 mm NB/150 mm NB*		
Pressure drop	*1.7 kPa*		

PART II: MECHANICAL ENGINEERING DESIGN

11.7   MECHANICAL ENGINEERING DESIGN REQUIREMENTS

The chemical engineering design establishes the basic after-cooler configuration and determines some key dimensions exactly. However, several parameters remain undetermined and there is insufficient information, at this stage, to construct the exchanger. The principal aspects of the design that have not yet been fixed include:

(a) materials of construction;
(b) vessel dimensions and wall thickness;
(c) joints (type and radiographic efficiency);
(d) insulation requirements;

(e) vessel supports;

(f) vessel foundation.

Finally, an engineering drawing must be produced so that fabrication and installation can proceed. Once the drawings have been completed, and all the relevant construction details are known, the exchanger can be costed to an accuracy of approximately 10%, so that vendors' quotes and estimates can be obtained.

11.8 MATERIALS OF CONSTRUCTION

Materials selection was based on the consideration of four main factors: resistance to phthalic and maleic anhydride vapours and condensate; strength; ease of fabrication; and low cost. Much of the vessel (both the shell and the tubes) will be in continuous contact with molten phthalic anhydride at moderate temperatures. Therefore, particular attention was given to corrosion resistance under those conditions.

Phthalic and maleic anhydride vapours are not particularly corrosive to most steels, but phthalic anhydride liquid especially when containing impurities will rapidly corrode many stainless steels. These include the most widely used (and least expensive) types, 304 and 316, at the operating temperatures present in the after-cooler. Ideally, the average corrosion rate should be less than 0.05 mm/year (Rabald, 1968) but there is a risk of crevice corrosion if liquid stagnation occurs. Thus, the sloped inserts in the base of the vessel are important. The diathermic oil used on the tube-side of the exchanger has little or no effect on stainless steel unless very high levels of impurities (especially sulphur) are present. The preferred material of construction for the after-cooler is type 317 stainless steel. The full material specification is given in Table 11.6.

The fabrication properties of stainless steel type 317 are considered satisfactory for the construction of the after-cooler, as only cold working and welding should be necessary. Hot working or machining could also be performed satisfactorily, if required. The mechanical properties of type 317 at the maximum operating temperature of 205°C are also satisfactory, and the tensile strength is only 5% lower at the maximum temperature than it is at ambient conditions. The increased cost per unit weight (compared with, say, type 304 or 316) will be offset by the improved strength which will reduce the minimum wall thickness.

Table 11.6 Material specification for the after-cooler.

Description	Austenitic stainless steel
ASTM designation	A240
Grade	317-00
Composition	0.08% C, 2.00% Mn, 0.045% P, 0.030% S, 0.75% Si, 18.0–20.0% Cr, 11.0–15.0% Ni, 3.0–4.0% Mo
Tensile strength (MPa)	515
Yield strength (MPa)	205
Heat treatment	Solution annealing 1040°C and water quenching

It may be possible to use a low-alloy chromium steel where the vessel walls are only exposed to phthalic anhydride vapours. However, testing would be required as the exact effects of crude phthalic anhydride vapours are not well documented. Regardless, the tubes and the base of the vessel should be constructed of stainless steel as they will be constantly exposed to the liquid product. Overall, the use of a single material, type 317 stainless steel, is considered safer and easier than other options.

Key Results

- The construction materials were selected for chemical resistance, and mechanical properties and cost considerations.
- Stainless steel, type 317, will be used to fabricate the after-cooler.
- The resistance to warm liquid phthalic anhydride was the most critical materials selection consideration.
- Liquid stagnation should be avoided at all times as it will increase the corrosion rate.
- The mechanical properties of type 317 stainless steel are satisfactory for the mechanical processes required during fabrication.

11.9   VESSEL DIMENSIONS

The vessel was designed to adequately house the tube bundle, to provide sufficient clearance above and below the bundle for the gases to distribute evenly and disengage from the condensate, and to allow effective drainage of liquid product. Many dimensions were determined during the chemical engineering design phase. The inlet and outlet nozzles on both sides of the exchanger, but especially the shell-side, were also determined during the chemical engineering design phase as they significantly affected the overall vessel design.

Unconventional vessel shapes (e.g. a box structure with short tubes) could have been used to more effectively house the required components of the exchanger and to meet the chemical engineering design requirements. However, this would require extra material at the construction stage (due to the weaker structural design) and would have added substantially to the overall cost. This is due to the difficulties in fabrication and the cost of completing a detailed mechanical analysis which would be required with an unconventional design. Therefore, a conventional cylindrical design was utilised.

Several vessel parameters can be calculated directly using the methods given in the Australian Standard AS1210 for Unfired Pressure Vessels, or the equivalent British Standard, BS5500. The wall thickness was calculated from equation (11.17) which is from AS1210, while the end thickness (flat, unstayed ends) was calculated from equation (11.18), below, which is from BS5500. The tube-plate thickness was determined from equation (11.19) which is also from BS5500.

Vessel walls:

$$t_w = \frac{PD}{2f\eta - P} \qquad\qquad (11.17)$$

Vessel ends:

$$t_e = C \times D \times (P/f)^{0.5} \qquad\qquad (11.18)$$

Tube plate:

$$t_p = C \times D \times (P/(\lambda \times f))^{0.5} \qquad\qquad (11.19)$$

where: t_w = thickness of vessel walls (mm); t_e = thickness of vessel ends (mm); t_p = thickness of tube-plate (mm); D = shell inside diameter or tube-plate diameter (mm); P = design pressure (MPa); f = tensile strength of material (MPa); C = a design constant dependent on the edge constraint; η = joint efficiency; λ = ligament efficiency.

The design pressure for the shell-side is 200 kPa which is equal to the reactor outlet pressure, and is well above the normal after-cooler operating pressure of 70 kPa. On the tube-side, the design pressure is 400 kPa, which is approximately 10% higher (in absolute terms) than the normal operating pressure. Both these pressures are relatively low so that, where no significant economic penalty exists, the vessel was designed to withstand pressures of up to 700 kPa. This is a typical design philosophy which increases the intrinsic safety level of a vessel. The normal differential pressure across the tube-plate is 290 kPa but it could be up to 400 kPa based on the design pressures. This higher value was used to determine the tube-plate thickness. Note that, in this case, a design pressure of 700 kPa adds significantly to the material requirements and cost of the vessel without improving the intrinsic safety accordingly.

The minimum wall thickness was found to be 2.8 mm. Steel plate is most readily available in nominal thicknesses of $\frac{1}{8}''$ (3.2 mm), $\frac{3}{16}''$ (4.8 mm) and $\frac{1}{4}''$ (6.3 mm), etc. The MAWP with a wall thickness of 4.8 mm, incorporating a corrosion allowance of 1.0 mm (based on a 20-year vessel life), was calculated to be 0.95 MPa (950 kPa). This is acceptable for the after-cooler considering pressure effects only. However, this plate thickness is relatively small for the dimensions of the after-cooler and could present welding difficulties if much thicker steel plate is used for the vessel ends. A heavier duty plate is preferred for practicality and overall vessel integrity, and a thickness of 12 mm ($\frac{1}{2}''$ nominal thickness) was selected.

The minimum allowable end thickness was calculated to be 44.9 mm. An end thickness of 50.8 mm was specified (the nearest standard thickness above the minimum calculated value) which resulted in a MAWP of 0.90 MPa (900 kPa). This is lower than the MAWP calculated for the vessels walls and is, therefore, the overall MAWP for the vessel.

The minimum required thickness of the tube-plates was found to be 74.7 mm, where the ligament efficiency was 21% and the design differential pressure was 0.4 MPa. A tube-plate thickness of 76.2 mm was specified which is well above the minimum allowable tube-plate thickness specified by AS1210 (20 mm). The fixed-head and floating-head of the after-cooler are detailed in Figure 11.6. The tubes should be fixed to the tube-plates by expansion rolling at both ends.

According to AS1210, the vessel should be constructed to Class 2A specifications which requires that the longitudinal welds should be double-welded butt-joints. The circumferential welds should be either double-welded butt-joints or single-welded butt-joints, with backing strips which remain in service. The

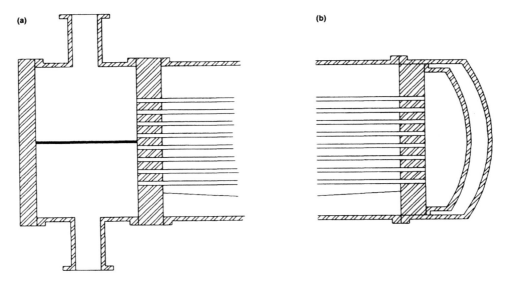

Figure 11.6   Tube bundle end detail: (a) fixed head; (b) floating head.

maximum joint efficiency for these welds is only 80%. For the specified plate thickness, double-welded, single 'V' butt-joints are recommended.

Key Results

- The thicknesses of the vessel walls, vessel ends and the tube-plates were calculated from the tensile strength of stainless steel type 317 (the principal fabrication material) according to the equations from AS1210 and BS5500 which are listed above.
- The vessel walls will be 12 mm thick, including a corrosion allowance of 1.0 mm.
- Flat, unstayed ends (50.8 mm thick) were specified.
- The tube plate thickness will be 76.2 mm. The tubes will be fixed to the tube plate by expansion rolling.
- Double-welded, single 'V' butt-joints are recommended for all welded joints.

11.10   INSULATION

Thermal insulation was selected on the basis of the shape of the vessel walls, the operating temperatures, and an economic thickness of material. Three basic forms of insulation are available: rigid blocks; flexible sheets (or blankets); and spray-on cement. The most suitable type for flat walled vessels is rigid (or semi-rigid) blocks, but flexible sheets are preferred for cylindrical vessels. Here, a combination of rigid blocks (for the vessel ends) and flexible sheets (vessel walls) were specified.

The insulation material must be selected for the maximum operating temperature of the vessel and based on the thermal conductivity of the insulation. At temperatures from 100°C to 200°C, fibreglass, mineral fibre and pearlite foam all have suitable properties. Mineral wool is widely available and economically attractive when compared with the other alternatives. Furthermore, it is available in rigid blocks *and* flexible blankets. To determine the economic thickness of insulation, a unit price (including protective jacket and installation costs) of $100 per 100 mm thickness per m^2 was assumed. A further fixed charge for installation of $500 was incorporated into the optimisation.

The value of energy was estimated as the value of medium-pressure steam less production costs, which was valued at $0.04 per kWhr. Other assumptions were a free convection coefficient of 25 W/m^2/°C (Kreith and Bohn, 1986), thermal conductivity of 0.034 W/m/°C, ambient temperature of 25°C and a three-year effective life for the insulation. Using these values, the optimum thickness of insulation was estimated to be approximately 100 mm. An optimisation procedure is described in Appendix B for the reactor design, and a similar method was applied here. A single layer of preformed rigid and flexible blocks of mineral wool, with sheet metal flashing to protect the insulation from mechanical damage and wear, should be installed on all exposed surfaces of the after-cooler. The insulation should be secured in an appropriate manner using facing tape and metal banding. The flashing should be secured with pins located approximately 250 mm from each edge, at approximately 350 mm centres both horizontally and vertically.

For a small capital outlay, significant energy savings can be achieved by insulating the vessel. The economic analysis was completed by assuming a three-year project life but, if the insulation is still in good condition after this time then it should remain in service. However, if it has become worn or damaged then its effectiveness can be reduced and it will be profitable to replace all or part of the insulation. The economic benefit of installing insulation is substantial and it should be viewed as an essential element of the overall construction process.

Key Results

- An economic thickness of thermal insulation was calculated for the operating conditions.
- Rigid and flexible blocks of mineral fibre should be fitted to the vessel for a minimum thickness of at least 100 mm to provide adequate thermal insulation.
- The insulation should be secured by metal flashing and appropriate fastening.
- The economic benefit from installing insulation is such that it should be viewed as an essential requirement.

11.11 SUPPORTS AND FOUNDATION

The vessel supports were designed to: fit the shape of the vessel; provide access beneath the vessel; and provide sufficient strength to withstand the removal of the tube bundle. Neither the earthquake-load nor the wind-load were considered significant. The plant site is within a geologically stable zone and the vessel only extends by a total of 1.8 m above ground level. The combined weight of the equipment and foundation already provide significant resistance to overturning without further strengthening. A support structure consisting of two concrete piers set on a concrete foundation block was selected for strength, construction simplicity and low cost. Two saddle supports will be constructed from 6 mm steel plate and fixed to the piers in order to secure the vessel. This configuration is shown in Figure 11.7. The saddle detail is given in Figure 11.8.

The piers require reinforcing in order to resist the force created by removal of the tube-bundle. Size 5 re-bars should be set vertically, and size 4 re-bars should be set horizontally in each pier (Schwartz, 1983). Each pier should clear the ground level by 800 mm to provide clearance beneath the vessel for connections and pipes. A width of 450 mm and a depth of 800 mm provides adequate stability and strength, and spreads the vessel load across 0.72 m² which is sufficient to adequately spread the compressive load on the foundation slab. The foundation slab should be set at least 300 mm below ground level and should extend 300 mm beyond the piers. The total length of the foundation slab should be 7.5 m.

The total weight of the vessel, the supports, the piers and the foundation block was estimated at 22,710 kg, of which the vessel contributes approximately 10,180 kg. The compressive stresses on the piers, the foundation and the soil with these loads are all less than 10% of the allowable loads.

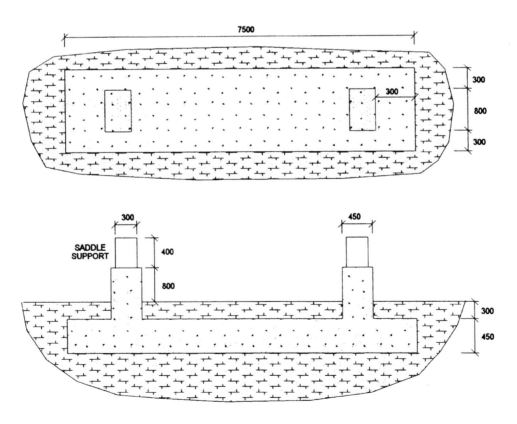

Figure 11.7 After-cooler supports and foundation: plan-view (above); side-view (below).

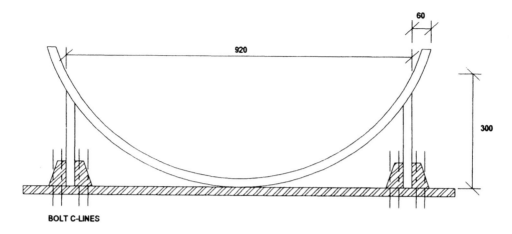

Figure 11.8 Saddle design for after-cooler support.

Key Results

- The after-cooler supports will consist of two steel saddles set on concrete piers which will be supported on a concrete foundation slab.
- The supports and foundation provide adequate strength against likely earthquake and wind loads and also against a bundle-pull.
- The compressive stresses on all components are much less than the allowable stresses.
- The weight of the vessel was estimated to be 10,180 kg, while the total weight of the vessel plus supports and foundation was estimated to be 22,710 kg.

11.12   COSTING

A factored cost estimate for the after-cooler (accurate to 15–20% without vendors' quotes for fabrication) was based on results obtained using the method of Purohit (1983) for shell-and-tube heat exchangers. This method accounts for economies of scale for larger exchangers and most of the standard TEMA exchanger configurations, and is summarised by equation (11.20), below. The calculated result was then extrapolated to 1997 prices using the Chemical Engineering Cost Index and converted into Australian dollars at the current prevailing exchange rate of A\$0.79 = US\$1.00.

$$b = \left[\frac{6.6}{1 - e^{[(7 - D_i)/27]}} \right] \times p \times f \times r \times A \qquad (11.20)$$

where: b = base cost for heat exchanger (1982 US\$); D_i = shell diameter (inch); A = heat transfer area (square feet); p = cost multiplier for tube layout and pitch; f = cost multiplier for front head; r = cost multiplier for rear head.

The base cost of the vessel was modified for the construction materials, and to provide an additional contingency of 10% for the condensate outlet nozzle (not allowed for in the base-case cost estimate). The final estimate of the total delivered cost of the after-cooler was A$41,000 with an uncertainty of at least 25%.

Several additional costs, including installation, foundations and supports, instrumentation and control hardware and miscellaneous other costs, will be incurred in addition to the delivered cost before the after-cooler is in place and operable. These are detailed in Table 11.7 and total approximately A$48,000. *Therefore, the estimated total installed cost of the after-cooler was A$89,000*. This estimate is considered to be accurate to only 25% without more detailed cost information from appropriate vendors.

Key Results

- The base cost of the after-cooler vessel was estimated in US dollars from a correlation for shell-and-tube exchangers and then adjusted for inflation, construction materials and the prevailing Australian dollar exchange rate.
- The total delivered cost was estimated at A$41,000 with additional costs of approximately A$48,000 for installation, supports and instrumentation, etc.
- *The estimated total cost of the after-cooler, installed and operable, was A$89,000, with an uncertainty of approximately 25%.*

Table 11.7   Additional costs for after-cooler.

Item	Rate	Cost
Installation	40% equipment cost	$16,400
Foundations	$270/m³, laid	$1500
Supports	—	$2500
Fixings	—	$500
Insulation	—	$2500
Process control	20% equipment cost	$9600
Inventory	10% equipment cost	$4800
Other expenses	30% equipment cost	$14,400
Total		$48,000

11.13 ENGINEERING SPECIFICATION

Mechanical Design for After-Cooler

Functional Description	*Desuperheater-cooler-condenser*		
Type of Exchanger	*Shell-and-tube*		
Orientation	*Horizontal*		
Operation	*Continuous*		
Design Code	*AS1210 / BS5500*		

Shell-Side		Tube-Side	
Type	*TEMA type E*	Type	*Plain, seamless*
Material	*Stainless steel, type 317*	Material	*Stainless steel, type 317*
Diameter	*1000 mm i.d.*	Number	*280*
Passes	*1*	Passes	*2*
Baffles	*5*	Length	*6.0 m*
Baffle Type	*Horizontal-cut, seg.*	Diameter	*25.4 mm o.d.*
Front Head	*Fixed (type A)*	Gauge	*14 BWG (medium)*
Rear Head	*Floating (type S)*	Pattern	*Square*
Insulation	*Mineral fibre (100 mm)*	Pitch	*32 mm centres*
Nozzles	*1 inlet x 760 mm,*	Nozzles	*1 inlet x 150 mm NB,*
	1 outlet x 760 mm,		*1 outlet x 150 mm NB*
	1 outlet x 40 mm NB		

Vessel		Supports and Foundation	
Wall Thickness	*12 mm*	Supports	*Steel saddles (2),*
Ends Thickness	*51 mm*		*concrete piers (2)*
Tube-plate Thickness	*76 mm*	Foundation	*Concrete slab*
Corrosion Allowance	*1.0 mm*		
Operating Weight	*10,180 kg*	Combined Weight	*12,530 kg*

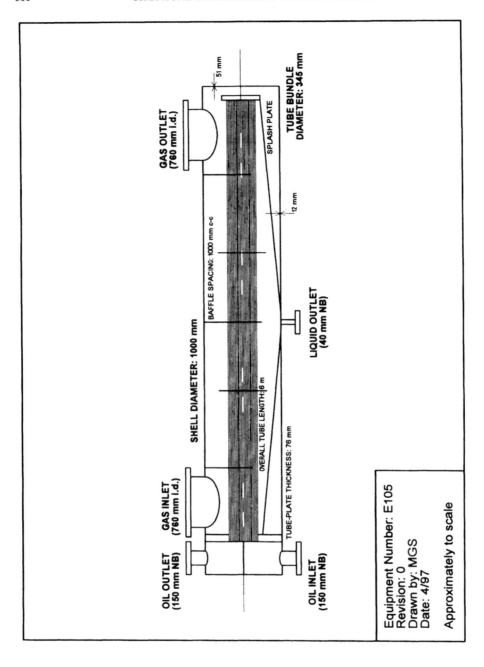

Figure 11.9(a) Cross-sectional view of the phthalic anhydride after-cooler.

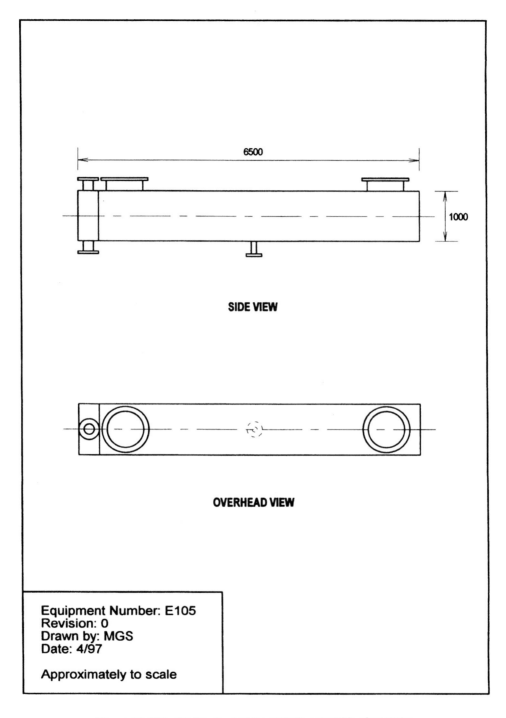

Figure 11.9(b) Profile views of the phthalic anhydride after-cooler.

PART III: OPERATIONAL CONSIDERATIONS

11.14 HAZOP ANALYSIS

A preliminary HAZOP analysis of the after-cooler was used to identify a range of precautionary measures which were, subsequently, incorporated into the design. The HAZOP tables (one for each stream associated with the after-cooler) are included in Appendix G. A final HAZOP must still be completed for the 'as-built' process before commissioning. The final HAZOP analysis should also be done by a multidisciplinary team, rather than an individual design engineer, in order to incorporate input from as many relevant sources as possible. The principal results arising from the preliminary HAZOP analysis are detailed in Sections 11.15 to 11.18 which discuss process hazards, safety, operability and environmental considerations.

11.15 PROCESS HAZARDS

There are essentially three types of streams associated with the after-cooler: reaction gases, containing phthalic and maleic anhydride vapours at moderate to high temperatures (135–205°C); diathermic oil at moderate temperatures (100–140°C); and crude phthalic anhydride condensate (98–99% pure). Each stream has the potential to be hazardous if they are accidentally released to the environment.

The principal process hazard associated with the gas streams is the possibility of an explosion occurring if the vessel shell or connecting pipes are breached. Phthalic anhydride forms explosive mixtures with air at concentrations of 1.5–10.5% and both the inlet and outlet gas streams contain sufficient phthalic anhydride to create a mixture within that range of concentrations. Two measures will be taken to counter-act this hazard: regular operator patrols of all gas lines will be instigated; and all sources of ignition will be prohibited from the site. Any leak will be clearly visible as hot phthalic anhydride vapours will sublime on contact with air and will resemble a white dust cloud. If a leak is detected, from either a vessel or attached pipe-work, an immediate emergency shut-down should be initiated.

Phthalic anhydride condensate is also potentially hazardous. Liquid phthalic anhydride is a severe irritant to the skin and moist tissue, including the eyes, nose and mouth. Contact with liquid phthalic anhydride can cause burns and induce diseases of the respiratory and digestive systems. Two sources of confusion are the cause of many accidents involving contact with phthalic anhydride. First, molten phthalic anhydride has the colour and viscosity of water. Second, the low thermal conductivity of the solid can insulate a pool of liquid from further solidification. As a result, hot liquid can be trapped below a layer of solid which then becomes treacherous, like thin ice. The surface solid is insufficiently strong to support the weight of a plant worker so that stepping on the hazard will release hot liquid product which may subsequently burn the worker.

Phthalic anhydride vapours which are released by the liquid are also toxic, and breathing apparatus should be used during the clean-up of any spills or leaks. The maintenance staff responsible for the control and clean-up of a spill, and all other workers, should be fully trained in the hazards associated with phthalic anhydride. Extreme care is required in handling any leak, vessel rupture or accidental spill.

The primary hazard associated with the diathermic oil streams is the potential for an employee to be burned by an unplanned release of oil from the system. Serious burns are possible at the temperatures used and it may be difficult to clean the hot, viscous liquid from an employee's skin or clothing. A major oil leak would warrant an emergency shut-down due to the loss of cooling. However, the after-cooler can be taken out of service for short periods, by transferring load to the switch-condensers and recycling reaction gases, so that a temporary repair can be affected in some circumstances.

All maintenance work should be carefully planned and monitored to ensure that the clearly dangerous practice of opening a 'live' line is avoided. During planned shut-downs, the integrity of all lines should be checked and gaskets and other sealing devices should be replaced to prevent possible accidents.

Key Results

- A leak from either the inlet or outlet gas streams has the potential to form an explosive mixture with air.
- A phthalic anhydride leak will initially be visible as a white dust cloud.
- Early detection of leaks and prohibition of all sources of ignition from the plant are vital.
- Accidental releases of molten phthalic anhydride represent a serious hazard to personnel. Operators should be trained in recognition and handling of liquid phthalic anhydride.
- The sudden release of hot diathermic oil could severely burn an employee who was accidentally exposed to the breach.
- Good practices during shut-downs and other maintenance activities are required to prevent accidents.

11.16 SAFETY

Safety will be a priority of the plant management. Effective safety policies are not only important to protect plant employees but have been shown to raise the productivity and effectiveness of employees. Safety has been considered at all stages of the design process and appropriate safety measures have been incorporated into the final design wherever they were required. Notably, care has been taken to minimise small operating problems and/or inconveniences so that they do not become the cause of future accidents.

Effective shut-down procedures, which are easy to implement, are essential to deal with emergency situations. The basic requirements for the after-cooler are: a bypass to allow the reaction gases to pass directly to the switch-condensers; and isolation valves around the unit. Both the bypass line and the isolation valves must be operable from the control room, and should be fitted with manual overrides so that they can still be operated during electrical or pneumatic failures. It will be possible to initiate the shut-down sequence automatically but there will be a series of checks that must be completed before process operation can recommence. It is not necessary to purge the vessel contents before an emergency shut-down as this can be done after the process is stopped.

A non-return valve should be installed in the liquid product line leaving the after-cooler. This prevents reverse flow into the vessel if there is a pressure build-up at or before the crude phthalic anhydride storage facilities. Intermediate storage tanks allow the reaction and recovery units (including the after-cooler) to be operated for several days during a shut-down of the purification section.

A regular maintenance program is essential to minimise process disruptions caused by equipment failure. A high process utilisation (defined as the percentage of time when the plant is fully operational) is important for plant profitability, but it is also effective in promoting good safety performance. Accidents are least likely when the plant is stable and accidents are most likely during disruptions which are frequently caused by equipment failures. In addition to the normal events and activities which occur during routine plant shut-downs, the rate of solid formation in the after-cooler should be investigated and recorded. The objective is to identify operating conditions that result in rapid solid formation which reduces the overall effectiveness of the after-cooler. The tube-bundle should also be cleaned regularly, to prevent unscheduled shut-downs and to maintain a high level of performance between successive shut-downs.

Key Results

- Safety will be a priority of the plant management.
- Shut-down procedures are designed for safety and speed.
- A feed bypass line (to divert reaction gases directly to the switch-condensers) and isolation valves, which are operable from the control room, will be installed on the after-cooler.
- The crude phthalic anhydride storage facilities will be protected by non-return valves in the condensate product line from the after-cooler.
- Regular maintenance is essential to maximise both plant utilisation and the plant safety performance.

11.17 OPERABILITY

The LAR process is sequential and all upstream activities have an impact on the operation of downstream units. Therefore, the crude phthalic anhydride yield from the after-cooler is dependent on the composition of the reactor product gas and, to a lesser extent, the outlet temperature from the gas cooler.

The phthalic anhydride recovery duty is shared between the after-cooler and the switch-condensers. The distribution of this duty can be varied (within limits) so that the after-cooler may be operated at a reduced loading or even removed from service for short periods. This facility provides flexibility but appropriate planning and fore-thought are required to maximise its value to the plant. Overuse will have a detrimental effect on the quality of crude product and the overall production rate.

The performance of the after-cooler must be monitored continuously to ensure optimal operation and also to detect reduced performance (which might be caused by fouling or excessive solid formation) and equipment failure. Samples of the gas streams and the liquid product should be taken regularly and analysed by the on-site laboratory. Several variables are important: the concentrations of phthalic and maleic anhydride must be measured to calculate the recovery yield; the water concentration in the condensate product must be measured to ensure that downstream acid formation is minimised; and the condensate purity should be measured, as a higher than normal concentration of by-products might indicate a possible blockage near the outlet.

Tight control of the after-cooler outlet temperature is important to prevent excessive formation of solid phthalic anhydride which has the potential to significantly reduce the heat transfer rate, and to block the condensate outlet. Ideally, only 45–52% of the phthalic anhydride in the reaction gases should be recovered in the after-cooler. During normal operation of the after-cooler, excess vapours will pass through the system to the switch-condensers. However, if the flow of liquid product becomes too great (blockages occurring or other problems), then part of the reaction gases can be vented directly to the switch-condensers.

Key Results

- Correct operation of upstream units is necessary to maintain an acceptable condensate yield from the after-cooler.
- All or part of the condensing duty can be transferred to the switch-condensers for short periods if required, but the overall crude product quality will be reduced.
- Regular sampling and analysis of gas and liquid streams should be employed to adequately monitor the performance of the unit.
- The after-cooler outlet temperature (and, therefore, the yield of phthalic anhydride condensate) should be continuously controlled to minimise the risk of solid formation in the vessel.

11.18 ENVIRONMENTAL CONSIDERATIONS

The after-cooler forms part of a closed system during normal process operation. There should be no release of any products (harmful or otherwise) to the environment unless an item of equipment fails and there is a breach of the vessel shell or a connecting pipe. There are no pressure relieving devices that vent directly to the atmosphere, or other possible sources of emissions. Before a shut-down, the whole process will be purged of hydrocarbons using nitrogen. The purged vapours will be flared or vented depending on the concentration of hydrocarbon present. Appropriate isolation equipment (block-and-bleed valves and bypass lines) are included on the after-cooler in order to control a process breach and limit its effect on the environment.

Three types of emissions are possible following a process breach: hydrocarbon vapour (including toxic phthalic anhydride and maleic anhydride vapours); crude phthalic anhydride condensate; and diathermic oil. The effects of these streams on personnel are discussed in Section 11.15 with respect to process hazards.

The environmental impacts of each of these streams relate primarily to the effect on people, and to the risk of an explosion.

The only significant source of noise from the after-cooler is the circulating coolant (diathermic oil) pump, which should be located in a specifically designated pump area where noise can be isolated from employees and neighbours (both residential and other industries) of the plant. The after-cooler itself will emit a low level of noise as vapour passes through the vessel at a relatively high speed, but will not create a noise hazard or disturbance.

The principal source of 'normal' emissions from the phthalic anhydride process is the waste gas stream from the switch condensers which is vented after being scrubbed to reduce the concentration of toxic components. The after-cooler has some impact on the composition of these waste gases as material not condensed in the after-cooler adds to the load on the switch condensers. Therefore, overall emissions from the plant are minimised by continual reliable operation of the after-cooler.

Key Results

- There are no emissions from the after-cooler or associated equipment during normal operation.
- A process breach has the potential to release toxic hydrocarbon vapour, liquid (crude) phthalic anhydride and/or diathermic oil.
- No significant sources of noise are present around the after-cooler.
- Continual, reliable operation of the after-cooler helps to minimise emissions from the plant.

11.19 CONTROL AND INSTRUMENTATION

The critical aspect of operation for the after-cooler is to accurately control the outlet temperature. This maximises the yield of condensate and minimises the formation of solids inside the vessel. If the outlet temperature is too high, insufficient phthalic anhydride is recovered and the switch condensers may be overloaded. If the outlet temperature approaches the melting point of phthalic anhydride ($130.8°C$), the condenser tubes may become coated with solid which will reduce the heat transfer rate and limit the throughput of reaction gases.

Only one primary control loop is required: temperature control of the outlet temperature. Both the coolant (diathermic oil) flow rate and the coolant temperature are available for manipulation by an automatic controller. The control scheme shown in Figure 11.10 was designed for disturbance rejection and set-point tracking so that very tight temperature control could be achieved.

Cascade control was used to improve the dynamic responsiveness of the control system, and was implemented by using the primary temperature controller (on the condensate outlet) output as the input to a secondary temperature controller (on the coolant inlet). Disturbances that affect the coolant inlet temperature are rejected by the secondary controller before they affect the primary control loop. The overall response is faster as the primary control loop can be

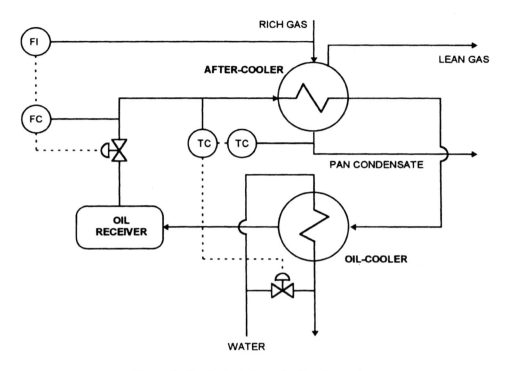

Figure 11.10   Control scheme for the after-cooler.

tuned quicker than would be possible without the cascade loop. Simple PI controllers are recommended for both control loops as the process response should be nearly linear for a reasonable range of temperatures around the optimum operating point.

Feed-forward control was also used to help reject disturbances that originate upstream of the after-cooler, and that would be perpetuated through the reaction gas inlet stream. The feed rate of the reaction gas inlet stream is monitored and the signal sent to a feed rate controller on the coolant circuit. A fixed ratio of coolant flow to gas flow is maintained by the controller (ratio control). Therefore, operator intervention is not required when the plant feed rate is changed, and the after-cooler will respond more smoothly.

Although not recommended, the heat transfer rate in the after-cooler could also be controlled by manipulating the condensate draw rate in order to vary the heat transfer area by covering and uncovering part of the tube-bundle. This type of control could be used to provide effective temperature control but would be slower to respond to disturbances in the coolant temperature and the gas inlet rate.

Temperature indicators should be installed at the inlet and outlet gas streams, and at several points along the liquid product line. Within the vessel, temperature readings allow the cooling duty to be calculated and provide continuous data regarding the performance of the after-cooler. Abnormally high temperatures indicate that the unit is unable to maintain the design service, and cleaning or a reduction in gas flow is required. As the molten phthalic anhydride product flows at just above its melting point (by about 5°C), the liquid lines must be continuously monitored to detect solidification and prevent potential blockages. The thermodynamic properties of solid phthalic anhydride are such that under normal conditions, a solid film will form on the pipe wall and liquid flow will continue inside. If a thick film is allowed to develop, it may become difficult to sustain normal operation. Low-level alarms, set at the melting point, should be connected to all temperature indicators in molten phthalic anhydride streams. Steam can then be used to melt the solid wherever it forms in the pipes between the after-cooler and the crude product storage facilities.

The coolant inlet will be regulated as part of the cascade control system but the outlet temperature does not need to be monitored continuously. However, a thermocouple should be provided so that the outlet temperature can be checked periodically to allow the after-cooler performance to be checked rigorously. This is useful to detect fouling or the excessive formation of solid on the tube surfaces.

The pressures inside the vessel are generally not critical but the shell-side pressure drop provides another indication of fouling inside the tube shell. Similarly, facilities are required to periodically check the tube-side pressure drop to detect fouling on the oil side.

Flow meters should be installed in the gas lines (inlet and outlet), liquid product line and the coolant line. The gas inlet flow and the coolant flow are required for the feed-forward controller on the coolant flow rate, but the measurement of all three streams is useful in monitoring the performance of the after-cooler and also determining the yield of phthalic anhydride from the overall process.

Key Results

- The primary controller will regulate the coolant inlet temperature in order to control the condensate outlet temperature.
- A secondary controller will be used to regulate the coolant inlet temperature in a cascade loop with the primary controller in order to improve the dynamic responsiveness of the system.
- A ratio control system will be used to regulate the coolant flow rate.
- The facility to measure pressures on the inlet and outlet of both the shell-side and the tube-side will be installed to detect fouling.
- Temperature indicators will be installed at the gas inlet and outlet and at several points along the condensate line.
- Flow meters will be installed to provide a continuous measure of the yield of condensate.

11.20 OPERATING CONSIDERATIONS

11.20.1 Operation Under Normal Conditions

The reactor product gases, containing approximately 10.5% phthalic anhydride vapour, are cooled by the gas cooler immediately downstream of the reactor, and then enter the after-cooler at about 205°C. The gases are cooled by contact with 25.4 mm Ø tubes which are kept below the condensing temperature by coolant (diathermic oil) which enters the exchanger at 100°C. As the temperature of the vapour decreases, the saturation pressure of phthalic anhydride vapours falls until it equals the partial pressure in the gas mixture. At this point, the temperature is approximately 155°C and the first droplets of liquid phthalic anhydride form and drop to the base of the after-cooler. The vapour temperature continues to fall via contact with the relatively cold tubes and the rate of condensation increases. Liquid phthalic anhydride with a purity of approximately 98.0% collects on the sloped base of the after-cooler and flows towards the centrally located condensate outlet nozzle. The gas leaves the after-cooler at approximately 136°C at which point it contains only about 50% of the phthalic anhydride that was present at the inlet.

Small amounts of maleic anhydride co-condense with phthalic anhydride but all of the other components, except water, are not present in sufficient quantity to condense at the temperatures present. A small amount of water condenses as its dew point exceeds the after-cooler outlet temperature. This is important as the water reacts quickly with liquid phthalic anhydride to form phthalic acid which reduces the overall purity of the phthalic anhydride product. Fortunately, only a very small quantity of water condenses and the acidification reaction is reversible.

Phthalic anhydride which remains in the vapour phase is recovered by sublimation in the switch-condensers which are located immediately downstream of the after-cooler. The products of both the after-cooler and the switch-condensers are transferred to intermediate storage before purification. Both the after-cooler and the switch-condensers expel heat to the circulating diathermic oil. This heat is recovered via heat exchange with feed-water for the HP steam generators which remove heat from the foundry salt which circulates through the reactor at high temperature.

A single operator would normally control both recovery units (the after-cooler and the switch-condensers) and all the associated equipment, including the oil-cooler(s). This is beneficial as it facilitates effective interaction of the two units, which is considered essential for optimal operation of the whole plant. A single operator is best placed to coordinate shifts in recovery load between the two units and to ensure that high purity phthalic anhydride is being recovered at all times. Frequent adjustments on the outlet temperatures of both units may be required to balance the recovery load and minimise the contamination of the phthalic anhydride product. Care will also be required to prevent solidification in liquid lines.

The control system will be capable of controlling much of the operation of the unit, but the operator has a responsibility to oversee all adjustments to the unit to prevent accidents from occurring. If control of the after-cooler is lost, the first action should be to temporarily transfer the load to the switch-condensers. If the problem persists, then it may be necessary to reduce the hydrocarbon feed rate to the reactor.

Key Results

- The reactor product gases are cooled from 205°C to 136°C in the after-cooler, by contact with tubes which are cooled by circulating diathermic oil. About 50% of the phthalic anhydride vapour which is present at the inlet is condensed.

- Crude molten phthalic anhydride is collected at the base of the after-cooler, with small quantities of impurities, and is then transferred directly to an intermediate storage tank.
- One operator is required to control both the after-cooler and the switch-condensers. Accurate control of the outlet temperature is critical.
- Abnormal situations should be controlled by either partially transferring the load to the switch-condensers or reducing the reactor feed rate.

11.20.2 Commissioning

Before the after-cooler is commissioned, several pre-commissioning checks and tests must be completed. First, the mechanical integrity of the complete vessel (including the internals, as far as possible) should be checked by visual inspection. Second, debris from the fabrication process should be flushed from the system to prevent corrosion or potential blockages in the tube-bundle. Third, a hydrostatic test should be conducted on both the shell-side and the tube-side.

The commissioning of the after-cooler should take place concurrently with the commissioning of the reactor, salt-cooler, gas-cooler and switch-condensers. The chief operator must effectively co-ordinate the activities in each area so that each item of equipment is ready for commissioning as the feed becomes available. The key steps in the commissioning procedure for the after-cooler are summarised in Table 11.8.

The first tasks for the operator responsible for the after-cooler are to establish cold circulation of diathermic oil through the after-cooler and water through the oil-cooler. The reaction gases will initially be very lean in phthalic anhydride, and there will be little or no flow of condensate from the after-cooler until the reactor operation is stabilised and has a significant volume of *o*-xylene in the feed. The

Table 11.8 Commissioning procedure for the after-cooler.

Step	Time	Action
1	0:00 hrs	Establish cold circulation of diathermic oil at 50% of design rate.
2	0:30 hrs	Estabilsh water flow through the oil-cooler at minimum rate.
3	1:00 hrs	Introduce reaction gases from the gas-cooler, and allow the hot gases to slowly heat up the exchanger until the outlet temperature reaches 120°C.
4	1:30 hrs	Open the condensate outlet line and start to collect condensate (as it appears) in a temporary storage vessel.
5	2:00 hrs	Set the primary temperature controller to automatic with a set point of 136°C. Set the oil-cooler control system to automatic. Set the coolant flow rate ratio controller to automatic.
6	4:00 hrs	Initiate a sampling program for the condensate product. Ensure the samples are analysed immediately and report results to the chief-operator.
7	10:00 hrs	Once a phthalic anhydride purity of 95% is reached, redirect the condensate product to a new storage tank for immediate processing in the purification section.
8	24:00 hrs	Commence a testing program to determine optimum operating conditions (especially the condensate outlet temperature).

hot gases should be used to heat up the after-cooler until the vapour outlet temperature approaches the design operating point (136°C). During this period, essentially all the phthalic anhydride (and any maleic anhydride which is present) will be recovered by the switch-condensers.

At high o-xylene concentrations in the reactor feed, the partial pressure of phthalic anhydride in the reaction gases will be sufficient for condensation to begin in the after-cooler. However, the composition of the initial condensate product will be unsuitable for collection in the intermediate storage tanks. A temporary collection vessel should be used until a satisfactory condensate purity is achieved. This product can be slowly reblended into the crude phthalic anhydride product after stable plant operation has been attained.

Once a flow of condensate from the after-cooler has been established, condensate samples should be taken every 30 minutes and immediately dispatched to the on-site laboratory for analysis. The results of these tests are important in the overall plant commissioning procedure and should be immediately reported to the chief operator. Its is important to quickly establish a balance in the product recovery distribution between the after-cooler and the switch-condensers.

Soon after operation has been successfully stabilised and the product quality is acceptable, a testing program should be established to determine the optimum operating conditions. Samples should be taken and operating temperatures (and pressures) recorded for various sets of operating conditions. The condensate outlet temperature is the most important variable in this process but it is also important to establish operability limits for all other variables, and to determine the relationship between the outlet temperature, the recovery yield and the product purity. This process should include testing the operation of the after-cooler at temperatures below the recommended (design) operating point, although this involves some risk as solidification of the condensate may occur. These tests will require close attention, particularly the condition of the liquid outlet and attached piping. Those areas should be monitored closely for several weeks after commissioning to assess the impact of the operating conditions being used.

Key Results

- The commissioning of the after-cooler should be concurrent with the commissioning of the reactor, gas-cooler and switch-condensers.
- Condensate product will not be produced until the reactor feed contains a high concentration of o-xylene, and the reactor operating conditions are close to design.
- Condensate should be collected in a temporary storage vessel until its quality is suitable for processing in the purification section of the plant.
- A testing program should be commenced soon after stable operation is reached in order to find the optimum operating conditions for the after-cooler.

11.20.3 Shut-Down and Start-Up

Two types of shut-down are considered: planned and emergency. During a planned shut-down, the reactor feed rate is slowly reduced to minimum. At that time, the after-cooler can be taken off-line by diverting all the reaction gases to the switch condensers. The low feed rate allows normal product recovery to be maintained using only the switch condensers. Once bypassed, the after-cooler can be drained of liquid product and then purged with nitrogen, if required.

The major steps involved in a shut-down of the after-cooler are detailed in Table 11.9. The coolant flow rate should be reduced to minimum as soon as the vessel is fully bypassed in order to prevent excessive solidification of condensate. However, coolant circulation should continue for at least four hours after bypassing to ensure that the phthalic anhydride vapour concentration has been reduced to a minimum. The isolation valves on the liquid product line and gas exit line should be left open until the vessel contents have been drained. Nitrogen purging would generally only be required for a vessel inspection, otherwise shutting all of the isolation valves completes the shut-down of the after-cooler.

An emergency shut-down is also initiated by diverting the flow of reaction gases to the switch-condensers. However, at high reactor feed rates, the switch condensers will be unable to effectively handle the entire gas flow and some phthalic anhydride will escape through the switch-condenser vents and will enter the scrubbing system. Under these circumstances, the reactor feed rate must be reduced immediately to prevent the scrubbing system from becoming overloaded, which would eventually result in the discharge of PAN and MAN vapours to the atmosphere.

If an emergency shut-down is the result of a failure unrelated to the operation of the after-cooler, the after-cooler can be drained and purged as if for a planned shut-down. However, if a failure in some part of the after-cooler is the cause of the shut-down, the vessel should be immediately isolated. Once the problem has been fully diagnosed with the vessel in this 'safe' condition, remedial action can be determined and implemented. If a blockage is suspected (either in the outlet line or the tube-bundle), the coolant flow should be reduced to the minimum safe flow rate to prevent excessive solidification of condensate. If a leak is detected in the outlet gas line, the coolant flow should be maximised to reduce the concentration of phthalic anhydride in the vapours which are escaping to atmosphere.

The after-cooler start-up procedure is similar to the commissioning procedure which was summarised in Table 11.8. Stable operation of the switch-condensers should slightly precede stable operation of the after-cooler as phthalic anhydride recovery will commence in the switch-condensers (due to the requirement for a high phthalic anhydride partial pressure in the after-cooler before condensation will occur). Initially, the reaction gases are used to heat up the after-cooler so that the outlet temperature can be stabilised near the desired operating point. With all

Table 11.9 Planned shut-down procedure for the after-cooler.

Step	Action
1	Divert reaction gases to the switch-condensers.
2	Reduce coolant flow rate to minimum.
3	Drain molten phthalic anhydride (condensate).
4	Shut isolation valves.
5	Purge vessel with inert gases (if necessary).

of the control systems operating automatically, the flow rate of condensate can build up until the operating conditions approach design. At full operating load, the total flow of reaction gases should pass through the after-cooler where approximately 50% of the phthalic anhydride vapours are recovered, before proceeding to the switch-condensers where the remaining PAN vapours are sublimed.

Key Results

- Shut-downs are initiated by diverting the flow of reaction gases to the switch-condensers.
- Isolation valves should be used to restrict the cause of emergency shut-downs. The after-cooler can be successfully shut-down without completely purging the vessel contents, if necessary.
- A planned shut-down will allow the vessel to be drained before being isolated.
- Start-up of the after-cooler follows the switch-condensers and is timed to commence as the reaction gas composition approaches the normal operating conditions.

11.20.4 Regular Maintenance

A full plant shut-down will be required every 18–24 months to allow the catalyst in the reactor to be replaced. Major maintenance jobs should all be scheduled during this time. Two to three weeks should be sufficient for a complete catalyst changeout and appropriate reactor inspections, and this is suspected to be the 'critical path' job. Therefore, all other jobs will have to be completed in this period.

After-cooler inspections should only be required at every second plant shut-down, although inspections are recommended for both of the first two shut-downs to assess the rate of scaling and fouling in the after-cooler and all other critical heat exchangers. The after-cooler should be fully cleaned (both tube-side and shell-side) whenever the tube-bundle is removed from the shell for inspection. Corrosion tests (thickness measurements) should also be completed at every shutdown for the first four years to assess the resistance to the PAN vapours and condensate. If necessary, samples of solid deposits should be taken for analysis. If some locations are found to be highly susceptible to fouling, extra baffles may need to be added to increase the fluid velocity.

It will sometimes be necessary to schedule a minor shut-down between the major shut-downs to complete smaller maintenance jobs which are critically affecting the operation of the plant. During these shut-downs, which should generally last less than a week, only the exchanger tubes need to be inspected (in-situ) unless definite evidence exists of fouling. High-pressure hoses and/or chemical cleaners should be used to accelerate the cleaning process on these occasions.

Key Results

- Major shut-downs will be required every 18–24 months to change the catalyst in the reactor. All planned maintenance activities should take place during planned major shut-downs.

- The after-cooler will initially be cleaned at each major shut-down. Thereafter, the tube-bundle will only be removed from the shell for a complete clean (shell-side and tube-side) at every second shut-down.
- Tube-side cleans of the after-cooler are possible during minor shut-downs without removing the tube-bundle.
- The vessel should be regularly checked for corrosion and deposits of condensed phthalic anhydride.

PART IV: CONCLUSIONS, RECOMMENDATIONS AND REFERENCES

11.21 CONCLUSIONS

Chemical Engineering Design

- The first phase of the design process was the selection of an appropriate exchanger type.
- The second phase was coolant selection.
- The third phase was an iterative determination of the exchanger configuration and its heat transfer area, using a specially developed computer model which was written in FORTRAN.
- The fourth phase of the design process was the calculation and evaluation of the remaining design parameters using various literature correlations.
- A standard exchanger design, a shell-and-tube condenser, was selected for the after-cooler.
- A configuration having condensation on the outside of horizontal tubes was selected on the basis of expected heat transfer coefficients, vapour–liquid disengagement, and ease of cleaning.
- Diathermic oil was the preferred coolant as it allows the inlet temperature to be tightly controlled so that condensing phthalic anhydride does not solidify.
- The use of diathermic oil makes the control of the coolant inlet temperature independent of other process units so that their operation is not compromised via heat integration.
- An auxiliary heat exchanger is required to exchange heat between the hot oil and coolant.
- The analysis of a desuperheater–cooler–condenser should be considered in two sections: with and without condensation.
- There are four heat transfer resistances occurring in the desuperheating section: coolant film, scale deposits, tube wall and gas phase.
- There are five heat transfer resistances occurring in the cooling–condensing section: coolant film, scale deposits, tube wall, gas phase and diffusion.
- A computer model was formulated to allow an iterative approach to the after-cooler design.
- The computer model calculates the required heat transfer area, the coolant flow rate and the condensate film temperature for a specified condenser geometry.
- The shell diameter and tube-side nozzle diameter were estimated by the program so that the exchanger geometry could be optimised.

- No correction was made for condensation in the calculation of the pressure drop as it accounts for only 5% of the total gas flow rate.
- The after-cooler tube diameter, gauge and length were selected from commonly available sizes.
- The heat transfer rate in the desuperheating zone was significantly lower than in the condensing zone.
- The shell diameter was selected to provide adequate clearance above and below the tube-bundle.
- A sloped base was installed to encourage the condensate to flow directly to the outlet nozzle in order to minimise the risk of corrosion.
- The vapour nozzles were sized smaller than the economic pipe diameter due to the relatively small exchanger shell diameter.
- The liquid-product outlet nozzle was sized to minimise the potential for pipe blockages caused by solidification.
- An impingement plate was specified due to the high vapour velocity at the inlet.
- Medium-gauge tubes were used to provide adequate corrosion resistance.
- A square tube-pattern layout with a standard pitch (32 mm centres) was used to simplify tube-bundle cleaning during maintenance.
- Pressure drops were calculated to be 1.0 kPa on the shell-side and 1.7 kPa on the tube-side.

Mechanical Engineering Design

- The construction materials were selected for chemical resistance, and from mechanical properties and cost considerations.
- Stainless steel, type 317, will be used to fabricate the after-cooler.
- Corrosion resistance to warm liquid phthalic anhydride was the most critical material consideration.
- Liquid stagnation should be avoided at all times as it will increase the corrosion rate.
- The mechanical properties of type 317 stainless steel are satisfactory for the mechanical processes required during fabrication.
- The vessel walls will be 12 mm thick, including a corrosion allowance of 1.0 mm.
- Flat, unstayed ends (50.8 mm thick) were specified.
- The tube plate will be 76.2 mm. The tubes will be fixed to the tube plate by expansion rolling.
- Double-welded, single 'V' butt-joints are recommended for all welded joints.
- The after-cooler supports will consist of two steel saddles set on concrete piers which will be supported on a concrete foundation slab.
- The supports and foundation provide adequate strength against likely earthquake and wind loads and also against a bundle-pull.
- The compressive stresses on all components are much less than the allowable stresses.
- The weight of the vessel was estimated to be 10,180 kg, while the total weight of the vessel plus supports and foundation was estimated to be 22,710 kg.
- The base cost of the after-cooler vessel was estimated in US dollars from a correlation for shell-and-tube exchangers and then adjusted for inflation, construction materials and the prevailing Australian dollar exchange rate.

- The total delivered cost was estimated at A$41,000 with additional costs of approximately A$48,000 for installation, supports and instrumentation, etc.
- The estimated total cost of the after-cooler, installed and operable, was A$89,000, with an uncertainty of approximately 25%.

Operational Considerations

- A leak from either the inlet or outlet gas streams has the potential to form an explosive mixture with air. Such a leak will initially be visible as a white dust cloud.
- Early detection of leaks and prohibition of all sources of ignition from the plant are vital for the safety of the plant and personnel.
- Accidental releases of molten phthalic anhydride represent a serious hazard to personnel. Operators should be trained in recognition and handling of liquid phthalic anhydride.
- The sudden release of hot diathermic oil could severely burn an employee who was accidentally exposed to the breach.
- Safe operating practices during shut-downs and other maintenance operations are required to prevent accidents.
- Safety will be a priority of the plant management.
- Detailed shut-down procedures should be designed for safety and speed.
- A feed bypass line (to divert reaction gases directly to the switch-condensers) and isolation valves, which are operable from the control room, will be installed on the after-cooler.
- The crude phthalic anhydride storage facilities will be protected by non-return valves in the condensate product line from the after-cooler.
- Regular maintenance is essential to maximise plant utilisation and for plant safety performance.
- Good operation of upstream units is necessary to maintain an acceptable condensate yield from the after-cooler.
- All or part of the condensing duty can be transferred to the switch-condensers for short periods if required, but the overall crude product quality will be reduced.
- Regular sampling and analysis of gas and liquid streams should be employed in order to adequately monitor the performance of the unit.
- The after-cooler outlet temperature (and, therefore, the yield of phthalic anhydride condensate) should be continuously controlled to minimise the risk of solid formation in the vessel.
- There are no emissions from the after-cooler or associated equipment during normal operation.
- A process breach has the potential to release toxic hydrocarbon vapour, liquid (crude) phthalic anhydride and/or diathermic oil.
- No significant sources of noise are present around the after-cooler.
- Continuous, reliable operation of the after-cooler helps to minimise emissions from the plant.
- The primary control system on the after-cooler will regulate the coolant inlet temperature in order to control the condensate outlet temperature.

- A secondary controller will be used to regulate the coolant inlet temperature in a cascade loop with the primary controller, in order to improve the dynamic responsiveness of the system.
- A ratio control system will be used to regulate the coolant flow rate.
- The facility to measure pressures on the inlet and outlet of both the shell-side and the tube-side will be installed to detect fouling.
- Temperature indicators will be installed at the gas inlet and outlet and at several points along the condensate line.
- Flow meters will.be installed to provide a continuous measure of the yield of condensate.
- The reactor product gases are cooled from 205°C to 136°C in the after-cooler, by contact with tubes which are cooled by circulating diathermic oil, in order to condense about 50% of the phthalic anhydride vapour which is present at the inlet.
- Crude molten phthalic anhydride is collected at the base of the after-cooler with small quantities of impurities and is then transferred directly to intermediate storage.
- One operator is required to control both the after-cooler and the switch-condensers. Accurate control of the outlet temperature is critical.
- Abnormal situations should be controlled by either partially transferring the load to the switch-condensers or reducing the reactor feed rate.
- The commissioning of the after-cooler should be concurrent with the commissioning of the reactor, gas-cooler and switch-condensers.
- Condensate product will not be produced until the reactor feed contains a high concentration of o-xylene and the reactor operating conditions are close to design.
- Condensate should be collected in a temporary storage vessel until its quality is suitable for processing in the purification section of the plant.
- A testing program should be commenced soon after stable operation is reached in order to find the optimum operating conditions for the after-cooler.
- Shut-downs are initiated by diverting the flow of reaction gases to the switch-condensers.
- Isolation valves should be used to restrict the cause of emergency shut-downs. The after-cooler can be successfully shut-down without completely purging the vessel contents, if necessary.
- A planned shut-down will allow the vessel to be drained before being isolated.
- Start-up of the after-cooler follows the switch-condensers and is timed to commence as the reaction gas composition approaches the normal operating conditions.
- Major shut-downs will be required every 18–24 months to change the catalyst in the reactor. All planned maintenance activities should take place during planned major shut-downs.
- The after-cooler will initially be cleaned at each major shut-down. Thereafter, the tube-bundle will only be removed from the shell for a complete clean (shell-side and tube-side) at every second shut-down.
- Tube-side cleans of the after-cooler are possible during minor shut-downs without removing the tube-bundle.
- The vessel should be regularly checked for corrosion and deposits of condensed phthalic anhydride.

11.22 RECOMMENDATIONS

The after-cooler design presented here is full and complete. Although no further work is required, the following areas should be reviewed if sufficient project resources are available:

- VLE data for the condensation of phthalic anhydride.
- Fouling coefficients should be compared with actual operating experience from other phthalic anhydride plants.

This design is NOT sufficient for construction of a phthalic anhydride after-cooler. The following items and tasks are still to be completed before construction can begin:

- Isometric drawings of the after-cooler, oil-cooler and all connecting equipment and pipes.
- Installed costs must be confirmed with suitable vendors.
- Contracts with construction company(s) must be negotiated and signed.
- The requisite funds for equipment construction, installation and commissioning must be secured.
- Relevant government approvals for construction must be obtained.
- The necessary engineering, operating and maintenance staff must be hired.

REFERENCES

ASTM Standards, *A240: Heat Resisting Chromium and Chromium–Nickel Stainless Steel Plate, Sheet and Strip for Pressure Vessels* (1989).

ASTM Standards, *A354: Quenched and Tempered Alloy Steel Bolts, Studs and Other Externally Threaded Fasteners* (1989).

Baillie, I., Estimating and Construction Costs, *The Architect*, 30(1), 41–47 (1990).

Brown, A.A., Tank Foundation Design, *Hydrocarbon Process.*, 53(10), 153–156 (1974).

Coulson, J.M. and Richardson, J.F., *Chemical Engineering Volume 1* (4th ed.), Pergamon, New York (1990).

DeRenzo, D.J., *Corrosion Resistant Materials Handbook* (4th ed.), Noyes Data Corp, New Jersey (1985).

Everett, A., *Materials*, Batsford, London (1981).

Fraas, A.P. and Necati Ozisik, M., *Heat Exchanger Design*, Wiley, New York (1965).

Gere, J.M. and Timoshenko, S.P., *Mechanics of Materials* (2nd ed.), Wadsworth International, Boston (1985).

Grant, I.D.R., Condenser Performance — The Effect of Different Arrangements For Venting Non-Condensing Gases, *Brit. Chem. Eng.*, 14(12), 1709–11 (1968).

Harker, J.H., Finding an Economic Pipe Diameter, *Hydrocarbon Process.*, 57(3), 74–76 (1978).

Kern, D.Q., *Process Heat Transfer*, McGraw-Hill, New York (1950).

Kreith, F. and Bohn, M.S., *Principles of Heat Transfer* (4th ed.), Harper & Row, New York (1986).

Mark, H.F. *et al.* (eds), Phthalic Acids and Other Benzene-polycarboxylic Acids, in (Kirk-Othmer) *Encyclopedia of Chemical Technology* (3rd ed.), Wiley, New York (1978–84).

McAdams, W.H., *Heat Transmission* (3rd ed.), McGraw-Hill, New York (1954).

McKetta, J.J. (ed.), Phthalic Anhydride, (McKetta) *Encyclopedia of Chemical Processes and Design*, Marcel Dekker, New York (1990).

MICA, *Commercial and Industrial Insulation Standards* (1979).

National Association of Corrosion Engineers, *Corrosion Data Survey* (6th ed.), NACE, Houston (1974).

Perry, R.H. and Green, D. (eds), *Perry's Chemical Engineers' Handbook* (6th ed.), Chapters 4, 10, 11, McGraw-Hill, New York (1984).

Peters, M.S. and Timmerhaus, K.D., *Plant Design and Economics for Chemical Engineers* (4th ed.), McGraw-Hill, New York (1991).

Purohit, G.P., Estimating Costs of Shell-and-Tube Heat Exchangers, *Chem. Eng. (N.Y.)*, 22 August, 56–67 (1983).

Rabald, E., *Corrosion Guide* (2nd ed.), Elsevier, Amsterdam (1968).

Schlunder, E.N., *Heat Exchanger Design Handbook*, Hemisphere Publishing, Dusseldorf (1983).

Schwab, R.F. and Doyle, W.H., Hazards in Phthalic Anhydride Plants, *Chem. Eng. Prog.*, 66(9), 49–53 (1970).

Schwartz, M., Supports for process vessels and storage equipment, *Chem. Eng. (N.Y.)*, 3 September, 119–123 (1983).

Shinskey, F.G., *Process Control Systems* (4th ed.), McGraw-Hill, New York (1996).

Sinnott, R.K., *Chemical Engineering Volume 6* (2nd ed.), Pergamon, Oxford (1993).

Standards Association of Australia, *AS1210:1989 Unfired Pressure Vessel Code*, Sydney (1989).

Standards Association of Australia, *AS1358:1989 Bursting Disks and Bursting Disk Devices*, Sydney (1989).

APPENDIX E.   CALCULATIONS FOR PHTHALIC ANHYDRIDE AFTER-COOLER DESIGN

E.1   PROVISIONAL AFTER-COOLER DESIGN

The initial design was based on seven assumptions:

1. equal recovery of the PAN vapours as liquid (after-cooler) and solid (switch-condensers);
2. liquid purity of 98% PAN;
3. condensate film temperature 1.5°C higher than the melting point;
4. sub-cooling is 20% of the overall ΔT;
5. *minimum* shell diameter is equal to the inlet nozzle diameter (1000 mm);

6. tube layout consists of 6.0 m × 25 mm diameter tubes on a 32 mm square pitch;
7. coolant temperature rise of 15–20°C.

The after-cooler outlet temperature was estimated to be approximately 136°C from vapour–liquid equilibrium considerations using assumptions (1) and (2), and the physical property database in HYSYS. The melting point of PAN at the inlet pressure is 130.8°C. The film temperature (i.e. the temperature of the liquid PAN on the surface of the tubes) was estimated to be 132.3°C using assumption (3). This defines the extent of sub-cooling as the difference between the equilibrium temperature (136°C) and the film temperature (132.3°C), and therefore the overall temperature difference as 18.5°C, using assumption (4). The coolant (diathermic oil) inlet temperature should, therefore, be around 120°C to provide the necessary ΔT. The number of tubes that could be fitted into the minimum shell diameter was estimated from assumptions (5) and (6) to be 729. The mass flow rate of coolant was estimated to be approximately 44 kg/s from assumption (7) and the known after-cooler duty of 1290 kW. With only a single tube-pass, the tube-velocity for this combination of mass flow rate and number of tubes is given by:

$$u = \frac{\dot{m}}{\rho \times N_T \times ((\pi/4)d^2)} \tag{E.1}$$

$$= \frac{44}{885 \times 729 \times ((\pi/4)0.0212^2)}$$

$$= 0.19 \, \text{m/s}$$

where: u = tube-side velocity (m/s); $\dot{m}$ = mass flow rate of coolant ≈ 44 kg/s; ρ = density of coolant at the operating temperature = 885 kg/m³; N_T = number of tubes per pass; d = tube inside diameter = 21.2 mm.

The calculated tube-side velocity is very low and not acceptable. Therefore, more than one tube pass must be used, and fewer than the 729 tubes must be used for a satisfactory design. Initially, two tube passes were preferred. The tube count was re-estimated using equations (E.2) and (E.3) below, an estimated overall heat transfer coefficient of 800 W/m²/°C (for condensing hydrocarbon vapours) and a log mean temperature difference (LMTD) of 10°C. The tube-count using this method was estimated to be 335, and this number was used as the starting point for the optimisation of the after-cooler.

$$A = \frac{Q}{U \times \Delta T} \tag{E.2}$$

$$N = \frac{A}{\pi \times D \times L} \tag{E.3}$$

where: A = heat transfer area (m²); Q = after-cooler duty = 1290 kW; U = estimated overall heat transfer coefficient = 800 W/m²/°C; ΔT = estimated LMTD = 10°C; N = number of tubes; D = tube outside diameter = 25.4 mm; L = tube length = 6.0 m.

E.2 SHELL-SIDE CROSS-FLOW AREA

The cross-flow area which is available for flow (i.e. not occluded by tubes) is a function of the tube pitch, the number of tubes in each row and the number of baffles:

$$A = \frac{(P-OD)}{P} \times N_{T/R} \times P \times \frac{L}{(B+1)} \qquad (E.4)$$

where: A = cross-flow area (m^2); P = tube pitch (mm); OD = tube outside diameter (mm); $N_{T/R}$ = number of tubes per row; L = tube length (m); B = number of shell-side baffles.

Initially, $P = 32.0$ mm, OD = 25.4 mm, $N_{T/R} \approx 13$, $L = 6.0$ m and $B = 1$. With this configuration, the shell-side cross-flow area is 0.26 m^2. However, equation (E.4) must be re-evaluated for each new after-cooler configuration.

E.3 PRESSURE DROPS

E.3.1 Tube-Side Pressure Drop

The tube-side pressure drop was estimated from the following correlations for the pressure loss in straight tubes (equation (E.5)), and entrance and exit losses (equation (E.6)):

$$\Delta P = f \times \frac{\rho \times u^2}{2} \times \left(\frac{L}{D}\right) \times N_P \qquad (E.5)$$

$$= 0.028 \times \frac{885 \times 0.40^2}{2} \times \left(\frac{6.0}{0.021}\right) \times 2$$

$$= 1.13 \text{ kPa}$$

$$\Delta P = 4 \times \frac{\rho \times u^2}{2} \times N_p \qquad (E.6)$$

$$= 0.57 \text{ kPa}$$

where: ΔP = pressure drop (Pa); f = friction factor = 0.028 at a Reynold's number of 56,000; ρ = coolant density = 885 kg/m^3; u = tube-side velocity = 0.40 m/s; L = tube length = 6.0 m; D = tube inside diameter = 0.021 m; N_P = number of tube-side passes = 2.

The overall tube-side pressure loss is given by:

$$\Delta P_{TOT} = \Delta P_1 + \Delta P_2 \qquad (E.7)$$

$$= 1.70 \text{ kPa}$$

where: ΔP_{TOT} = total tube-side pressure drop (kPa); ΔP_1 = pressure loss due to the tubes = 1.13 kPa; ΔP_2 = pressure loss due to entrances and exits = 0.57 kPa.

E.3.2 Shell-Side Pressure Drop

The shell-side pressure drop in a shell-and-tube heat exchanger can be calculated using a correlation from Kern (1950):

$$\Delta P = \frac{f \cdot G_s^2 \cdot D_s \cdot (N+1)}{2 \cdot \rho \cdot g \cdot D_e \cdot \phi} \qquad (E.8)$$

$$= 0.76\,\text{kPa}$$

where: f = friction factor (determined graphically from the Reynold's number and surface roughness) = 0.0026×144; G_s = mass flux of the condensing vapours $\approx$ 290 kg/m²/s; D_s = shell inside diameter = 1.00 m; N = number of baffles = 5; ρ = density of the condensing vapours $\approx$ 1.3 kg/m³; D_e = effective tube diameter = 0.030 m; ϕ = number of shells = 1.

The entrance and exit losses were estimated to be approximately 0.2 kPa (equivalent to two velocity heads), but could not be determined exactly due to the complex exchanger geometry around the nozzles. Therefore, the total shell-side pressure drop was estimated to be 1.0 kPa.

E.4 MECHANICAL DESIGN

E.4.1 Wall Thickness

The wall thickness was specified in accordance with AS1210 for a Class 1 pressure-vessel, using an equation based on circumferential stress:

$$t = \frac{P \cdot D}{2 \cdot f \cdot \eta - P} \qquad (E.9)$$

$$= 2.8\,\text{mm}$$

where: t = minimum wall thickness (mm); D = shell inside diameter = 1000 mm; P = design pressure = 0.7 MPa; f = tensile strength of stainless steel, type 317 at the operating temperature = 125 MPa; η = joint efficiency = 100%.

A corrosion allowance of 1.0 mm (equivalent to 0.05 mm for 20 years) was added to the minimum thickness required for adequate mechanical strength (2.8 mm) in order to determine the minimum plate thickness for the vessel walls. The next thicker standard size for steel plate is 4.8 mm (nominal thickness: $\frac{3}{16}''$), and this thickness was specified for the after-cooler shell. The maximum allowable working pressure (MAWP) for the after-cooler was determined from:

$$P = \frac{2 \cdot f \cdot \eta \cdot t}{D+t}$$

$$= 950\,\text{kPa}$$

where: t = (actual thickness − corrosion allowance) = 3.8 mm; and other variables are defined above.

E.4.2 Head Covers

Flat, unstayed head covers were used for both ends of the after-cooler. The minimum thickness of the head covers (end-plates) can be calculated from equation (E.10), which is from the British Standard BS5500:

$$t = C \times D \times \left(\frac{P}{f}\right)^{0.5} \tag{E.10}$$

$$= 44.9 \, \text{mm}$$

where: t = minimum required thickness (mm); C = a constant from BS5500 = 0.6 for welded plate; D = diameter of the end-plate = 1000 mm; P = design pressure = 0.7 MPa; f = tensile strength of the end-plate material = 125 MPa. Steel plate with a thickness of 50.8 mm (nominal thickness: 2″ was specified). The corresponding MAWP was calculated to be 900 kPa. Therefore, the overall MAWP for the after-cooler is also 900 kPa.

E.4.3 Tube-Plate Thickness

The minimum required tube-plate thickness was calculated in accordance with BS5500 from:

$$t = C \times D \times \left(\frac{P}{\lambda \times f}\right)^{0.5} \tag{E.11}$$

$$= 74.7 \, \text{mm}$$

where: t = minimum required thickness (mm); C = a constant from BS5500 = 0.6 for bolted joints; D = diameter of the tube-plate = 1000 mm; P = maximum differential pressure across the tube-plate = 0.4 MPa; λ = ligament efficiency = 21% for the pitch and tube diameter in the after-cooler (see BS5500); f = tensile strength of the tube-plate material = 125 MPa. A tube-plate thickness of 76.2 mm was selected (nominal thickness: 3″). Although designed from BS5500, this thickness is also in accordance with AS1210 which specified a minimum tube-plate thickness of 20 mm, but which provides an alternative method of calculation.

E.4.4 Nozzle Diameters

The process connections (inlets and outlets) were sized using the Harker equation (1978) for the optimum pipe (and nozzle) diameter, modified for SI units:

$$D_{opt} = 8.41 \frac{W^{0.45}}{\rho^{0.31}} \tag{E.12}$$

where: D_{opt} = pipe diameter (mm); W = mass flow rate (kg/hr); ρ = fluid density (kg/m^3).

Generally, the nozzle was sized as the standard pipe size nearest to the optimum, but both the gas inlet and outlet were sized below the optimum to

Table E.1 Reactor nozzle diameters.

Stream	W(kg/hr)	ρ(kg/m³)	D_{opt}(mm)	D
Rich gas inlet	49,650	1.55	953	760 mm
Lean gas outlet	47,120	1.13	1026	760 mm
Liquid product outlet	2530	1178	31.9	40 mm NB
Diathermic oil inlet	62,840	885	148	150 mm NB
Diathermic oil outlet	62,840	885	148	150 mm NB

conserve space and reduce the number of tube rows which would have to be removed to accommodate the larger nozzles. The optimum pipe diameters and selected sizes for all the inlet and outlet streams are given in Table E.1.

E.5 SUPPORTS

E.5.1 Weight of the Vessel

The components of the total weight of the after-cooler are summarised in Table E.2. A steel density of 7850 kg/m³ was assumed for all metal components. Two types of impact load were considered: general impacts (e.g. accidents, etc.); and a bundle-pull which resulted in an impact equivalent to 50% of the weight of the bundle. However, the combined weight of impact loads, live loads and the process streams was determined to be less than the weight of water which would be present during a hydrostatic test. Therefore, only the hydrostatic test condition weight was evaluated.

E.5.2 Saddles and Piers

The after-cooler will be supported and secured at two points using saddle supports mounted on concrete piers. An evenly distributed load would results in equal forces on each saddle and pier, equal to:

$$F = \frac{m \times g}{2} \tag{E.13}$$

$$= \frac{10,180 \times 9.81}{2}$$

$$= 49.9 \, \text{kN}$$

where: F = force on each saddle and pier (N); m = total weight of the after-cooler, during the hydrostatic test = 10,180 kg.

Each saddle support should be able to withstand the full load due to the direct weight of the vessel. Therefore, the cross-sectional area of each support can be

Table E.2 After-cooler weight during hydrostatic test.

Component	Weight (kg)
Shell, including miscellaneous fittings	1480
Tube-plate	770
Tubes	2030
Baffles and other internals	200
Unsupported piping	450
Insulation	150
Water	5100
Total	10,180

found from a simple structural analysis:

$$CSA = \frac{F \times \cos(\theta)}{\sigma} \tag{E.14}$$

$$= \frac{49{,}900 \times \cos(35°)}{55}$$

$$= 744\,mm^2$$

where: CSA = cross-sectional area of the saddle support (mm^2); F = force on the support (N); σ = compressive strength of the support material $\approx$ 50% tensile strength = 55 MPa; θ = reaction angle; $\cos(\theta)$ = (shell outside diameter/distance between supports). The required cross-sectional area can be comfortably satisfied with 6 mm plates which extend over the length of the saddle (300 mm).

The load on each pier is dependent only on the maximum weight of the vessel. The compressive stress on a pier which is 800 mm deep and 450 mm wide (see Figure 11.7) is given by:

$$\sigma = \frac{F}{0.800 \times 0.450} \tag{E.15}$$

$$= 0.14\,MPa.$$

The maximum allowable compressive stress on concrete is approximately 5.0 MPa so that the specified design is more than adequate. However, each pier should be reinforced in the vertical plane to resist the overturning moment resulting from a bundle pull. Size 5 re-bars, set 75 mm from the edges of each pier, are recommended.

E.6 FOUNDATION

E.6.1 Footing

The maximum compressive stress on the concrete footing is equal to the compressive load calculated for the piers. Therefore, no further structural checks are required for the concrete footing as the compressive load is significantly less than the allowable limit.

E.6.2 Soil Stresses

The weight of the saddle supports, the concrete piers and the foundations was estimated to be 12,530 kg using a concrete density of 2400 kg/m³ and a steel density of 7850 kg/m³. The maximum compressive stress on the underlying soil was calculated from the total weight (vessel plus supports and foundation) and the foundation area, and it was found to be only 21 kPa. The soil bearing capacity of the sandy soil in the region should be around 300 kPa (to be confirmed by geological testing). Therefore, no additional strengthening is required.

APPENDIX F. FORTRAN PROGRAM FOR PHTHALIC ANHYDRIDE AFTER-COOLER SIMULATION

```
C       Program to calculate the heat transfer area for After
C       Cooler in phthalic anhydride process (LAR).
C       Desuperheating condenser with non-condensables.
C
C-------------------------------------------------------------
C
C       Desuperheating
C       Four resistances:
C           1. Convection through coolant
C           2. Scale resistance
C           3. Conduction through tube wall
C           4. Gas phase resistance
C
C       Condensation
C       Five resistances:
C           1. Convection through coolant
C           2. Scale resistance
C           3. Conduction through tube wall
C           4. Condensate film resistance
C           5. Vapour diffusion through gas phase
C
C-------------------------------------------------------------
C
C       Inputs:
C       1. Tube diameters (inside and outside)
C       2. Tube length
C       3. Total number of tubes
C       4. Tube pitch
C       5. Number of baffles
```

```
C      6. Tube-side velocity
C      7. Inlet coolant temperature
C
C      Outputs:
C      1. Temperature profiles through gas, condensate film and
C         diathermic oil
C      2. Heat transfer area required
C      3. The actual area as a fraction of the required area
C      4. The coolant flow rate required
C      5. Estimates of the shell diameter and tube-side nozzle
C         diameter
C
C----------------------------------------------------------
C
       REAL DENSV(9), CPV(9), VISV(9), KV(9),
      .           DENSF(9), CPF(9), VISF(9), KF(9),
      .           DENSW, CPW, VISW, KW, KST, OD, ID, PITCH,
      .           AREAG, TEMP(9), FLOWG(9), Q(8), TF(9), TW(9),
      .           QTOT, FLOWW, AREAW, DTW, VELW,
      .           RE, PR, NU, SC, KG, UT(9), LH,
      .           H1, H2, H3, H4, H(9), RD, U, HTOT, LMTD, LMUT,
      .           MA, MB, VA, VB, LHS, RHS,
      .           PRESS, PF, PS, PSF, PV, PGLN, DT1, DT2,
      .           SA, HTAREA(8), AREA, AREA2, LEST, LENGTH
       INTEGER I, J, K, NP, NT, NBAF
C
C      Physical properties for reaction gases and condensate
C      film.
       DATA  DENSV/1.086,1.182,1.186,1.180,1.180,1.172,1.169,
      .1.166,1.154/
       DATA CPV/1.105,1.088,1.087,1.086,1.086,1.085,1.085,
      .1.085, 1.084/
       DATA VISV/2.46E-5,2.31E-5,2.29E-5,2.26E-5,2.25E-5,
      .2.24E-5, 2.22E-5,2.22E-5,2.21E-5/
       DATA KV/0.0350,0.0326,0.0323,0.0321,0.0320,0.0319,
      .0.0318,0.0317,0.0316/
       DATA DENSF/1161,1161,1161,1167,1170,1174,1177,1179,
      .1181/
       DATA CPF/1.708,1.708,1.708,1.698,1.692,1.686,1.680,
      .1.676,1.673/
       DATA VISF/1.05E-3,1.05E-3,1.05E-3,1.12E-3,1.17E-3,
      .1.22E-3,1.27E-3,1.31E-3,1.34E-3/
       DATA KF/0.114,0.114,0.114,0.115,0.116,0.116,0.117,
      .0.117,0.117/
C
C      Oil Properties.
       DENSW = 885
```

```
          CPW = 1.945
          VISW = 3.5E-3
          KW = 0.114
C
C         Water Properties.
C         DENSW = 960
C         CPW = 4.26
C         VISW = 2.2E-4
C         KW = 0.68
C
C         Other Miscellaneous Properties.
          KST = 25.0
          LH = 485000
          MA = 148
          MB = 29
          VA = 140
          VB = 29.5
C
          DATA TEMP/205,160,155,150,146,143,140,138,136/
          DATA Q/591.6,112.9,188.5,104.8,99.8,95.2,60.6,36/
          DATA FLOWG/13.489,13.489,13.409,13.180,13.061,12.952,
         .12.853, 12.793,12.756/
C
C         Inputs.
          PRINT*,'        Enter the tube outside diameter (mm) : '
          READ*,OD
          OD = OD/1000
          PRINT*,'        Enter the tube inside diameter (mm) :   '
          READ*,ID
          ID = ID/1000
          PRINT*,'        Enter the tube length (m) :             '
          READ*,LEST
          PRINT*,'        Enter the total number of tubes :       '
          READ*,NT
          PRINT*,'        Enter the number of tube passes :       '
          READ*,NP
          PRINT*,'        Enter the tube pitch (mm) :             '
          READ*,PITCH
          PITCH = PITCH/1000
          PRINT*,'        Enter the number of baffles :           '
          READ*,NBAF
          PRINT*,'        Enter the tube-side velocity :          '
          READ*,VELW
          PRINT*,'        Enter the oil inlet temperature (C) :   '
          READ*,TW(9)
C
C         Initialize variables
```

```
         QTOT = 0.0
         AREA = 0.0
         RD = 3.5E-4
C
C        Calculations
C
C        Step 1 - Coolant temperature profiles
         AREAW = (NT/NP) * (3.14159 * ID * ID/4)
         FLOWW = AREAW  * VELW * DENSW
         DO 10 J = 8,1,-1
         DTW = Q(J) / (FLOWW * CPW)
         TW(J) = TW(J+1)+DTW
         QTOT = QTOT + Q(J)
   10    CONTINUE
C
C        Step 2 - Coolant film coefficient
         RE = DENSW * ID * VELW / VISW
         PR = 1000 * CPW * VISW / KW
         NU = 0.027 * 0.8 * PRRE * 0.333
         H1 = KW**NU / ID
C
C        Step 3 - Tube wall coefficient
         H2 = 2 * KST / (OD * LOG(OD/ID))
C
C        Step 4 - Gas side coefficients
         PRESS = 1.35
         DO 100 I = 1,9
C
C        Both sections (gas phase resistance)
         AREAG = (PITCH-OD) * (NT/2))**0.5 * LEST / (NBAF+1)
         RE = DENSV(I) * FLOWG(I) / (AREAG * VISV(I))
         PR = 1000 * CPV(I) * VISV(I) / KV(I)
         NU = 0.40 * RE**0.60 * PR**0.36
         H(I) = NU * KV(I) / OD
C
C        Condensing section (two resistances)
         IF (I .GE. 2) THEN
         DIFF = 1.0E - 7 * TEMP(I)**1.75 * (1/MA+1/MB**0.5 /
        .(PRESS * (VA**0.333 + VB**0.333)**2)
         SC = VISV(I) / (DENSV(I) * DIFF)
         KG = H(I) * (PR/SC**0.667 / (CPV(I)*1000)
         TF(I) = TEMP(I) - 0.25 * (TEMP(I) - TW(I))
         DELTF = 2.5
   20    PV = (101.325/760) * 10**8.022 - 2868.5/(TEMP(I)+273))
         PS = (1.35 * 101.325) - PV
         PF = (101.325/760) * 10**8.022 - 2868.5/(TF(I)+273))
         PSF = (1.35*101.325) - PF
```

```
        PGLN = (PS - PSF)/LOG(PS/PSF)
        H4 = 1.25 * 0.725 * (KF(I)**3 * DENSF(I)**2 * LH * 9.81/
       .(OD * VISF(I)*(TF(I) - TW(I)))**0.25
        HTOT = 1 / (ID/OD/H1 + RD + 1/H2 + 1/H4)
        LHS = H(I)*(TEMP(I)- TF(I)) + KG * LH * (PV - PF)/PGLN
        RHS = HTOT * (TF(I) - TW(I))
        TF2 = (LHS/HTOT) + TW(I)
             IF (ABS(TF2-TF(I)) .GT. 0.001 * TF(I)) THEN
                  IF (LHS .GT. (LHS+RHS)/2) THEN
                       TF(I) = TF(I) + DELTF
                  ELSE
                       TF(I) = TF(I) - DELTF
                  ENDIF
                DELTF = DELTF/2
                  IF (DELTF .LT. 0.0001) DELTF = 2.0
                       GOTO 20
                  ENDIF
                UT(I) = (LHS + RHS)/2
             ENDIF
  100   CONTINUE
C
C       Step 5 - Heat transfer area
        DO 200 I = 1,8
        IF (I .EQ. 1) THEN
C
C           Desuperheating section
            H3 = (H(1) + H(2))/2
            U = 1 / (ID/OD/H1 + RD + 1/H2 + 1/H3)
            DT1 = TEMP(1) - TW(1)
            DT2 = TEMP(2) - TW(2)
            LMTD = (DT1 - DT2) / LOG(DT1/DT2)
            HTAREA(I) = 1000 * Q(I) / (LMTD * U)
        ELSE
C
C           Condensing section
            LMUT = (UT(I) - UT(I+1)) / LOG(UT(I)/UT(I+1))
            HTAREA(I) = 1000 * Q(I) / LMUT
        ENDIF
  200   CONTINUE
C
C       Step 6 - Heat transfer area
        DO 30 J = 1,8
        AREA = AREA + HTAREA(J)
   30   CONTINUE
C
C       Short cut calculations
        U = 1 / (ID/OD/H1 + 1/H2 + RD + 1/H3)
```

```
      DT1 = TEMP(1) - TW(1)
      DT2 = TEMP(9) - TW(9)
      LMTD = (DT1 - DT2) / LOG(DT1/DT2)
      AREA2 = 1000 * QTOT / (LMTD * U)
C
C     Output Results
C
      PRINT*,'------------------------------------------------'
      PRINT*,
      PRINT*,'          Heat transfer... '
      PRINT*,
      PRINT*,'              HEAT (kW)            AREA (m²) '
      QTOT = 0
      AREA = 0                                         •
      DO 50 K = 1,8
      QTOT = QTOT + Q(K)
      AREA = AREA + HTAREA(K)
      PRINT4,QTOT,AREA
  50  CONTINUE
      PRINT*,
      PRINT*,'------------------------------------------------'
      PRINT*,
      PRINT*,' Temperature profiles (based on counterflow)... '
      PRINT*,
      PRINT*,'      GAS          PAN          OIL'
      PRINT2,TEMP(1),' - ',TW(1)
      DO 40 K = 2,9
      PRINT3,TEMP(K),TF(K),TW(K)
  40  CONTINUE
      PRINT*,
      PRINT*,'------------------------------------------------'
      PRINT*,
      PRINT1,' The required heat transfer area is : ',AREA,' m2'
      PRINT5,' The actual area is ',LEST/LENGTH * 100,'% of the
     .required area.'
      PRINT*,
      PRINT6,'        The required coolant flow rate is : ',
     .(AREAW * VELW * 3600),' m3/hr'
      PRINT6,'    The estimated tube bundle diameter is : ',
     .(PITCH * 1000 * NT**0.5) + 25,'mm '
      PRINT6,' The estimated coolant nozzle diameter is : ',
     .(4/3.1416 * AREAW * VELW /2.5)**0.5 * 1000,'mm'
      PRINT*,
      PRINT*,' Using short cut calculations...   '
      PRINT1,' The required heat transfer area is : ',AREA2,' m2'
      PRINT*,
      PRINT*,'------------------------------------------------'
```

```
      PRINT*,
    1 FORMAT (A44,F6.1,A4)
    2 FORMAT (9X,F5.1,11X,A5,12X,F5.1)
    3 FORMAT (9X,F5.1,11X,F5.1,12X,F5.1)
    4 FORMAT (15X,F6.1,14X,F6.2)
    5 FORMAT (A26,F5.1,A23)
    6 FORMAT (A50,F5.1,A7)
C
  300 CONTINUE
C
      STOP
      END
```

APPENDIX G. HAZARD AND OPERABILITY STUDY FOR PHTHALIC ANHYDRIDE AFTER-COOLER

Table G.1 Reactor product gases from gas-cooler at 49.65 T/hr and 205°C.

Key Word	Deviation	Possible Causes	Consequences	Action Required
NONE	No flow	1. Flow stopped upstream.	Process stops. Inconvenience but no hazard.	(a) Warnings should be installed upstream.
		2. Isolation valve fails shut.	As for 1. Valve may overheat and/or rupture.	(b) Install low pressure warning and manual override on valve.
		3. Line breakage.	As for 1. Release of explosive mixture to atmosphere.	(c) Regular operator patrol of all lines. Emergency shutdown procedures.
MORE OF	More flow	4. Reactor operating at above rated capacity.	Increased duty on cooling system.	(d) Ensure over-capacity of oil supply.
	More temperature	5. Upstream cooling equipment working ineffectively.	As for 4.	Covered by (d).
		6. Reaction favouring complete oxidation.	Low concentration of PAN, low flow of liquid product.	(e) Good operation of reactor. Regular renewal of catalyst.
LESS OF	Less flow	7. Reduced feed.	Reduced steam production from reactor.	No action required.
		8. Leaking flange.	As for 7. Release of explosive mixture to atmosphere.	Covered by (c).
	Less temperature	9. Excess cooling upstream.	Maybe small quantity of liquid in feed.	No action required.
	Less pressure	10. Partial line blockage.	Reduced throughput.	(f) Install bypass line or clear blockage. Maintenance to follow.
		11. Isolation valve partially open.	Rich gases bypass after-cooler.	Close valve and repair, if required.
PART OF	Low PAN concentration	12. Reaction favouring complete oxidation (catalyst spent).	Reduced flow of liquid PAN.	Covered by (e).
OTHER	Maintenance	13. General equipment failure or catalyst changeover in reactor.	Process stops.	(g) Good practices in construction and operation. Ensure shutdown and start-up procedures are well detailed.

Table G.2 Liquid phthalic anhydride (crude) to storage at 2.53 T/hr and 136°C.

Key Word	Deviation	Possible Causes	Consequences	Action Required
NONE	No flow	1. Blockage at outlet.	Vessel fills with liquid, possibly over-pressuring joints.	(a) Regular sampling and analysis of liquid.
			Liquid carried out with vapour.	(b) Shut down and maintenance.
		2. No PAN in feed.	Reduced condensation from feed. Reduced pressure in line.	(c) Good operation of reactor. Regular renewal of catalyst.
		3. Valve fail shut.	As for 1. Valve may overheat and/or rupture.	(d) Install manual override on valve.
MORE OF	More flow	4. High concentration of PAN in feed.	Increased risk of line or vessel blockage.	(e) Good operating procedures. Limit upstream operations to design capacity.
	More temperature	5. Inadequate cooling.	Liquid may start to vaporise again. Two phase flow.	(f) Control diathermic oil flow based on feed rate and composition.
	More pressure drop	6. Partial blockage in line or vessel.	As for 4.	Covered by (a) and (b).
LESS OF	Less flow	7. Low concentration of PAN in feed.	Process inconvenience but no hazard.	Covered by (c).
		8. Partial blockage at outlet.	As for 4. Liquid build-up in vessel.	Covered by (a) and (b).
	Less temperature	9. Excessive cooling.	Oil may start to condense forming acid from anhydride.	Covered by (f).
AS WELL AS	Low purity of crude PAN	10. Excessive by-products from reactor.	Increased duty in purification units.	Covered by (c).
REVERSE	Reverse flow of liquid PAN	11. Crude storage vessel full and over pressured.	Vessel fills with liquid, possibly over-pressuring joints. Liquid carried out with vapour. No condensation from feed.	(g) Install high level alarm in crude storage vessel.
OTHER	Maintenance	12. General equipment failure or catalyst changeover in reactor.	Process stops.	(h) Good practices in construction and operation. Ensure shutdown and start-up procedures are well detailed.

Table G.3 Lean gases to switch condensers at 47.12 T/hr and 136°C.

Key Word	Deviation	Possible Causes	Consequences	Action Required
NONE	No flow	1. Isolation valve fails shut.	Vessel may become over pressured. Valve may overheat and /or rupture.	(a) Install pressure relief system including warnings and alarms. (b) Install manual override on valve. (c) Emergency shutdown.
		2. Line or vessel breakage.	Process stops downstream. Release of explosive mixture to atmosphere.	
MORE OF	More flow	3. No condensation of PAN.	Switch Condensers overloaded.	(d) Install control system to ensure temperature is matched to flow and composition. (e) Good operating procedures. Limit upstream operations to design capacity.
		4. Increased feed.	Risk of equipment or line failure increased.	
	More temperature	5. Inadequate cooling.	As for 3.	**Covered by (d) and (e).**
	More pressure drop	6. Partial blockage in vessel or line.	As for 4.	**Covered by (e).** (f) Implement regular maintenance.
LESS OF	Less flow	7. Leaking flange.	Release of explosive mixture to atmosphere.	Covered by (c).
		8. Reduced feed.	No hazard.	No action required.
		9. Gas leaving with liquid.	Contamination of crude PAN.	(g) Install vent from liquid line to switch condenser feed line.
	Less temperature	10. Excessive cooling.	Condensation of moisture in line with the potential formation of phthalic acid.	Covered by (d).
PART OF	Low oil concentration.	11. Oil condensation in vessel.	Phthalic acid formed.	Covered by (d).
AS WELL AS	High PAN concentration	12. Reduced condensation in vessel.	As for 3.	Covered by (d) and (e).
OTHER	Maintenance	13. General equipment failure or catalyst changeover in reactor.	Process stops.	(h) Good practices in construction and operation. Ensure shutdown and start-up procedures are well detailed.

Table G.4 Cold diathermic oil at 71 m^3/hr and 100°C.

Key Word	Deviation	Possible Causes	Consequences	Action Required
NONE	No flow	1. Pump failure.	No condensation from reaction gases.	(a) Install spare pump in parallel. Maintenance.
		2. Valve shut incorrectly.	As for 1.	(b) Install manual override on valve.
		3. Line rupture.	As for 1. Release of 'hot' oil to atmosphere.	(c) Emergenecy shutdown. Maintenance.
MORE OF	More temperature	4. Increased duty upstream.	Reduced heat transfer, less condensation.	(d) Regulate diathermic oil flow with heat duty.
			Increased duty on switch condensers.	(e) Ensure over-capacity of diathermic oil is available.
LESS OF	Less flow	5. Pump working below rated capacity.	As for 4.	Covered by (a).
		6. Leaking flange or pipe.	As for 4. Release of 'hot' oil to atmosphere.	(f) Temporary maintenance if possible, otherwise emergency shutdown.
		7. Controller malfunctioning or incorrectly tuned.	As for 4.	(g) Good monitoring and operating techniques.
				(h) Deactivate controller if necessary.
	Less temperature	8. Winter conditions.	Improved heat transfer — over cooling.	Covered by (d).
		9. Reduced duty upstream.	As for 8.	Covered by (d).
AS WELL AS	Air	10. Suction at vent.	Pockets of air in condenser — reduced heat transfer.	(i) Vents installed correctly.
	Suspended solids	11. Contamination of supply.	Increased fouling rate.	(j) Install filter if oil quality is inconsistent.
				(k) More frequent exchanger cleaning.
OTHER	Maintenance	12. General equipment failure or catalyst changeover in reactor.	Process stops.	(h) Good practices in construction and operation. Ensure shutdown and start-up procedures are well detailed.

Table G.5 Hot diathermic oil at 71 m³/hr and 138°C.

Key Word	Deviation	Possible Causes	Consequences	Action Required
NONE	No flow	1. No flow into vessel.	No condensation from reaction gases.	(a) Recycle gases through switch condensers until flow can be restored.
		2. Vessel tubes or walls ruptured.	Oil contacts reaction gases with possible reaction. Process stops.	(b) Product contaminated. Stop flow to storage to prevent contamination of good product. (c) Emergency shut-down.
MORE OF	More temperature	3. Increased heat duty.	Reduced duty on steam generators.	(d) Regulate diathermic oil flow with heat duty.
LESS OF	Less temperature	4. Reduced heat duty. 5. Increased flow of oil into vessel. 6. Fouling reducing heat transfer rate.	Increased duty on steam generators. Increased pumping duty. Heat transfer rate falls slowly. Condensation duty transferred to switch condensers.	Covered by (d). (e) Select pump with inbuilt over-capacity. (f) Shutdown and clean tubes.
OTHER	Maintenance	7. General equipment failure or catalyst changeover in reactor.	Process stops.	(g) Good practices in construction and operation. Ensure shutdown and start-up procedures are well detailed.

FINAL COMMENTS

At this stage of the design project the student probably writes an appropriate conclusion and submits the report. In an industrial situation there are one thousand and one tasks still to do. The next stage would be a thorough and accurate costing of the entire project and submission of detailed quotations for all aspects of equipment, design, construction, etc. If the project is to proceed, company policy decisions will have to be made by senior management — including project approval and obtaining of finance. The decisions made in the design stages are frequently modified as the project proceeds and the calculations are revised. The iterative design process continues. The start-up and commissioning stages are considered in the design process, as are procedures for plant shut-down and detailed maintenance programmes. Plant modifications should only be performed with reference to the final design study. In this respect plant design is a truly dynamic activity and a final solution is never really obtained.

A student design project can only provide an approximation to industrial design work. However, it should be possible to consider the nature of the design process and various aspects realistically. Student design projects are restricted only by time. In industry the same constraint applies with the additional consideration of economics.

Action: Review the design project, identify aspects that should be performed in greater detail if more time was available.

How does this academic exercise differ from an industrial design study?

INDEX